Edited by
A. C. Sigleo and A. Hattori

MARINE AND ESTUARINE GEOCHEMISTRY

Library of Congress Cataloging in Publication Data
Main entry under title:

Marine and estuarine geochemistry.

Bibliography: p.
Includes index.
1. Chemical oceanography—Congresses. 2. Geochemistry—Congresses. 3. Estuarine oceanography—Congresses.
I. Sigleo, A.C. (Anne C.) II. Hattori, Akihiko, 1926–
GC110.M36 1985 551.46′01 85-13127
ISBN 0-87371-007-X

LEWIS PUBLISHERS, INC.
121 South Main Street, Chelsea, Michigan 48118

PRINTED IN THE UNITED STATES OF AMERICA

PREFACE: PURPOSE AND SCOPE

As the world's terrestrial environments become further populated, marine and estuarine ecosystems are being increasingly impacted by anthropogenic activities, particularly by the disposal of waste products. To predict the effects of these activities, it is necessary to understand fundamental marine and estuarine processes. Both directly and indirectly the chapters in this volume address this issue, integrating concepts and analytical techniques from chemistry, biochemistry, geochemistry and oceanography.

Topics in organic and inorganic geochemistry, as well as data on nutrient cycling, are represented in this volume. Studies of specific biogenic compounds are complemented by studies of the distributions and fate of anthropogenic PCBs (polychlorinated biphenyls), silicones and the pyrolyzate styrene. Many of the chapters emphasize the need for in-depth measurements over annual cycles to establish baseline biogenic inputs. Even in the deep sea, the flux of organic matter in the water column, and its composition, are subject to seasonal fluctuations of primary productivity in surface waters. For long-range resource planning in estuaries, the results from San Francisco Bay (USA) serve to illustrate the value of data over decades which include the maxima in interannual variability (i.e., wet and dry years).

A fundamental problem in geochemistry is the presence of natural organometallics in the environment. Actual measurements of carbon-metal compounds, and their distribution and production in marine and estuarine environments, are reported in this book. Of additional interest, the chapter on methyl iodide suggests a mechanism for producing many of the organometallics by marine algae. The chapters on metal speciation, along with those on organometallics, are of particular interest to biologists and toxicologists because it is the form, or species, of the metal, rather than simply its concentration, which makes a metal bioavailable, toxic, or harmless.

Geographically, these chapters cover the Pacific, Atlantic and Antarctic oceans, and major estuaries from Tokyo Bay and the Keum Estuary (Korea) in Asia to San Francisco Bay, Chesapeake Bay, and the St. Lawrence Estuary in North America. Studies include transport processes, and nutrient and metal distributions by depth as well as regionally. Nutrient cycling and the mass balance of essential elements such as carbon and nitrogen have been considered using a wide range of analytical methods and state-of-the-art sampling techniques. To determine the effects of anthropogenic activities on marine and estuarine environments, these processes must continue to be studied and adequately understood.

This work is the peer reviewed proceedings of the symposium on Marine and Estuarine Geochemistry, International Chemical Congress of Pacific Basin Societies (Honolulu, Hawaii, Dec. 16–21, 1984). The importance of the topic is reflected by the large number of contributed papers from both sides of the Pacific. The symposium was sponsored by the Chemical Society of Japan, the American Chemical Society, and the Chemical Institute of Canada. We thank these societies for their support and the opportunity to convene an international symposium.

A. C. Sigleo
U.S. Geological Survey
Reston, Virginia
USA

A. Hattori, Director
Ocean Research Institute
University of Tokyo
Tokyo, Japan

CONTENTS

Organometallics and Trace Metals

ISOTOPE AND ORGANIC GEOCHEMISTRY

CHAPTER 1

VARIATION OF ^{15}N NATURAL ABUNDANCE OF SUSPENDED ORGANIC MATTER IN SHALLOW OCEANIC WATERS

Toshiro Saino and Akihiko Hattori
Ocean Research Institute, University of Tokyo
1-15-1, Minamidai, Nakano-ku, Tokyo, 164 Japan

ABSTRACT

Natural abundances of ^{15}N in suspended particulate organic nitrogen (PON) were determined at 5 or 10 m intervals in the shallow layer at a station in the northwestern North Pacific (45°N,160°E) in summer. An ammonium maximum was observed in the bottom of the euphotic layer, and a nitrite maximum appeared below the ammonium maximum layer. The ^{15}N natural abundance of suspended PON exibited a minimum of -1.5 per mil at slightly above the ammonium maximum, and then increased with depth to 4.8 per mil at 80 m. In the euphotic layer, suspended PON is enriched in ^{15}N. Variation in the natural abundance of ^{15}N in suspended PON is interpreted in terms of nitrogen cycling processes. We infer that during POM decay, the isotope fractionation in deamination, followed by the uptake of ^{15}N enriched ammonium (produced by ammonium oxidizing bacteria) by the nitrifiers and other bacteria is the primary cause of ^{15}N enrichment in POM.

INTRODUCTION

Particulate organic matter (POM) plays a central role in the process of vertical transport of organic matter in the ocean. The isotopic abundances of biophilic elements in various fractions of biological materials are determined by the isotope fractionation in the biochemical reactions during its formation and degradation. In the open ocean where the supply of organic matter from precipitation or river runoff is not substantial, POM is produced mainly by phytoplankton primary production. The POM in the shallow euphotic layer is in a dynamic balance between its gain by production and loss by degradation or sinking. In the deeper aphotic layer, POM is decomposed during sinking to the seafloor.

We previously reported the first vertical profile of ^{15}N natural abundance of PON in the sea [1], and discussed the potential usefulness of ^{15}N natural abundances in studying the vertical transport processes of organic matter in the sea. In order to correlate the data on ^{15}N natural abundances of PON with

MARINE AND ESTUARINE GEOCHEMISTRY, Sigleo, A. C., and A. Hattori (Editors)

nitrogen cycling in the sea, however, it is necessary to collect an adequate number of data sets from various sea areas where some processes in the cycle of nitrogen are prominent, while others are not.

In the northern North Pacific and the Bering Sea in summer, an ammonium maximum layer is developed at the bottom of the euphotic zone [2]. Below the ammonium maximum a nitrite maximum is found. We present here the first observation of a closely spaced vertical profile of the ^{15}N natural abundance of PON, together with various environmental parameters in shallow waters at a station in the northwestern North Pacific in summer.

MATERIALS AND METHODS

The location of the sampling station is shown in Fig. 1. Seawater samples were collected with 23 ℓ Niskin bottles from depths of 20, 30, 35, 40, 45, 55, 60, 65, 70, 75, and 80 m.

Temperature was measured by reversing thermometers and salinity by an inductive salinometer. Dissolved oxygen was determined by the Winkler's method [3]. Ammonium, nitrite, and nitrate were determined using a Technicon Autoanalyser [4] within 30 minutes after the sampling. A 250 mℓ aliquot was filtered through $MgCO_3$ coated Whatman GF/C filter for determination of chlorophyll *a* by fluorometry [5].

Approximately 20 ℓ of seawater were filtered through Whatman GF/F filters (47 mm; precombusted 450 ℃, 4h) for PON analysis.

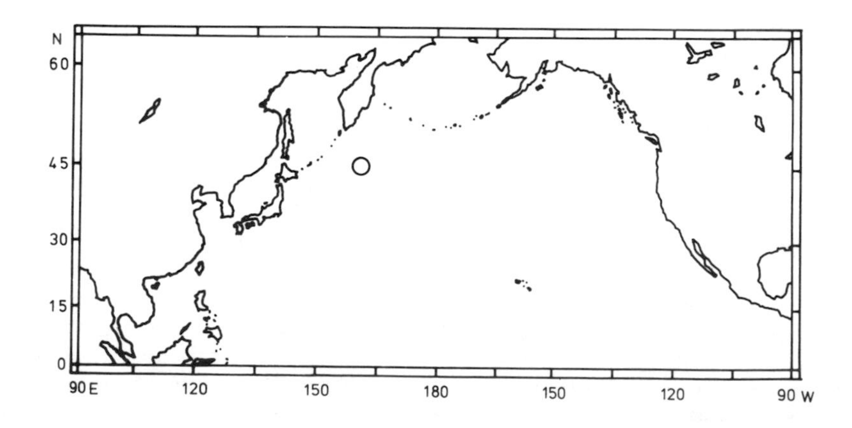

Fig. 1 Location of the sampling station. Water sampling was made at 05:05 on 19 August 1983.

Filters with collected POM were stored frozen and brought back to the laboratory. Before analysis the filters were freeze dried. PON concentration and its ^{15}N abundance were measured by the method of Wada et al. [6] using a Hitachi RMU-6 Mass spectrometer for isotope ratio measurements. The ^{15}N abundance is expressed as per mil deviation from standard air nitrogen:

$$\delta^{15}N \text{ (per mil)} = [(^{15}N/^{14}N)_{smpl}/(^{15}N/^{14}N)_{std} - 1] \times 1000.$$

RESULTS

Vertical profiles of ^{15}N natural abundance of PON and its concentration are shown in Figure 2 together with temperature, salinity, σ_t, per cent saturation of dissolved oxygen, ammonium, nitrite, nitrate, chlorophyll *a* and phaeopigments concentrations. A strong pycnocline appeared at the 20-60 m depth interval. The mean density gradient in the pycnocline was 0.0237 m^{-1}. A subsurface ammonium maximum layer (>2.4 μM) was observed at 35-50 m zone, and a nitrite maximum (1.22 μM) at 70 m. Nitrate concentration increased with depth. Dissolved oxygen was supersaturated above 40m and undersaturated below 45 m. Particulate materials (PON, chlorophyll *a*, and phaeopigments) decreased sharply between 20-30 m and gradually below 30 m. A small concentration maximum of PON was present at 45 m. Ammonium and PON concentrations correlate well with each other below 30 m. The $\delta^{15}N$ of PON at 20 m was 2.3 per mil. The value decreased with depth to a minimum (-1.5 per mil) at 35 m and then increased with depth to 4.8 per mil at 80 m. The euphotic depth (1% light depth) was approximately 25 m.

DISCUSSION

Characteristics of nitrogen cycling processes

A schematic model of nitrogen cycling in this sea area is presented in Figure 3. PON is defined as nitrogen retained on glass fiber filters (Whatman GF/F) after filtering 23 ℓ of seawater collected with Niskin bottles. The PON is composed of phytoplankton, micro-zooplankton, bacteria, and detritus. It contains only a small amount, if any, of large size particles such as fecal pellets, amorphous organic aggregates known as marine snow, or zooplankton itself [7]. The dissolved organic nitrogen (DON) consists of organic materials such as proteins, peptides, amino acids, nucleic acids and other constituents of organisms or their derivatives not retained on the filter used. In Figure 3 we assumed that the production of DON is mainly mediated by breakdown

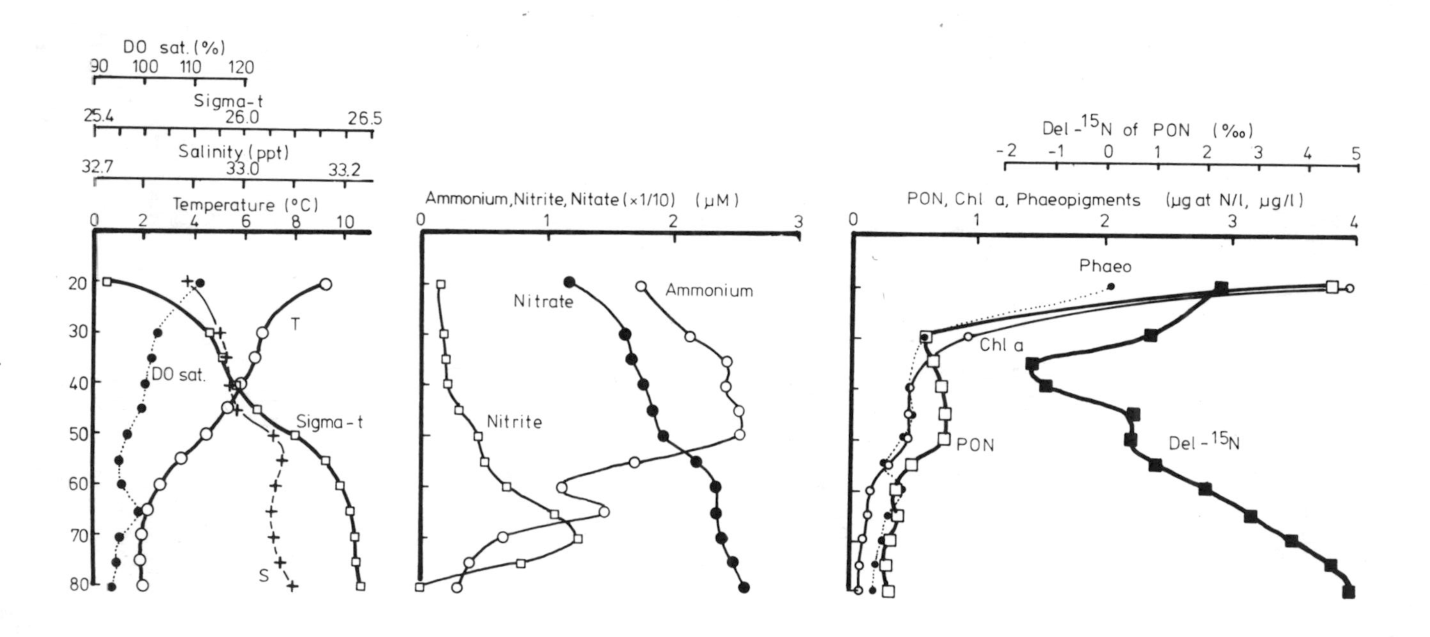

Fig. 2 Closely spaced vertical distribution of PON concentration and $\delta^{15}N$ of PON (right) together with temperature, salinity, σ_t, per cent oxygen saturation (left), ammonium, nitrite, nitrate (middle), chlorophyll *a*, and phaeopigments (right).

of organisms by zooplankton grazing or bacterial attack. In this process some portion is liberated as ammonium, some as DON, and the rest remains as PON. Ammonium is produced by decomposition of PON (or DON) by zooplankton and/or bacteria. Deamination is the most probable enzymatic reaction leading to ammonium liberation. Consumption processes of ammonium are uptake by phytoplankton either in the light or in darkness, uptake by bacteria, or oxidation to nitrite by a group of nitrifying bacteria. Nitrite is an intermediate of nitrification; it is produced by ammonium oxidation and consumed by nitrite oxidation to nitrate. Nitrite can also be taken up by phytoplankton or produced by phytoplankton during nitrate reduction. Nitrate is taken up by phytoplankton and produced by the bacterial nitrification.

A concentration maximum of ammonium occurred at 35-50 m (Fig. 2). It is within a pycnocline and corresponds to the zone where dissolved oxygen is slightly lower than 100%. Similar ammonium maxima were observed widely in the northern North Pacific and the Bering Sea in summer [2]. We explained that the high concentration of ammonium in the maximum layer is regenerated from PON that is produced in the surface layer using nitrate transported to the euphotic zone by mixing in winter. The same explanation can be applied in the present study.

A nitrite maximum layer appeared below the ammonium maximum layer. If a steady state exists in the distribution of nitrite there must be a source of nitrite in the maximum layer. A steady state occurs at least on a time scale of a week, since we observed the same vertical distribution patterns for various parameters in all water sampling casts (25 casts) from shallow depths near this

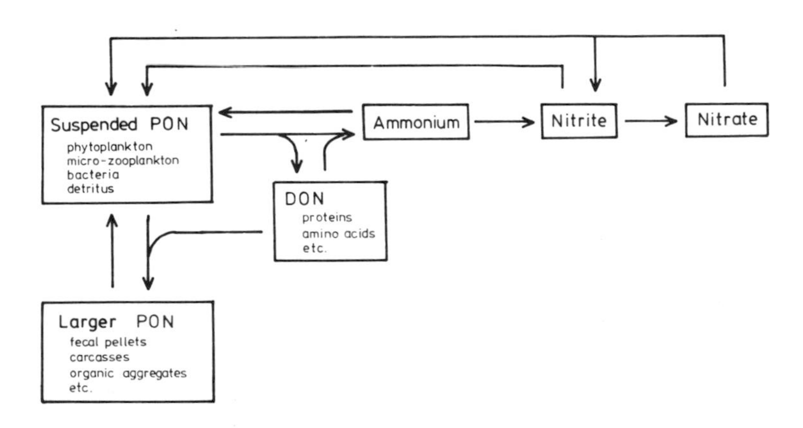

Fig. 3 Schematic model of nitrogen cycling in the sea. Nitrogen fixation is neglected.

station for one week. Olson [8] proposed that differential photoinhibition of the two nitrification processes, ammonium oxidation and nitrite oxidation, is mainly responsible for formation of the primary nitrite maximum. Our results can be explained by his hypothesis; ammonium oxidation is inhibited in the shallow depths (<40 m), and nitrite oxidation is inhibited above 70 m, and neither ammonium and nitrite oxidation processes are inhibited below 70 m. The midday maximum light intensitis at 40 and 70 m are estimated as 5×10^{13} and 2×10^{11} quanta cm^{-2} sec^{-1}, respectively. The light intensity for inhibition of the two processes of nitrification is approximately two or three orders of magnitude lower than those given by Olson [9].

Temperature and nitrate concentration have a linear relationship from 20 m to 70 m (Fig. 4). This indicates that nitrate is not produced or consumed in the depth zone of 20-70 m. It is well known that high concentrations of ammonium supress nitrate uptake by phytoplankton [10], and that the light dependence of nitrate uptake is much more significant than it is for ammonium uptake [11]. The production of nitrate is probably supressed above 70 m due to photoinhibition of nitrification.

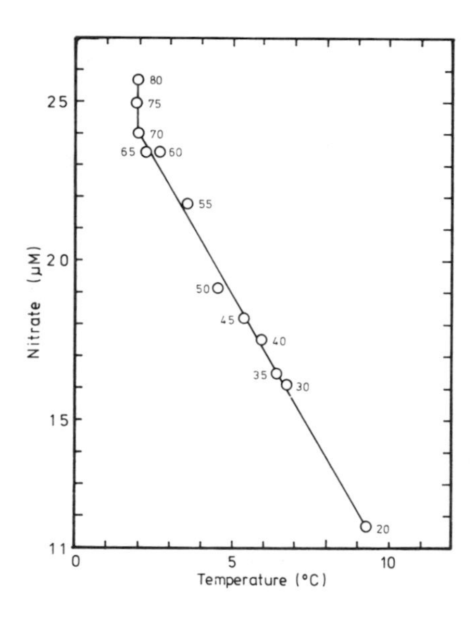

Fig. 4 Temperature-nitrate diagram for the station.

Variation of $\delta^{15}N$ in PON

Based on the above considerations of nitrogen cycling at this station, the shallow water column can be divided into 4 regimes by depth. Photosynthetic oxygen production exceeds respiratory oxygen consumption above 40 m. Ammonium serves as a primary nitrogen source for phytoplankton growth. Net production of ammonium occurs in 45-50 m layers where ammonium regeneration from PON exceeds ammonium uptake by phytoplankton. In the depth range from 50-70m, ammonium regeneration and oxidation of ammonium by ammonium oxidizing bacteria are predominant processes in nitrogen cycling. Below 70 m, ammonium is further oxidized to nitrate by nitrifying bacteria.

We further simplify the schematics shown in Fig. 3 as

$$\text{PON} \underset{k_3}{\overset{k_1}{\rightleftharpoons}} \text{Ammonium} \xrightarrow{k_2} \text{Nitrite}, \qquad (1)$$

where k_1, k_2, and k_3 are the first order rate constants for ammonium production by decomposition, ammonium oxidation by nitrification, and ammonium uptake by phytoplankton or bacteria. In a steady state, $[\text{ammonium}]/[\text{PON}] = k_1/[k_2 + k_3]$. The ammonium:PON ratio remains constant within the 30 to 65 m zone,

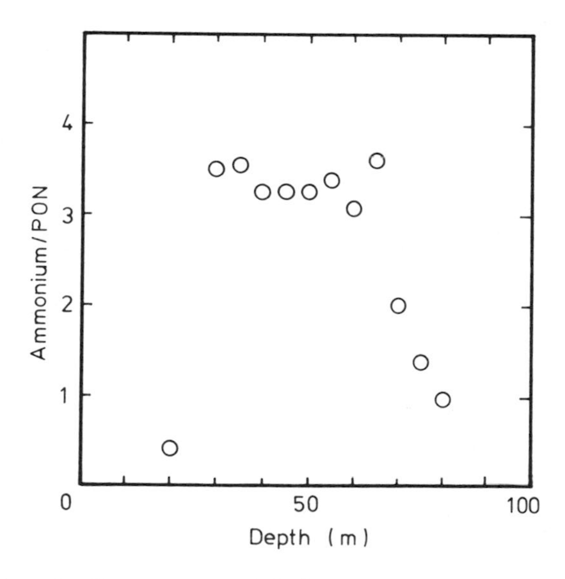

Fig. 5 Variation of PON/ammonium ratio with depth

and decreases above or below that depth interval (Fig. 5). This indicates that the same mechanism of consumption and production as depicted most simply by equation (1) dominates a steady state distribution of ammonium within the depth range of 30 to 65 m. Deviation from the trend can be explained by enhanced ammonium uptake (k_3) by phytoplankton at 20 m, and by the onset of nitrite oxidation below 70 m.

DON concentration in shallow waters of the studied area exceeds 10 μg atoms ℓ^{-1}, and several times higher than PON [12]. The close correlation observed between ammonium and PON implies that DON contribution to ammonium production is not significant compared with PON. Nitrogen isotope fractionation in the transformation process of PON to DON [13] is too small to explain the observed variation of $\delta^{15}N$ of PON. We infer ammonium derived mainly from PON.

The PON concentration is high when the density gradient is high (Fig. 6). This suggests that the main part of PON consists of

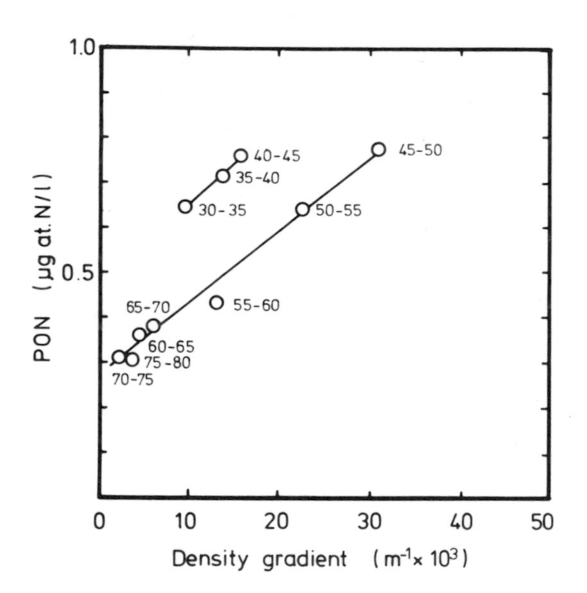

Fig. 6 Relationship between density gradient and PON concentration. Density gradient was calculated as difference of σ_t between the sampling interval $[\sigma_t(z_1) - \sigma_t(z_2)]/(z_1 - z_2)$. Mean PON concentration between the sampling depths, $[PON(z_1) + PON(z_2)]/(z_1 + z_2)$, are plotted against the density gradient. Data from 20-30 m are off scale.

small particles which have a slow sinking velocity comparable to or less than the water turbulence. These small particles are likely to be produced in the pycnocline by breakdown of larger particles transported from shallow layer or produced biologically *in situ* (Fig. 3). Chlorophyll *a* and PON are correlated well with each other within the depth range 40-80 m; [Chl. *a*; $\mu g\ \ell^{-1}$] = 0.18 + 1.28x[PON; μg atom ℓ^{-1}], r=0.98 (n=9). This indicates that the composition of POM does not change from 40 to 80 m. On the other hand, there exists a discontinuity at 45 m in PON-density gradient relationship (Fig. 6); above and below that depth, two parallel regression lines can be drawn. This discontinuity depth coincides with the depth above which ammonium starts to decrease, and 100% oxygen saturation and a step in the $\delta^{15}N$ profile are observed. The occurrence of PON production due to the growth of phytoplankton or bacteria are significant above the discontinuity. Phytoplankton production near the bottom of the euphotic zone is certainly light limited. Nitrogen isotope fractionation during nitrate uptake is substantial under the light limited conditions [14]. The occurrence of isotope fractionation in ammonium uptake under light limited conditions is highly possible at the bottom of the euphotic zone, since ammonium concentrations are high at this station.

The decrease of nitrate concentration in the surface mixed layer starts concomitantly with the beginning of phytoplankton blooming and water column stratification. Since nirate concentration is not a limiting factor of nitrate uptake during blooming of phytoplankton at this station (Fig. 2), isotope fractionation will occur during nitrate uptake. This causes an enrichment of ^{15}N in the remaining nitrate above the pycnocline, and an enrichment of ^{14}N in produced PON. Field and laboratory data support this assumption. Figure 7 shows the variation of ^{15}N abundance in sea surface PON with ambient nitrate concentration. Surface PON has a $\delta^{15}N$ number of nearly zero when the ambient nitrate concentration is high. The $\delta^{15}N$ number of PON is inversely correlated with surface nitrate concentration. The $\delta^{15}N$ number of surface PON in nitrate depleted water is approximately 6 per mil, close to the observed number for deep water nitrate in the Pacific [15,16]. In this oceanic area the concentration of surface nitrate before the onset of blooming can be estimated from the concentration of nitrate in the subsurface temperature minimum layer [2,17].

The ammonium concentration in the subsurface maximum layer is dependent on the amount of nitrate removed from the surface water above the pycnocline as shown in a previous report (Fig. 7 in [2]). Therefore, ammonium in the maximum layer is likely be derived from PON produced in the surface mixed layer utilizing nitrate. Ammonium in its concentration maximum layer will have a $\delta^{15}N$ number similar (in case of no isotope fractionation during ammonium regeneration) to or lower (in case of some isotope frac-

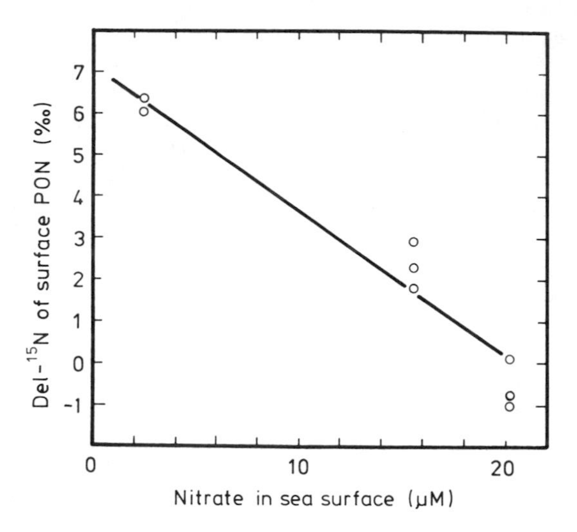

Fig. 7 Relationship between δ^{15}N of surface PON and ambient concentration of nitrate. Samples were collected on the Hakuho Maru Cruise KH-78-3 at Station 3 (11 July '78,47°57'N; 176°27'E, nitrate=20.2μM), Station 4 (15 July,'78, 53°34'N; 177°10'E, nitrate=15.6μM and Station 8 (18 July '78, 58°00'N; 175°06'W, nitrate=2.5μM) in the northern North Pacific (Sta. 3) and the Bering Sea (Stas. 4 and 8).

tionation during ammonium regeneration) than isotopically light PON transported from the shallow layer. If PON is again reproduced using ammonium in that layer, it will be isotopically much lighter than the ammonium because of probable isotope fractionation in the uptake process.

In the surface mixed layer after the water column stratifies, PON is produced using both nitrate from the deep layer and ammonium from its concentration maximum layer, and nitrification processes are supressed by sunlight. The PON, DON and ammonium turn over rapidly within the surface mixed layer and some portions of large sized PON particles are lost by sinking. The δ^{15}N number of PON will be mainly determined by the source nitrate and ammonium transported from depth, and by the relative contribution of these nitrogen sources to primary production. The enrichment of ^{15}N along the food chain [18] might cause some variation in the δ^{15}N of PON as defined previously because larger organisms of higher trophic levels are not involved in the PON. The δ^{15}N profile of PON above its minimum is explained by a combination of these effects.

The decomposition of PON prevails in the water column below 45 m. The $\delta^{15}N$ number of PON increases with depth. The same trend was invariably seen in the shallow vertical profiles of PON $\delta^{15}N$ ([1] and our unpublished data). Nitrite is probably the end product of oxidative degradation processes within the 45-70 m zone, and nitrate below 70 m. It is possible that isotope fractionation occur during each step of the oxidative decomposition of PON leading to nitrate formation, as shown for ammonium regeneration [19], and ammonium oxidation [13]. We cannot as yet specify the specific step, among the oxidative degradation steps of PON, responsible for ^{15}N enrichment in PON with depth. Nitrite decreased rapidly below 70 m to almost zero at 80 m, but no abrupt change in the $\delta^{15}N$ profile at 70 m was observed (Fig. 2). This indicates that nitrite oxidation during nitrification is not responsible for the enrichment of ^{15}N in PON. We infer that in the oxidative decomposition of POM the deamination followed by the uptake of ^{15}N enriched ammonium, produced by nitrifying (ammonium oxidizing) bacteria, by nitrifiers themselves and other bacteria is the primary cause of ^{15}N enrichment in POM.

CONCLUSIONS

Variation in the natural abundance of ^{15}N in suspended PON is interpreted in terms of nitrogen cycling processes. It is shown that the biological decomposition of PON during vertical transport is responsible for the observed characteristic features such as the subsurface ammonium and nitrite maxima. The $\delta^{15}N$ of PON thus provides a novel parameter for the nitrogen cycling study. The ^{15}N natural abundance data as such are not sufficient enough to quantify each process causing variation of ^{15}N abundance, but they can offer unperturbed and at learst qualitative information on the processes responsible for ^{15}N variation under *in situ* conditions. Most data on nitrogen cycling have been gathered by use of ^{15}N tracer techniques, but the technique in principle perturbs natural conditions. The natural abundance of ^{15}N therefore provides a complementary approach to elucidate cycling of nitrogen in the sea.

ACKNOWLEDGEMENTS

We thank Prof. John J. Goering, Univeristy of Alaska and Visiting Professor of University of Tokyo, for reviewing the manuscript. Thanks are also due to Captain, officers and crew of R/V Hakuho Maru, University of Tokyo, and scientist aboard the Cruise KH-83-3. We are grateful to Drs. E. Wada and I. Koike for reading the manuscript. This work was partly supported by grant

#58740253 to TS from the Ministry of Education, Science and Culture, Japan.

REFERENCES

[1] Saino,T. and A. Hattori. "^{15}N natural abundance in oceanic suspended particulate matter," *Nature.* 283: 752-754 (1980).
[2] Saino, T., H. Otobe, E. Wada and A. Hattori. "Subsurface ammonium maximum in the northern North Pacific and the Bering Sea in summer," *Deep-Sea Research.* 30:1157-1171 (1983).
[3] Japan Meteorological Agency (ed). "Manual for Oceanographic Observation," Oceanographical Society of Japan. 429 pp. (1970; in Japanese).
[4] Strickland, J.D.H, and T.R. Parsons. "A Practical Handbook of Seawater Analysis," *Bulletin of the Fisheries Research Board of Canada.* 167:310 pp (1972).
[5] Holm-Hansen, O., C.J. Lorenzen, R.W. Holmes and J.D.H. Strickland. "Fluorometric determination of chlorophyll," *Journal du Conseil, Conseil permanent International pour l'Exploration de la Mer.* 30:3-15 (1965).
[6] Wada, E., T. Tsuji, T. Saino and A. Hattori. "A simple procedure for mass spectrometric microanalysis of ^{15}N in particulate organic matter with special reference to ^{15}N tracer experiments," *Analytical Biochemistry.* 80:312-318 (1977).
[7] McCave, I.N. "Vertical flux of particles in the Ocean," *Deep-Sea Research.* 22:491-502 (1975).
[8] Olson, R.J. "^{15}N Tracer studies of the primary nitrite maximum," *Journal of Marine Research.* 39:203-226 (1981).
[9] Olson, R.J. "Differential photoinhibition of marine nitrifying bacteria: a possible mechanism for the formation of the primary nitrite maximum," *Journal of Marine Research.* 39:227-238 (1981).
[10] McCarthy, J.J., W.R. Taylor and J.L. Taft. "Nitrogenous nutrition of the plankton in the Chesapeake Bay. 1. Nutrient availability and phytoplankton preference," *Limnology and Oceanography.* 22:996-1011 (1977).
[11] Slawyk, G. "^{13}C and ^{15}N uptake by phytoplankton in the Antarctic Upwelling area: Results from the Antiprod I Cruise in the Indian Ocean Sector," *Australian Journal of Marine and Freshwater Research.* 30:431-448 (1979).
[12] Hattori, A (ed.). "Preliminary Report of The Hakuho Maru Cruise KH-71-3 (IBP Cruise)," Ocean Research Institute, University of Tokyo, 69 pp. (1973).
[13] Miyake, Y. and E. Wada. "The isotope effect on the nitrogen in biochemical oxidation-reduction reactions," *Records of Oceanographic Works in Japan.* 11:1-6 (1971).

[14] Wada, E. and A. Hattori. "Nitrogen isotope effects in the assimilation of inorganic nitrogenous compounds by marine diatoms," *Geomicrobiology Journal*. 1:85-101 (1978).

[15] Cline, J.D. and I.R. Kaplan. "Isotope fractionation of dissolved nitrate during denitrification in the eastern tropical Pacific Ocean," *Marine Chemistry*. 3:271-299 (1975).

[16] Reid, J.L. "Northwest Pacific Ocean waters in winter," The Johns Hopkins Oceanographic Studies, No 1. The Johns Hopkins University Press, Baltimore, 96 pp.

[17] Miyake,Y. and E. Wada. "The abundance ratio of $^{15}N/^{14}N$ in marine environments," *Records of Oceanographic Works in Japan*. 9:37-53 (1967).

[18] Minagawa, M., and E. Wada. "Stepwise enrichment of ^{15}N along food chains: Further evidence and the relation between $\delta^{15}N$ and animal age," *Geochimica et Cosmochimica Acta*. 48:1135-1140 (1984).

[19] Macko, S.A., and M.L.F. Estep. "Microbial alteration of stable nitrogen and carbon isotopic composition of organic matter," *Annual Report of the Director, Geophysical Laboratory, Carnegie Institution*, 1982-1983:394-398 (1983).

CHAPTER 2

AMINO ACID VARIATIONS IN MARINE PARTICLES DURING SINKING AND SEDIMENTATION IN HARIMA-NADA, THE SETO INLAND SEA, JAPAN

Shigeru Montani and Tomotoshi Okaichi
Department of Agricultural Chemistry
Kagawa University
Miki, Kagawa, 761-07, Japan

ABSTRACT

The amino acid compositions of suspended matter, sinking particles and surface sediments in Harima-Nada, the Seto Inland Sea, Japan, were investigated together with those of a cultured phytoplankter, *Skeletonema costatum*, and natural population of a red tide flagellate *Chattonella marina*. On a dry weight basis, amino acid concentrations varied from 7.5 to 45 mg/g for suspended matter, 13 to 92 mg/g for sinking particles and 5.2 to 8.3 mg/g for surface sediment. The downward fluxes of amino acids at 5 m above the sea bottom were estimated to be 21 to 37 % of primary production in the surface layer. The amino acid composition of the suspended matter differed significantly either from that of the sinking particles collected by sediment traps or from that of the phytoplankton. Sulfur-containing and aromatic amino acids were depleted in the suspended matter. The results suggest the occurrence of rapid diagenesis of protein during the sinking and on the surface sediment.

INTRODUCTION

There is good evidence that the marine sediments play an important role in nutrient supply to the overlying water [1]. Organic matter is produced by photosynthesis of phytoplankton in the euphotic zone. A part of the phytoplankton is consumed by herbivorous zooplankton within the water column, and then sinks directly to the bottom.

There are a great deal of data on the spatial distribution of particulate organic carbon and nitrogen in coastal waters [2] and on sinking particles [3-12] but the information is still limited on the molecular composition of particulate organic matter.

Amino acids, the building blocks of protein molecules, form the largest reservoir of organic nitrogen in most organisms. Their composition may change along the transformation processes mediated by various organisms.

The present study investigates the variations of amino acid composition in various marine particles to obtain the insight into the transformation processes occurring in coastal sea. The particulate samples were collected in Harima-Nada, the Seto Inland Sea, Japan. Harima-Nada region is a good fishing area for

MARINE AND ESTUARINE GEOCHEMISTRY, Sigleo, A. C., and A. Hattori (Editors)

Crustacea (crab and shrimp), benthic fishes (sand launce and flounder) and migrating fishes (anchovie and yellowtail) as well as a spawning and nursery area for anchovie. The proteinaceous materials on the sea bottom serve possibly as an important food source for benthic fauna either directly, or indirectly through herbivorous animals.

EXPERIMENTAL

Sinking particles were collected at four stations (Figure 1, Table 1) using a sediment trap system. Water depths at Stations, 9, 64, 15 and 21 were 13, 17, 42 and 22 m, respectively. The sediment trap system consisting of an aluminium frame and two sets of paired polyvinylchloride cylinders, 45 cm high and 16 cm across (Figure 2) was used. The trap sylinder contained a 5 cm deep 1 cm x i cm grid at the bottom. The trap system was kept for 20 to 45 h at 5 m above the sea bottom. After retrival of the trap, sinking particles were collected by centrifugation at 3000 rpm for 15 min, freeze-dried and pulverized.

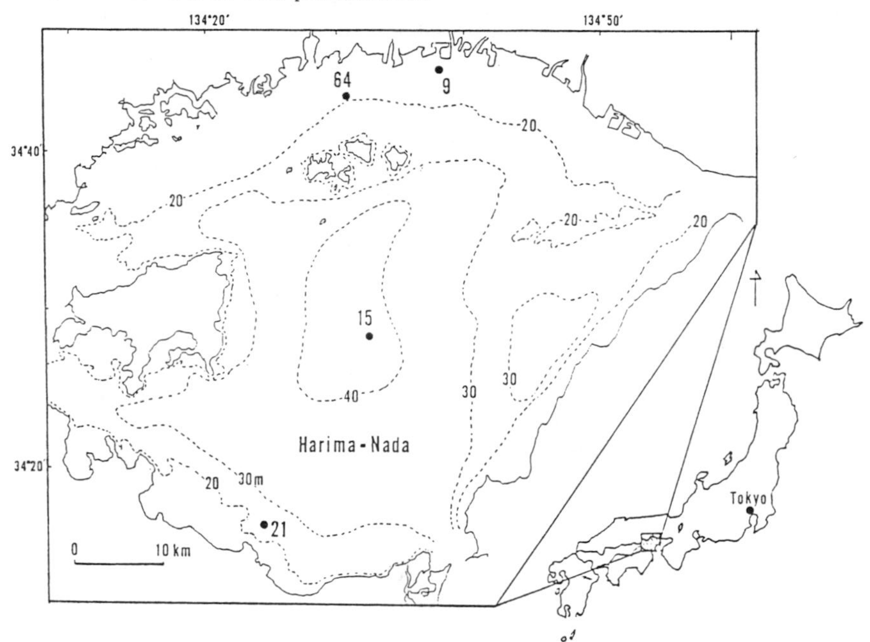

Figure 1. Location map of sampling stations in Harima-Nada, the Seto Inland Sea, Japan.

A portion of the trap sample was filtered through a Millipore HA filter, and the filter with residue was rinsed with a 3.3 % ammonium carbonate solution. The filter was freeze-dried, kept in a silica gel desiccater, and weighed later. Another portion of the trap sample was filtered through a pre-combused (450°C, 3 h)

Whatman GF/C glass-fiber filter. The filter was freeze-dried and stored for later determination of particulate organic carbon and nitrogen.

Table I. Deployment of sediment traps during 1982 and 1983. The station location is shown in Figure 1.

Date	Station	Trap
18 - 19 June 1982	21	I
19 - 20 June 1982	9	A
31 July - 1 August 1982	15	G
2 - 3 August 1982	9	B
30 June - 1 July 1983	21	J
1 - 2 July 1983	9	C
27 - 28 July 1983	64	D
8 - 10 August 1983	64	E
25 - 26 August 1983	64	F
26 - 28 August 1983	15	H

Water samples were collected from various depths by lo liter Van Dorn bottles, and suspended matter was collected by filtration through a pre-combusted, pre-weighed GF/C filter. The filters were rinsed with distilled water, dried and later weighed ashore.

Skeletonema costatum was cultured in 10-liter tanks fitted with fluorescent lamps. The culture was maintained bacteria free. Natural population of *Chattonella marina* was sampled with a Van Dorn bottle when its red tide ($\gg$1000 cells/ml) appeared in the northern part of Harima-Nada in the middle of July, 1983, and collected on the glass fiber filter.

Organic carbon and nitrogen were determined with a Yanagimoto MT-2 CHN analyzer or a Yanagimoto MT-500 CN analyzer, using 0.05 to 1.0 g of the samples which had been exposed to HCl fume. The analytical error was $\pm$ 1 % for organic carbon and $\pm$ 2 % for organic nitrogen.

The filters containing particulate organic matter were placed in 20 ml of 6N HCl and hydrolyzed, in the presence of added n-Leu (internal standard), for 22 h at 110°C under N_2 atomosphere. The hydrogen chloride was doubly distilled before use. The hydrolysate were filtered to remove glass fibers, and the filtrates were dried on a rotary evaporator at 50°c. The residue was redissolved in 20 ml of 0.01N HCl. This material was applied to the top of a 10 mm x 250 mm column of Dowex 50W X8 cation exchange resin which had been washed successively with 2N NaOH, and 2N HCl and then rinsed to neutrality with doubly-distilled water. The charged column was washed three times with 5 ml each of water, and the amino acids were eluted with 100 ml of 2N NH_4OH. The eluates were evaporated to dryness under vacuum. The residue was redissolved in few ml of 0.01N HCl and then amino acid composition was analyzed by a Hitachi KLA 5 liquid chromatograph. Reproducibility with standard mixtures of amino acids was within error range of 1.6 - 9.8 %, except valine, which error was 11.5 %.

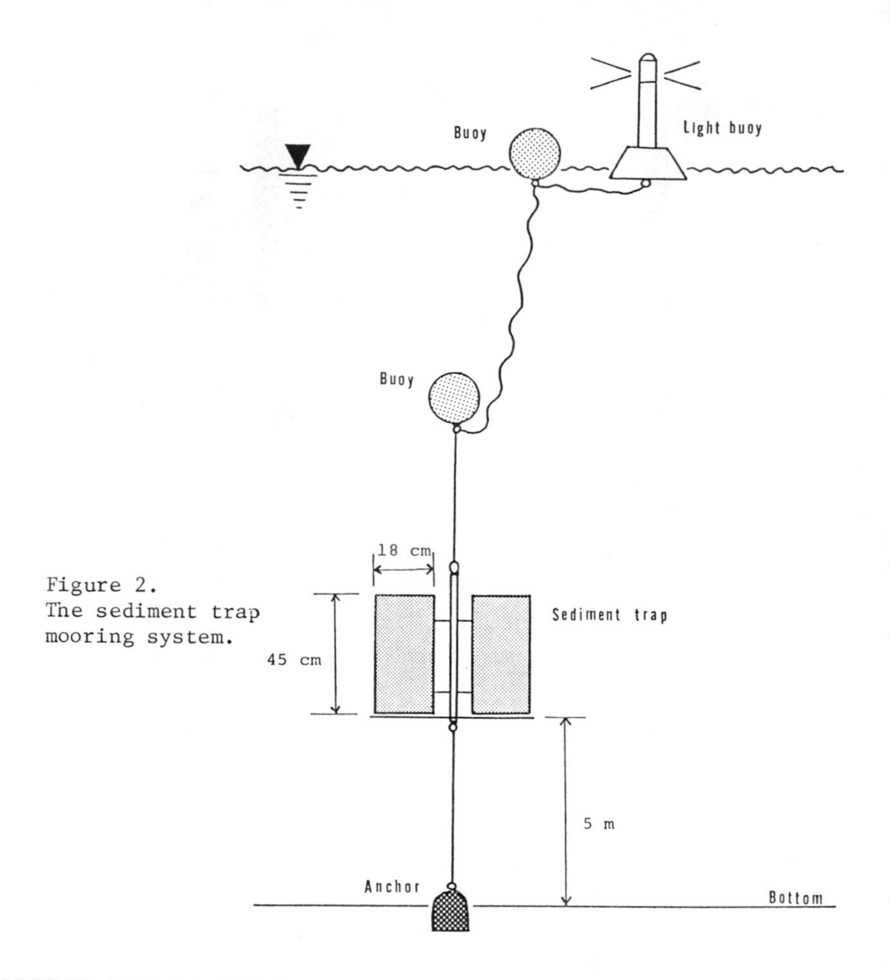

Figure 2.
The sediment trap mooring system.

RESULTS AND DISCUSSION

Distribution and sources of amino acids in marine particles.

The total mass fluxes varied between 3.56 and 24.0 g/m^2/d at the four stations (Figure 3). The mean fluxes of particulate organic carbon and nitrogen were 477 mgC/m^2/d and 80 mgN/m^2/d, respectively. Figure 3 also shows amino acid contents of sinking particles and vertical fluxes of amino acids calculated from these values and the total mass fluxes. Except for the high value of 92.3 mg/g in trap C at Station 9, the amino acid content of the sinking particles feel in a relatively narrow range of 13 to 34 mg/g. Two groups, A and B, can be distinguished (Figure 3): group A with a lower flux and a higher organic C content (Traps A, D, E, F, G and J) and group B with a higher flux and lower organic C content (Traps B, H and I).

The amounts of amino acids in the phytoplankton (*C. marina* and *S. costatum*), suspended matter and surface sediments are shown

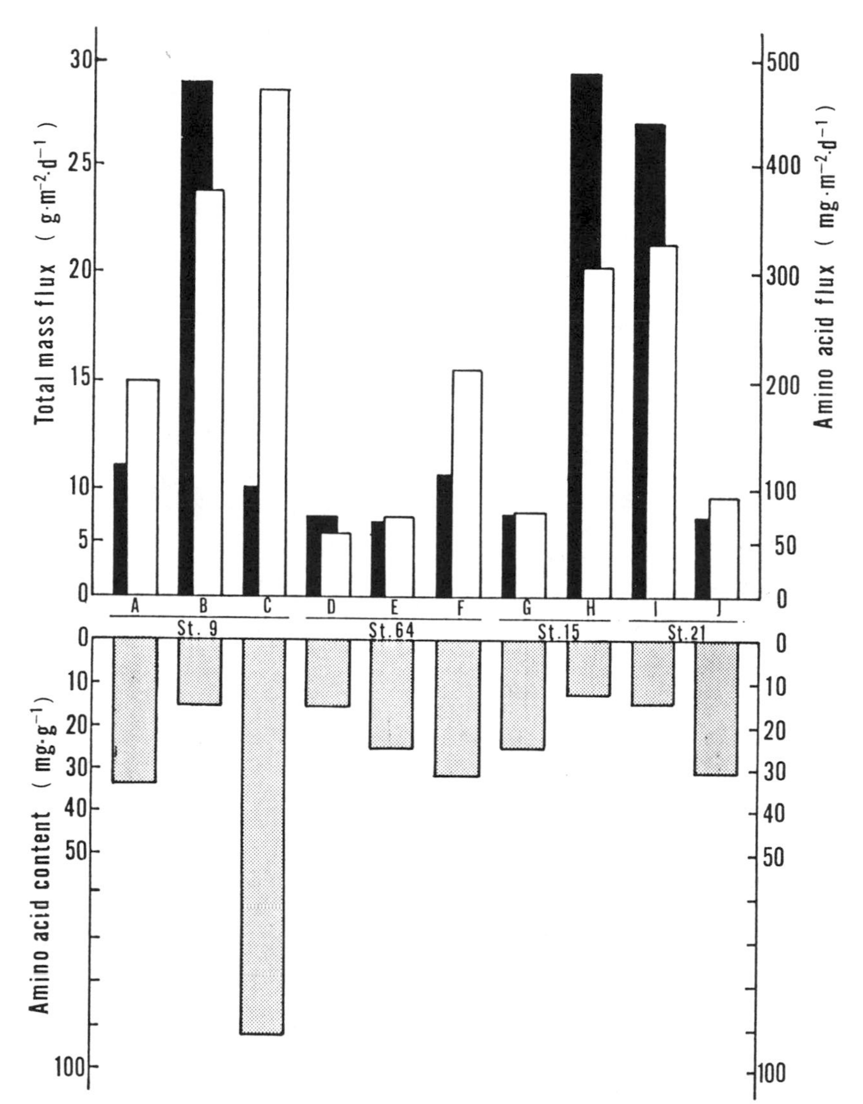

Figure 3. Fluxes (top) and contents (bottom) of amino acids obtained in sediment trap experiments. Total mass fluxes are shown by black bars and amino acid fluxes by white bars. See Table 1 for sampling date.

in Figure 4. Higher amounts of amino acids were often found in the suspended matter. The suspended matter appears to be mainly composed of phytoplankton debris.

The contents of amino acids in the surface sediments ranged between 5.2 and 8.3 mg/g. The values are higher than those reported with other coastal sediments [e.g. 13-15]. The higher amino acid contents probably reflect the higher primary productiv-

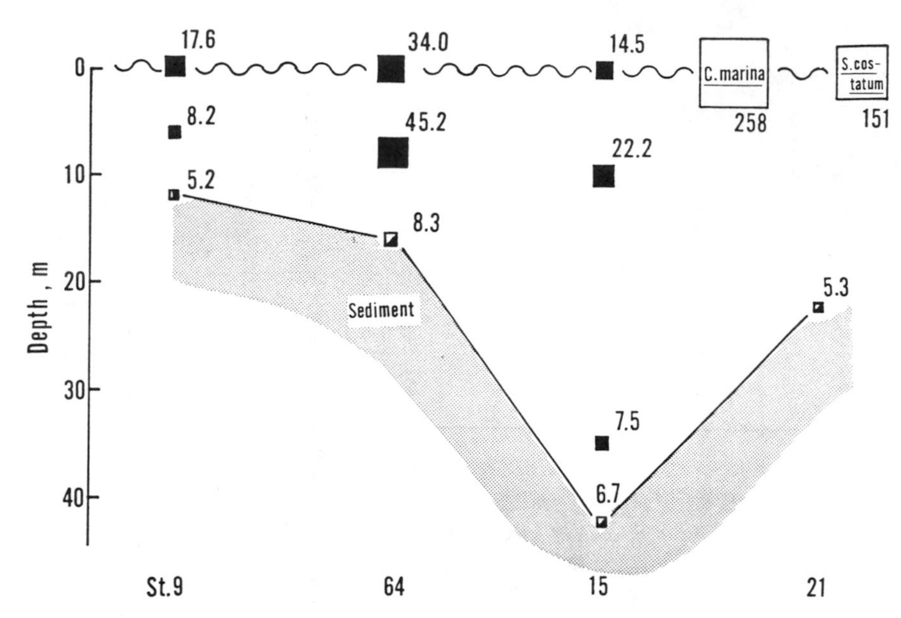

Figure 4. Amino acid contents (mg/g) of suspended matter (■) and surface sediments (◪) in Harima-Nada, the Seto Inland Sea, and of phytoplankton, *Skeletonema costatum* and *Chattonella marina* (□).

ity in this study area. The amino acid contents of the surface sediments were lower than the suspended matter and the sinking particles.

The contribution of amino acid carbon and amino acid nitrogen to total organic carbon and total organic nitrogen in sinking particles ranged from 17 to 44 % (average, 21.7±7.8%) and 27 to 71 % (average, 40.1±13.9%), respectively (Figure 5). These percentages do not vary significantly with time and place.

Numerous zooplankton fecal pellets were observed in sinking particles under the light microscope. The reported sinking rates of fecal pellet vary from 15 to 941 m/d [16-20]. The present data strongly suggest that fecal pellets produced by zooplankton are transported rapidly to the sea floor in the study area and provide a fresh source of amino acids.

The percentages of amino acids carbon and nitrogen relative to the organic carbon and nitrogen in the suspended matter (27±7, 58±12 %) were slightly greater than those in the sinking particles, suggesting that the suspended matter largely consists of phytoplankton debris

Changes in amino acid composition during sinking and sedimentation

Figure 6 shows relative abundance of individual amino acids in the suspended matter, sinking particles and surface sediments at Station 9 of Harima-Nada together with that in *S. costatum*.

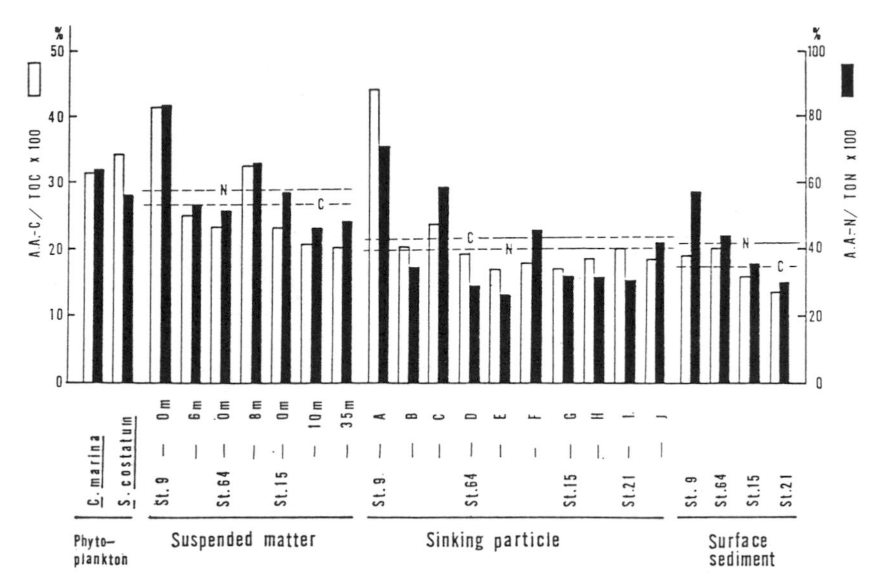

Figure 5. Percentages amino acid carbon (A.A.-C) and amino acid nitrogen (A.A.-N) relative to total organic carbon(TOC) and total organic nitrogen (TON) of suspended matter, sinking particles and surface sediments in Harima-Nada, the Seto Inland Sea, and of phytoplankton.

For convenience, amino acids were separated into acidic, basic, hydroxyl, neutral, aromatic and sulfur-containing amino acid groups (Figure 7). The following order of abundance was commonly observed: neutral acidic hydroxyl basic aromatic S-containing. The similarity in the amino acid compositions among the suspended matter, sinking particles and surface sediments suggests that these particles have a common source.

A close examination of the amino acids composition, however, discloses some differences. Basic amino acids (Arg, Lys and His) are depleted in the surface sediments. Aromatic (Phe, Tyr) and S-containing (Cys, Met) amino acids are less abundant in the sinking particles than in phytoplankton. The sinking particles would consist of slightly older materials which contain zooplankton and/or fish fecal pellets.

Amino acid compositions of the suspended matter can be characterized by the deficit of aromatic amino acids and S-containing amino acids. The occurrence of bacterial degradation of particulate organic matter during sinking is suggested. Actually, the aromatic amino acid in the suspended matter decrease with water depth (Figure 7; St.15). The aromatic amino acids can be used as an indicator to assess "age" or "freshness" of the particulate matter.

The quantitative study of the amino acid composition of suspended matter has been discussed by introducing the concept of essential amino acid index [21].

Figure 8 illustrates the amino acid compositions in the

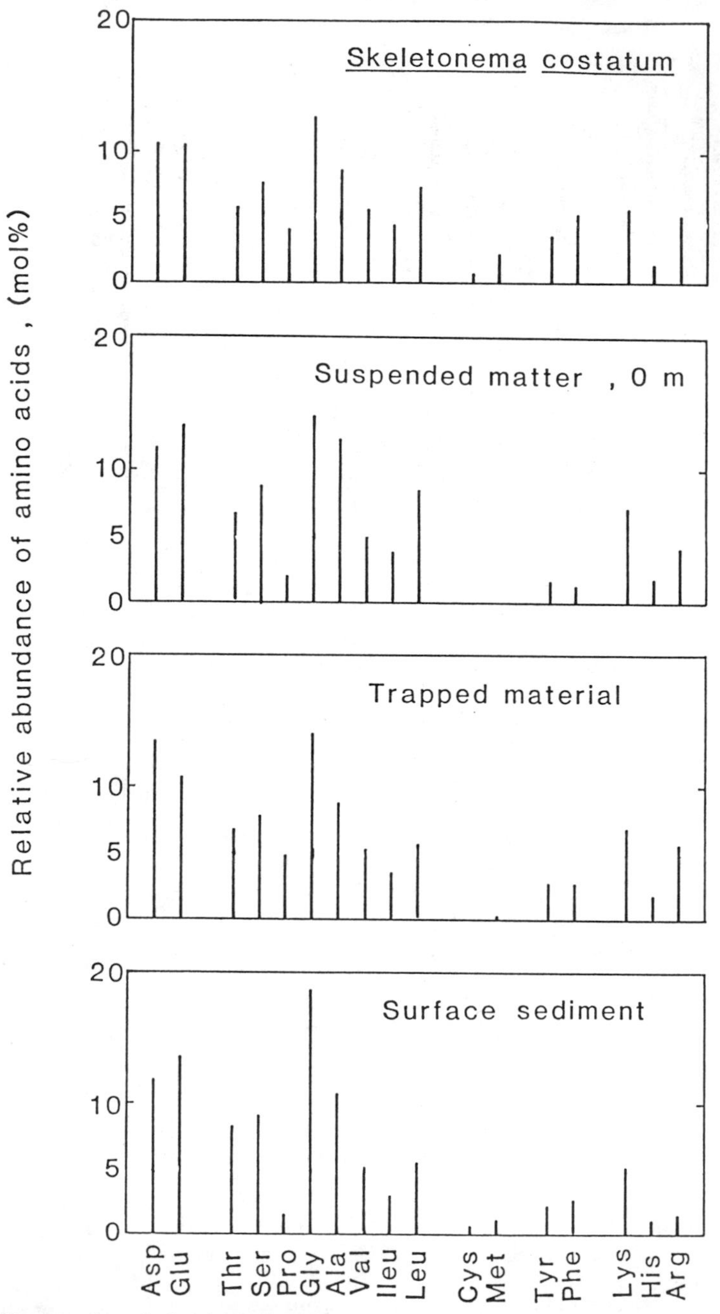

Figure 6. Relative abundances of amino acids in suspended matter, sinking particles and the surface sediments at Station 9, and in *Skeletonema costatum*

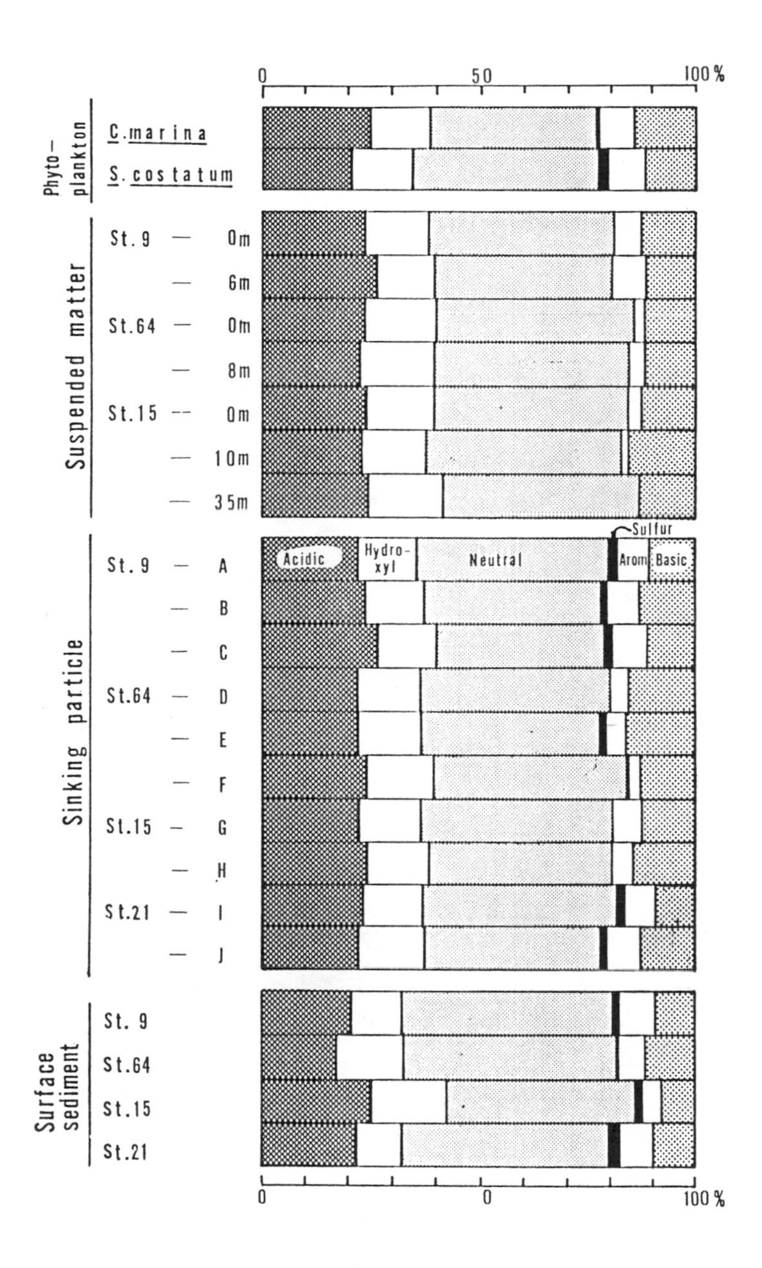

Figure 7. Relative abundances of acidic, hydroxyl, neutral, sulfur containing, aromatic and basic amino acids in suspended matter, sinking particles and surface sediments in Harima-Nada, the Seto Inland Sea, and in phytoplankton.

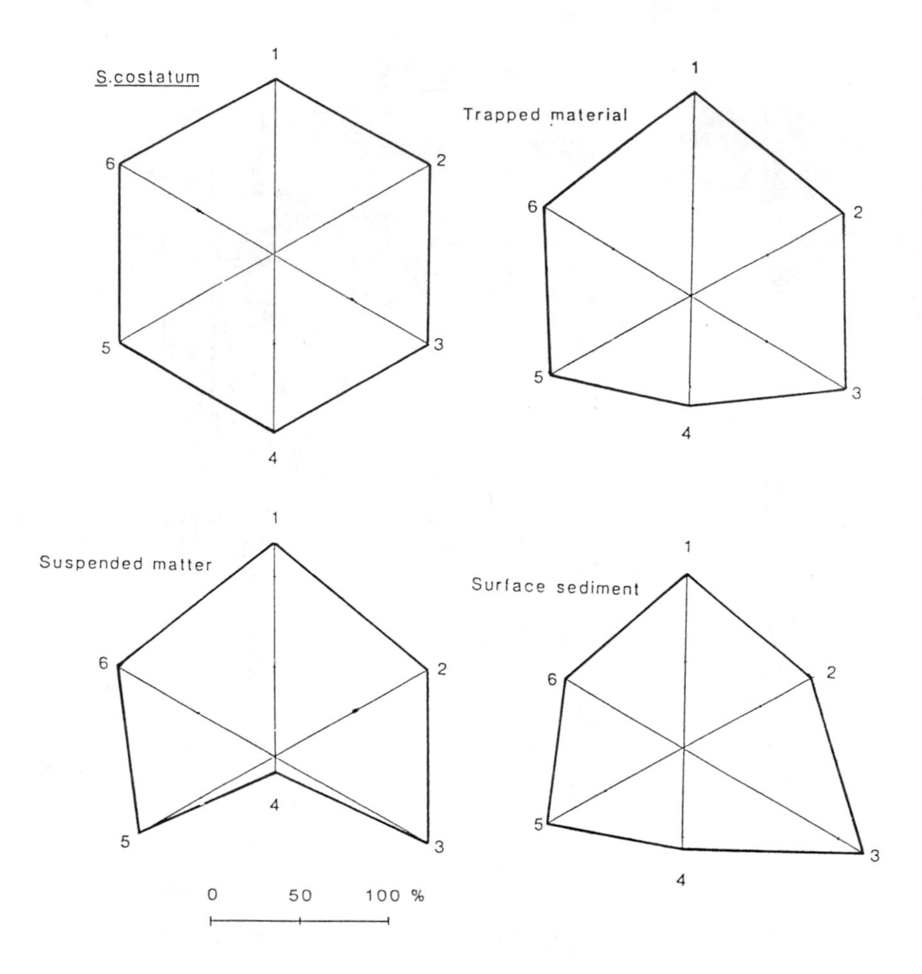

Figure 8. Abundances of acidic, hydroxyl, neutral, sulfur containing, aromatic and basic amino acids in suspended matter, sinking particles and surface sediments in Harima-Nada, the Seto Inland Sea, relative to those in *Skeletonema costatum*. 1, acidic, 2, hydroxyl, 3, neutral, 4, sulfur containing, 5, aromatic, and 6, basic.

phytoplankton, suspended matter, sinking particles and surface sediments using hexagonal diagrams. In these diagrams, the abundances of six amino acid groups in the suspended matter, sinking particles and surface sediments relative to those in *S. costatum* are given as the distances from the center. The similarity between the phytoplankton and sinking particles again suggests that the sinking particles mainly consist of fecal pellets produced by zooplankton from phytoplankton.

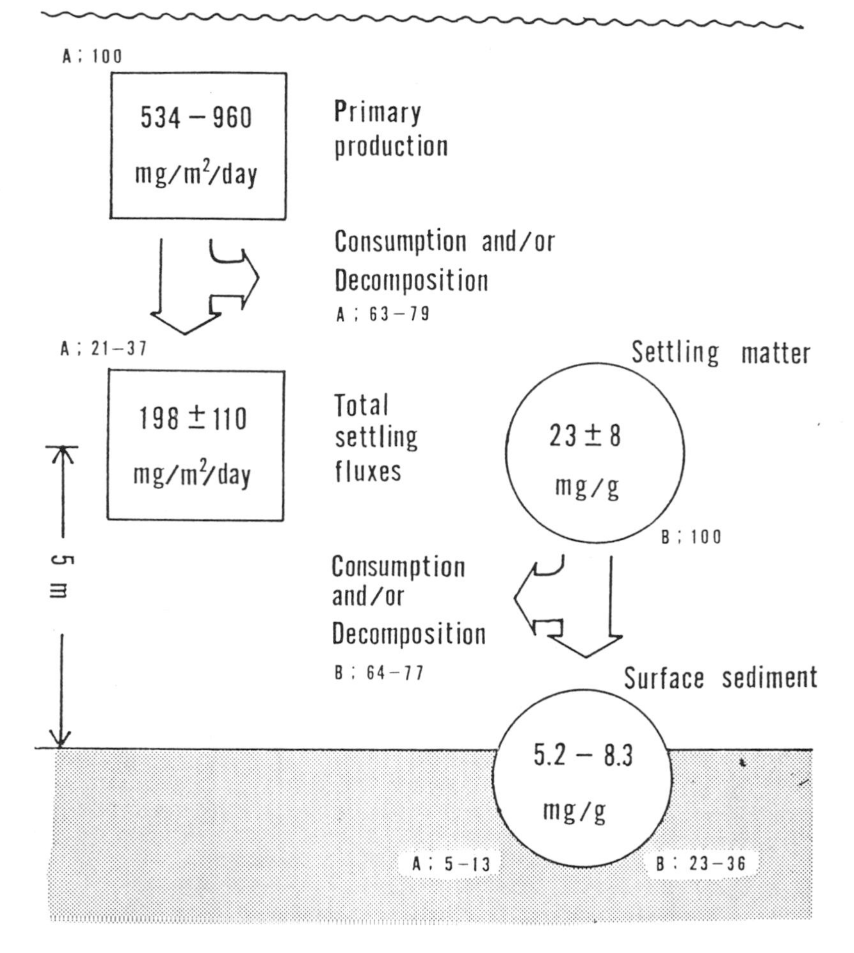

Figure 9. Amino acid budget in Harima-Nada, the Seto Inland Sea, Japan, as expressed in terms of dry weight.

Amino acid budget in Harima-Nada in summer.

Presented in Figure 9 is an estimate of amino acid budget in Harima-Nada based on our data. Primary productivity in Harima-Nada has been reported to be 0.54 to 0.97 $gC/m^2/d$ [22]. Particulate amino acid production is calculated using these values and amino acid contents in phytoplankton. A large fraction (63-79%) of produced amino acids is consumed and/or decomposed during sinking down to 5 m above the sea bottom. The benthic organisms thus receive a supply of fresh proteinaceous matter (amino acids) of $\sim$200 $mg/m^2/d$. Of this supply, 64 - 77 % or 15 - 25 % of primary production is further decomposed and/or consumed on or near the bottom surface. Only 5 - 13 % of the amino acids porduced by primary production is buried in the sediment.

ACKNOWLEDGMENTS

The autnors wish to thank Prof. A.Hattori of the Ocean Research Institute, University of Tokyo, for his critical reviewing manuscript. We are greatly indebted to Dr.A.C.Sigleo of U.S. Geological Survey, for kindly advice.

REFERENCES

1. Zeitzschel,B. "Sediment-water interactions in nutrient dynamics, " in *Marine Benthic Dynamics*, K.R.Tenore and B.C.Coull, Eds. (University of South Carolina Press, 1980), pp. 195-218.
2. Yanada,M. and Y.Maita, "Production and decomposition of particulate organic matter in Funka Bay, Japan," *Mar. Chem.* 13: 181-194 (1978).
3. Crisp,P.T., S.Brenner, M.I.Venkatesan, E.Ruth and I.R.Kaplan. "Organic chemical characterization of sediment-trap particles from San Nicholas, Santa Barbara, Santa Monica and San Pedro Basins, California," *Geochim. Cosmochim. Acta* 43: 1791-1801 (1979).
4. Prahl,F.G. and R.Carpenter. "The role of zooplankton fecal pellets in the sedimentation of polycyclic aromatic hydrocarbons in Dabob Bay, Washington," *Geochim. Cosmochim. Acta* 43: 1959-1972 (1979).
5. Deuser,W.G. and E.H.Ross. "Seasonal changes in the flux of organic carbon to the deep Sargasso Sea," *Nature* 283: 364-365 (1980).
6. Wakeham,S.G., J.W.Farington, R.B.Gagosian, C.Lee, H.Debaar, G.E. Nigrelli, B.W.Tripp, S.O.Smith and N.M.Frew. "Fluxes of organic matter from a sediment trap experiment in the equatorial Atlantic Ocean," *Nature* 286: 798-800 (1980).
7. Knauer,G.A. and J.H.Martin. "Primary production and carbon-nitrogen fluxes in the upper 1500 m of the North West Pacific," *Limnol. Oceanogr.* 26: 181-186 (1981).
8. Gagosian,R.B., S.O.Smith and G.E.Nigrelli. "Vertical transport of steroid alcohol and ketones measured in a sediment trap experiment in the equatorial Atlantic Ocean," *Geochim. Cosmochim. Acta* 46: 1163-1172 (1982).
9. Lee,C. and C.Cronin. "The vertical flux of particulate organic nitrogen in the sea: decomposition of amino acids in the Peru upwelling area and the equatorial Atlantic," *J.Mar. Res.* 40: 227-251 (1982).
10. Repeta,D.J. and R.B.Gagasian. "Carotenoid transformations in coastal marine waters," *Nature* 295: 51-54 (1982).
11. Lee,C. and S.G.Wakeham and J.W.Farrington. "Variations in the composition of particulate organic matter in a time-series sediment trap," *Mar. Chem.* 13: 181-194 (1983).
12. Davis,J.M. and R.Payne. "Supply of organic matter to the sediment in the northern North Sea during a spring phytoplankton bloom," *Mar. Biol.* 78: 315-324 (1984).

13. Mopper, K. and E.T. Degens. "Aspects of the biogeochemistry of carbohydrates and proteins in aquatic environments," in *Technical Report, Woods Hole Oceanogr. Inst.*, (Woods Hole, Mass. Reference No.72-68, 1972).
14. Rosenfeld, J.K. "Amino acid diagenesis adsorption in nearshore anoxic sediments," *Limnol. Oceanogr.* 24: 1014-1021 (1979).
15. Montani, S., Y. Maita and S. Fukase. "Flux of nitrogen compounds in coastal marine sediment and pore water (Funka Bay, Hokkaido)," *Chem. Geol.* 30: 35-45 (1980).
16. Smayda, T.J. "Some measurements of the sinking rate of fecal pellets," *Limnol. Oceanogr.* 14: 621-625 (1969).
17. Fowler, S.W. and L.F. Small. "Sinking rates of euphausiid fecal pellets," *Limnol. Oceanogr.* 17: 293-296 (1972).
18. Wiebe, P.H., S.H. Boyd and C. Winget. "Particulate matter sinking to the deep-sea floor at 2000 m in the Tongue of the Ocean, Bahamas, with a description of a new sedimentation trap," *J. Mar. Res.* 34: 341-354 (1976).
19. Turner, J.T. "Sinking rates of fecal pellts from the marine copepod *Pontella meadii*," *Mar. Biol.* 40: 249-259 (1977).
20. Honjo, S. and M.R. Roman. "Marine copepod fecal pellets; production, preservation and sedimentation," *J. Mar. Res.* 36: 45-57 (1978).
21. Okaichi, T. "Significance of amino acid composition of phytoplankton and suspensoid in marine biological production," *Bull. Jap. Soc. Sci. Fish.* 40: 471-478 (1974).
22. Endo, T., H. Koyama and H. Imabayashi. "Phytoplankton in Harima-Nada in relation to oceanographic conditions, particularly nutrient and physicological properties of sea water," in *Fuundamental Studies of the Effects of Marine Environment on the Outbreaks of Red Tide (1980-1981)*. (Report of Research Project, Grant in Aid for Scientific Research B148-R14-8, 1982) pp.81-92.

CHAPTER 3

STABLE ISOTOPE AND AMINO ACID COMPOSITION OF ESTUARINE DISSOLVED COLLOIDAL MATERIAL

A.C. Sigleo
U.S. Geological Survey
431 National Center
Reston, Virginia 22092 U.S.A.

S.A. Macko
Department of Earth Sciences
Memorial University of Newfoundland
St. John's, Newfoundland, Canada

ABSTRACT

Samples from the Patuxent Estuary, Maryland, were collected in surface-water (0.5 m depth) and deep-water (0.5 m above the sediment) pairs during a period of high productivity in July and a period of low productivity in October-November, 1983. Concentrations of total dissolved free amino acids (DFAA) were lower (0.06 to 1.9 μM) than those of dissolved combined amino acids (0.6 to 20 μM). Carbon isotope ratios in colloidal samples averaged -24.8 per mil $\delta^{13}C$ throughout the estuary for both seasons. In the colloidal material ^{15}N was enriched in the surface samples (10.8 per mil average $\delta^{15}N$) relative to deep samples (8.5 per mil average $\delta^{15}N$), whereas particulates (>0.4 μm) from the same stations were uniformly depleted in ^{15}N in surface waters relative to deep-water samples. All of the samples showed a strong seasonality, with higher values for ^{15}N in summer when the deep waters were anoxic. During periods of higher summer productivity, remineralized benthic ammonium as the primary source of nitrogen may be responsible for the ^{15}N enrichments observed in summer samples.

INTRODUCTION

An essential component of estuarine and coastal marine productivity is the recycling of nutrients, particularly those containing nitrogen [1,2]. Nitrogen is commonly the limiting nutrient in coastal United States waters, and a careful understanding of the sources and processes involving nitrogen is required to predict the environmental response to various anthropogenic activities [3,4]. A potential means of tracing nitrogen processes and sources is the distribution of the natural ratio of the two stable isotopes, ^{14}N and ^{15}N [5-7]. Nitrogen isotope ratios in marine and estuarine environments range from -3 per mil $\delta^{15}N$ for some plankton to more than 18 per mil $\delta^{15}N$ for more complex aquatic organisms [8].

The distribution of nitrogen isotope ratios in a well studied estuary where benthic metabolism and seasonal nutrient cycles are

MARINE AND ESTUARINE GEOCHEMISTRY, Sigleo, A. C., and A. Hattori (Editors)

well documented was investigated in the present study. Benthic remineralization of organic matter contributes significantly to the nutrient budget in this estuary [2]. Of the total particulate nitrogen deposited annually in these sediments, Boynton et al. [2] estimated that approximately 34% was returned to the water column as ammonium, 24% was denitrified, and 41% was stored as particulate nitrogen.

Previous studies have examined nitrogen isotopes in aquatic suspended particulate matter, plankton and other aquatic organisms, sediments, and the nutrients nitrate, nitrite, and ammonium [5-14]. In the present study natural abundances of nitrogen isotopes were determined for the first time in the colloidal fraction of dissolved organic matter (1.2 nm to 0.40 um). In addition, nitrogen isotope data for particulates (>0.4 um) from the same samples, along with carbon isotopic data are reported.

The primary organic nitrogen compounds in the biosphere are combined amino acids, which occur as enzymes, peptides, and proteins in cells and extracellular products. Free amino acids are individual monomers formed by the hydrolysis of peptides and proteins from enzymes and bacterial activity, or as plankton intra- and extracellular products. In Patuxent River waters, combined amino acids plus combined ammonium can account for up to 100% of the colloidal nitrogen [15]. The colloidal fraction is a potentially reactive and labile part of the aquatic system due to its large surface area and high nutrient value. As the major organic nitrogen component, amino acid compositions also were determined for these samples.

EXPERIMENTAL METHODS

Water samples were collected at 0.5 m depth and at approximately 0.5 m from the bottom at three stations in the Patuxent River, Maryland, USA, on 21 July 1983 and 2-4 November 1983 (Fig.1). A second sample at Station 2 was collected the shallows on 26 October 1983. Two surface samples were collected near Stations 2 and 3 in April 1983. The stations were selected to obtain a range from fresh to brackish water across the turbidity maximum. The Patuxent, with a discharge of 1.4 to 370 x 10^8 L/day, is a major tributary to Chesapeake Bay. Land use in the Patuxent watershed is primarily underdeveloped forests (50% of the area), farms (35%) and suburbs (15%). The primary industrial development in the Patuxent watershed is a large power plant located near Station 2.

The samples were prefiltered through 0.4 um Nucleopore[1] filters to remove suspended sediment and microorganisms. Dissolved free amino acids (DFAA) were analyzed directly with no preconcentration within 24 hours after collection. Combined

---

1. Use of brand names in this chapter is for identification purposes only and does not constitute endorsement by the U.S. Geological Survey.

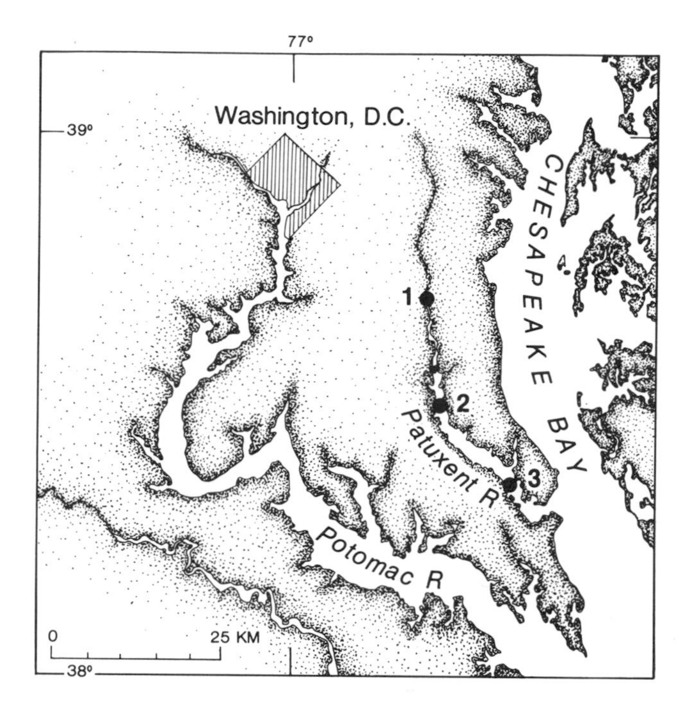

Figure 1. Sample location map

amino acids were determined in 2 ml aliquots which were freeze-dried and hydrolzyed in 6 N HCl at 155°C for 15 minutes in a nitrogen atmosphere. The hydrolozates were evaporated to dryness in a stream of N_2 to remove the HCl, and then brought to 1 ml volume with triple distilled H_2O. Prior to injection, both samples and standards were derivatized by combining 0.2 ml of sample with 0.05 ml of o-phthaldialdehyde, prepared according to Hare [16]. After 2 minutes, the solution was injected onto a 10 cm column containing 3 um C-18 resin (Ranin), and analyzed by reverse phase HPLC [17].

After filtering, samples (10 L) for isotopic analyses were concentrated and desalted using an Amicon hollow-filter cartridge (H1P5, 5,000 approximate molecular weight cutoff). This size range has been defined as a colloidal fraction [15]. The ultrafiltration step was necessary to obtain sufficient desalted sample for isotopic analysis. Particulate fractions (>0.40 um) of the same samples were collected from the Nucleopore filters and lyophilized. All samples were tested with 30% HCl for the possibility of carbonate. Samples were then combusted at 900°C for 1 hour by a modified Dumas technique in sealed quartz tubes according to the method of Macko et al. [18]. The N_2 and CO_2 gases obtained were isolated from the combustion products and analyzed in a Micromass 903E (V.G., Ltd.) instrument. On the basis of prior replicate analyses of samples, the reproducibility in combustion and measurement is within ±0.2 per mil [18]. All data are presented as per mil $\delta^{15}N$ values using atmospheric nitrogen as the standard, or per mil $\delta^{13}C$ using the Chicago PDB carbonate, where

$$\delta^{15}N = [(^{15}N/^{14}N)_{samp}/(^{15}N/^{14}N)_{air} - 1] \times 1000$$
$$\delta^{13}C = [(^{13}C/^{12}C)_{samp}/(^{13}C/^{12}C)_{PDB} - 1] \times 1000$$

The organic nitrogen content of the colloidal material was calculated from the ion intensity of the gas in a calibrated volume of the mass spectrometer, and the organic carbon content was determined on a calibrated manometer.

Temperature, pH, dissolved oxygen and conductivity were determined shipboard with a Hydrolab. Salinity was calculated from conductivity measurements. Total organic carbon (TOC) in unfiltered water, and dissolved organic carbon (DOC) in filtered water were measured with an Oceanography International analyzer using the persulfate method of Menzel and Vaccaro [19].

RESULTS

During the summer months, the lower Patuxent is a stratified estuary with a salt wedge forming in the deeper waters. Oxygen becomes depleted in the deeper waters due to the decay of organic matter from the overlying photic zone (Table I).

The TOC and DOC contents of the surface water ranged from 2.5 to 6.4 mg/L and were higher during the summer by up to 3 mg/L. The deep water samples generally showed less seasonal variation (the high value, 9.1 mg/L, may have resulted from sediment resuspension). The carbon content of the colloidal material recovered varied from 11 to 23% by weight. If compared with previous carbon values for colloidal material from the same river, these values are low, particularly for summer samples which previously showed a range of 26% to 38% carbon [15]. The nitrogen content (1.4 to 4.1%) was correspondingly low compared to previous values (2.7% to 5.9%). It is not clear if these samples are different, or if the differences are analytical, since the previous data were obtained from a CHN analyzer.

The total dissolved free amino acids (DFAA) ranged from 0.3 to 1.9 uM in the July surface waters, and from 0.1 to 0.2 uM in the deep waters (Table II). The high value in the surface waters corresponds with a phytoplankton bloom, although there was also active productivity at the other two stations. The constant low values in deep water samples suggest that these DFAA were consumed as rapidly as they were produced. The most abundant DFAA in the Lower Marlboro surface sample was glutamic acid, which was also the most abundant DFAA in the deep sample from Marsh Point. For the other surface samples, and two of the deep water samples, serine was the most abundant DFAA, followed by glycine. In the samples where serine and glycine were the most

Table I. Water Composition and Physical Parameters

Station	pH	Temp °C	Dissolved Oxygen[a]	Salinity g/kg	Colloidal Material Recovered[a]	Organic Carbon[a] TOC	DOC
PATUXENT RIVER 7/21/83							
Lower Marlboro(1)							
0.5 m depth	6.7	29.3	5.5	0.6	3.8	6.4	5.1
6.7 m depth	6.6	29.2	4.8	0.6	1.9	6.7	4.6
Marsh Pt (2)							
0.5 m depth	7.9	29.4	9.1	0.8	1.9	5.2	4.0
9.5 m depth	6.9	24.8	0.1	13.2	0.8	2.6	-
St. Leonard Ck(3)							
0.5 m depth	7.9	28.9	7.7	9.2	1.8	6.2	3.4
4.0 m depth	7.0	24.4	0.1	16.4	1.0	3.4	2.8
PATUXENT RIVER 10/26/83[b], 11/2/83[c] and 11/4/83[d]							
Lower Marlboro(1)[c]							
0.5 m depth	-	11.6	8.3	0.7	3.0	4.8	3.4
6.5 m depth	-	11.7	8.1	0.7	4.0	9.1	3.5
Marsh Pt (2)[d]							
0.5 m depth	-	13.8	9.4	13.3	1.9	3.0	2.5
9.5 m depth	-	14.3	8.7	15.0	1.0	3.7	2.3
Marsh Pt(2a)[b]							
0.5 m depth	-	14.9	12.7	14.2	1.3	3.0	2.5
2.0 m depth	-	14.0	9.1	14.1	1.4	2.7	2.5
St Leonard Ck(3)[d]							
0.5 m depth	-	14.9	12.7	14.2	1.3	3.0	2.5
4.0 m depth	-	14.0	9.1	14.1	1.4	2.7	2.5

a) Values in milligrams per liter.

abundant DFAA, these two amino acids accounted for 50 to 75% of the total.

The total DFAA concentrations in the October-November samples were uniformly low (0.06 to 0.2 µM). In these samples either phosphoserine or serine was generally the most abundant DFAA, although glutamic acid, alanine, and glycine were also high (Table III). The sum of serine plus phosphoserine comprised 20 to 54% (46% average) of the total in these samples. Because phosphoserine may be associated with certain bacteria [20], these results may indicate a higher level of heterotrophic activity in the late fall samples, or the dominance of a different species.

The dissolved combined amino acids (DCAA) were more abundant than the DFAA by up to a factor of 50 (average 5.4), and in the summer decreased downstream from 20 µM to 2.2 µM in the surface waters, relative to 4.4 µM to 2.9 µM in the deep water samples. These values are slightly higher than those determined previously for colloidal amino acids (1.0 to 5.0 µM) isolated by ultrafiltration [15], suggesting that up to 50% of the DCAA may

Table II. Dissolved Free Amino Acids (DFAA) for 21 July 1983

Amino Acid	Lower Marlboro		Marsh Point[1]		St. Leonard	
	0.5 m	6.7 m	0.5 m	9.5 m	0.5 m	4.0 m
Aspartic Acid	85	8	112	47	70	-
Asparagine	tr	tr	28	tr	8	7.7
Glutamic Acid	97	8	69	73	28	35
Threonine	-	-	97	-	41	-
Serine	60	41	514	34	231	51
Glycine	42	44	358	20	185	30
Alanine	21	15	173	-	88	12
Valine	-	-	73	-	32	-
Methionine	-	-	72	-	-	-
Isoleucine	-	-	54	-	50	-
Leucine	-	-	67	-	36	-
Tyrosine	-	-	73	-	21	-
Phenylalanine	-	-	35	-	13	-
Lysine	-	-	72	-	-	-
Histidine	-	-	88	-	34	92
Arginine	-	-	11	-	11	-
Total DFAA	307	114	1868	174	840	220

Values in nanomoles per liter. 1. Deep Channel

Table III. Dissolved Free Amino Acids for October-November 1983

Amino Acid	Lower Marlboro		Marsh Point[1]		Marsh Point[2]		St Leonard	
	0.5 m	6.7 m	0.5 m	9.5 m	0.5 m	2.0 m	0.5m	4.0m
Aspartic Acid	4	5	18	12	15	7	14	11
Glutamic Acid	3	8	35	14	21	12	24	9
Threonine	-	-	8	9	6	9	10	-
Serine	15	16	21	45	17	14	42	8
Phosphoserine	36	35	11	21	15	18	11	25
Glycine	16	12	12	25	24	21	45	6
Alanine	25	20	26	27	26	28	33	-
Valine	10	8	-	-	-	-	18	-
Methionine	-	-	-	-	-	-	11	-
Isoleocine	-	-	5	8	-	-	6	-
Leucine	-	-	3	5	-	-	10	-
Tyrosine	-	-	3	3	1	7	5	-
Phenylalanine	-	-	-	-	-	-	14	-
Histidine	-	-	21	25	-	6	12	-
Arginine	-	-	6	3	-	5	9	3
Total DFAA	110	104	169	197	125	127	264	61

Values in nanomoles per liter. 1. Deep Channel 2. Shallows

be less than 5,000 molecular weight, (i.e., not trapped by the ultrafilter). For the October-November samples, the total DCAA concentrations were lower, ranging from 2 µM upstream to 0.6 µM in the brackish surface waters, and from 3.7 µM to 0.8 µM for the deep waters (Fig.2). The slightly higher values for some deep stations in late fall may be evidence for increased heterotrophic activity. In the summer surface samples glutamic acid or serine were the most abundant DCAA, whereas in the summer deep water samples and in the late fall samples, the neutral amino acids glycine, alanine, B-alanine and valine were generally more abundant, as illustrated by the data from Marsh Point in Figure 3. Phosphoserine, which may be prevalent in DFAA, is converted mostly to serine during the DCAA hydrolysis step. These variations in DFAA relative abundance are similar to those reported for different seasons [21, 22] and for DCAA during different stages of a phytoplankton bloom [23].

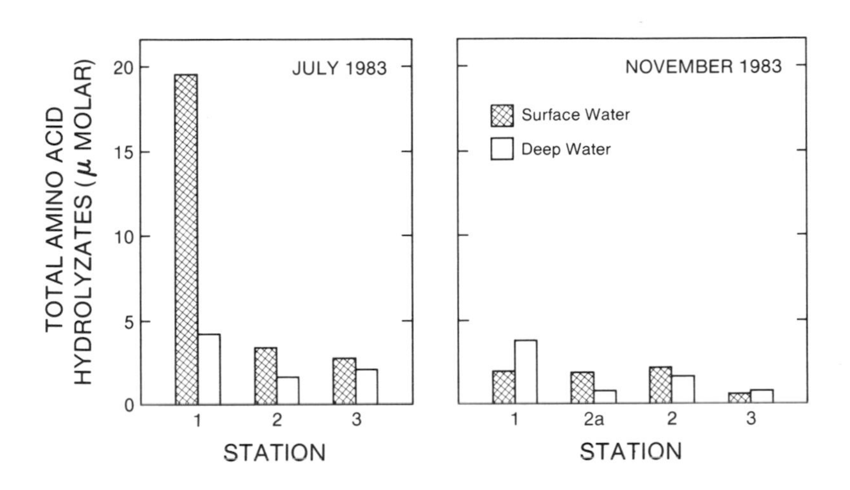

Figure 2. Comparison of July and November combined amino acid totals in surface and deep water samples at Lower Marlboro (1), Marsh Point (2 and 2a) and St. Leonard Creek (3)

Carbon isotope ratios for colloidal samples were relatively constant (-24.8% average $\delta^{13}C$) throughout the estuary for both seasons (Table IV). The carbon isotopic composition of the particulates is more variable than that of the colloidal samples. There is an increase in $\delta^{13}C$ from -24.0 per mil to -21 per mil in July and from -25.8 to -21 per mil in late fall from fresh water (Station 1) to brackish water (Station 3).

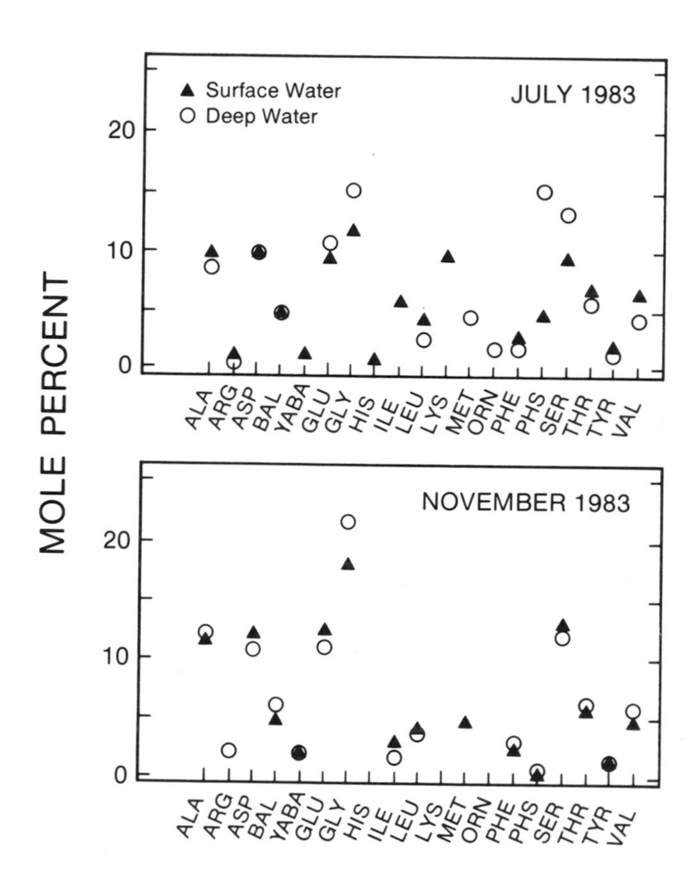

Figure 3. Individual combined amino acid abundances in mole percent for surface and deep water samples at Marsh Point deep channel, Station 2. The amino acids from left to right are alanine, aspartic acid, B-alanine, γ-amino butyric acid, glutamic acid, glycine, histidine, isoleucine, leucine, lysine, methioine, ornithine, phenylalanine, phosphoserine, serine, threonine, tyrosine, and valine.

Table IV. Patuxent River Carbon Isotope Data (per mil $\delta^{13}C$)

	Colloidal Material		Particulate Material	
	7/21/83	11/2-4/83	7/21/83	11/2-4/83
Lower Marlboro(1)				
0.5 m	-24.8	-24.6	-23.9	-25.5
6.7 m	-25.6	-24.4	-22.1	-25.8
Marsh Point(2)				
0.5 m	-24.9	-24.7	-23.6	-25.8
9.5 m	-24.0	-24.9	-22.0	-25.3
Marsh Point(2a)				
0.5 m		-24.6		-25.7
2.0 m		-24.9		-24.5
St. Leonard(3)				
0.5 m	-25.0	-24.7	-21.0	-23.2
4.0 m	-23.5	-25.8	-21.4	-21.9

The colloidal ^{15}N concentrations were uniformly depleted in deep samples (average 8.5 per mil) relative to surface samples (average 10.8 per mil, Table V). The differences between surface and deep water samples were greater in July (3.3 per mil) compared with the October-November data (average difference 1.5 per mil). The slightly enriched deep water ^{15}N values in the fall may reflect a late summer input of plankton debris. The particulate ^{15}N values, however, are consistently lower than the values for colloidal samples (Fig. 4). In addition, the surface values are uniformly lower than the deep particulate values by about 1 per mil. The average difference between particulate and colloidal fractions in surface waters is 4.1 per mil, whereas in deep waters the average difference is 2.0 per mil.

Table V. Colloidal Nitrogen Isotope Data (per mil $\delta^{15}N$)

	7/21/83	11/2-4/83
Lower Marlboro(1)		
0.5 m	10.40	10.10
6.7 m	8.27	9.09
Marsh Point(2)		
0.5 m	10.90	10.90
9.5 m	6.87	9.82
Marsh Point(2a)		
0.5 m		11.30
2.0 m		9.23
St. Leonard Ck(3)		
0.5 m	9.46	11.50
4.0 m	6.92	9.49

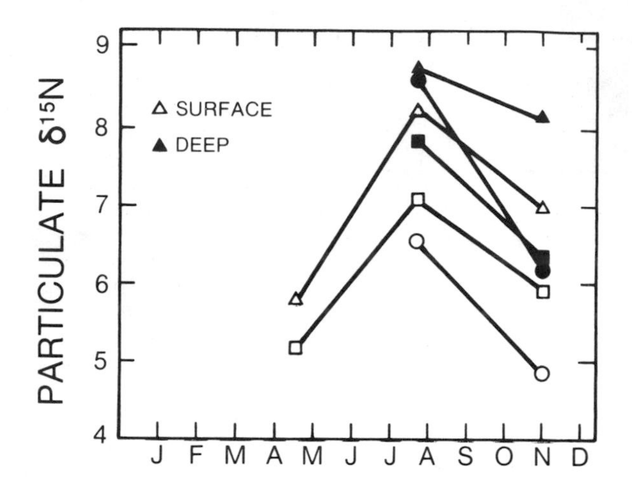

Figure 4. Particulate nitrogen isotope data by month for Lower Marlboro (O), Marsh Point (□), and St. Leonard Creek (Δ).

DISCUSSION AND CONCLUSIONS

The primary sources of DOC in marine and estuarine waters include extracellular release of photosynthetic products by algae, heterotrophic release of enzymes and adhesive materials, zooplankton grazing, and hydrolysis of particulate planktonic debris [24-29]. A number of the extracellular products have been identified, including glycolic acid, carbohydrates, sugar alcohols, sugar phosphates, lipids, free amino acids and proteins (including enzymes). Additional input into the dissolved-colloidal pool occurs during zooplankton feeding, and the zooplankton themselves excrete products which contain soluble components [30,31]. The major process which limits, or controls, DOC concentrations in water is heterotrophic uptake and utilization [24, 27-31].

The range in DFAA and DCAA both in total amount and relative composition suggests a complex succession of phytoplankton and heterotrophic activity with respect to season and location within the estuary. These variations are comparable to the seasonal and successional fluctuations which have been observed in other areas [21-23]. For example, in the Irish Sea, total DCAA concentrations showed major highs of about 1.15 uM in January and late July, during and following periods of high plankton productivity. Low concentrations of 0.1 uM were reported for February, April, and November [21]. Total DFAA concentrations (0.1 to 0.2 uM) were less variable throughout the year, as would be predicted when heterotrophic biomass and activity correlated with phytoplankton extracellular release [21,30]. In other words, the DFAA were utilized at about the same rate as they were formed by extracellular release, or by the hydrolysis of DCAA and particulates.

Much greater seasonal differences were reported in a shallow estuary in Denmark where DFAA concentrations of 0.7 to 2.5 uM occurred in spring and fall, and lower concentrations (0.2 uM) in summer and winter [22]. In addition, the mean turnover times of the five most abundant amino acids, serine, glutamic acid, glycine, ornithine, and alanine, varied from 2 hours in September to 41 hours in December, indicating a reduction of heterotrophic activity in winter [22]. These DFAA concentrations, among the highest reported, are comparable to those of the Patuxent surface waters which varied from 0.1 uM to 1.9 uM.

The greatest range in DCAA concentrations (1.0 to over 8.5 uM) and compositions occurred during a phytoplankton bloom in the North Sea [23]. The succession of species during the bloom was correlated with variations in DCAA composition. Glutamic acid was highest during the early stages of the phytoplankton bloom, histidine toward the end, and aspartic acid, glycine, alanine and lysine peaked after the bloom. Serine was present in large amounts throughout the time interval studied [23].

The DCAA totals for the Patuxent surface waters (4.6 uM average) are generally a factor of 2 higher than the sum of hydrolyzable amino acids in colloidal material isolated by ultrafiltration in a previous study [15]. These results suggested that approximately 50% of the DOC was concentrated by ultrafiltration. The DCAA abundances are more variable than those determined in colloidal material [15], possibly due to the greater range of DFAA in the present samples. In colloidal material, glycine and alanine each comprise approximately 13% of the sample on a molar basis, and aspartic and glutamic acids about 11% each. Cultured phytoplankton, containing little or no detritus, have aspartic-glutamic acid contents of 12 mole percent each, and glycine and alanine contents of 11% each [15]. The slight shift in acidic versus neutral amino acid abundances in all field samples suggests that degradation begins rapidly after the cell dies, perhaps simply with cell lysing.

Previous studies indicate that, with respect to amino acid concentrations and composition, the data from the Patuxent estuary are consistent with measured values from other areas where plankton productivity and extracellular release have been correlated with heterotrophic utilization. In addition, the DCAA reported here are similar in composition to previously determined amino acid abundances in colloidal material from the same estuary [15].

Isotopic Compositions

Variations in organism isotope ratios are related to the primary food source, which in turn is controlled in an estuary by several processes including river flow, primary productivity, and benthic regeneration of nutrients [13]. The particulate data for the Patuxent are comparable to literature carbon isotopic ratios of -26 per mil to -24 per mil for freshwater and estuarine sediments, particulates, and plankton. Higher values (near -21 to -18 per mil) are generally reported for marine sediments and particulates [6,7,34,35]. In the Patuxent, carbon isotopic ratios in the particulates increase from -23.9 per mil $\delta^{13}C$ at the freshwater end of the estuary to -21.0 per mil in the saline region in the summer, and from -25.8 per mil to -21.9 per mil in the fall.

The shift in carbon isotope ratios with increasing salinity has been described frequently in terms of a two member mixing model in which water containing terrestrial inorganic carbon is diluted by water containing marine inorganic carbon [6,7,36]. When simple dilution is the only process involved, the ^{13}C values exhibit a straight line relationship with salinity. However, when in situ processes are significant in comparison to flow rates, nonlinear distributions may result [33-36].

For Chesapeake Bay, Spiker [36] found that the $\delta^{13}C$ of total inorganic carbon increased from -9.5 per mil at the head of the Bay (Susquahanna River) to +1.5 per mil at the mouth [36]. The mid-Bay was significantly enriched in ^{13}C relative to the value predicted by a simple mixing model, indicating that processes other than dilution were significant. Furthermore, the ^{13}C decreased with depth in the water column to -14 per mil. These data were interpreted as an indication that up to 70% of the inorganic carbon was derived from the remineralization of benthic organic matter [36]. In the Patuxent, a subestuary of Chesapeake Bay, neither particulate nor colloidal carbon isotope ratios show a linear relationship with salinity (Table IV, Fig. 5), indicating that local processes, such as benthic remineralization and nutrient recycling have a substantial effect on these isotope ratios.

The nitrogen isotope data for particulates show trends similar to those of particulate carbon. Specifically, the order of increasing ^{15}N enrichment is from freshwater to saline, the heavier isotope is enriched in deep samples relative to samples from surface waters and the isotope ratios do not show a linear relationship to salinity (Fig. 5). Surface particulate samples were uniformly more depleted in ^{15}N than samples from deeper waters. Saino and Hattori [10] also observed an increase with depth in ^{15}N enrichment in particulate matter from the Indian Ocean. The nitrogen isotope ratios for particulate organic nitrogen (PON) in the euphotic zone, where the production of PON exceeded its degradation, were similar to those for phytoplankton. The increase of ^{15}N enrichment with depth was due to bacterial degradation [10]. This explanation is also a reasonable one for Patuxent particulate variations with depth.

The data for nitrogen isotopes show a pronounced seasonality (Fig. 4). The low values for the two April samples could reflect an input of nitrate from fertilizer, which has an isotopic value near zero [6]. Alternatively, the difference in nitrogen isotope ratios for different seasons may be the result of planktonic utilization of a nitrogen source from other processes as discussed below. In the water column, from winter to late spring there is

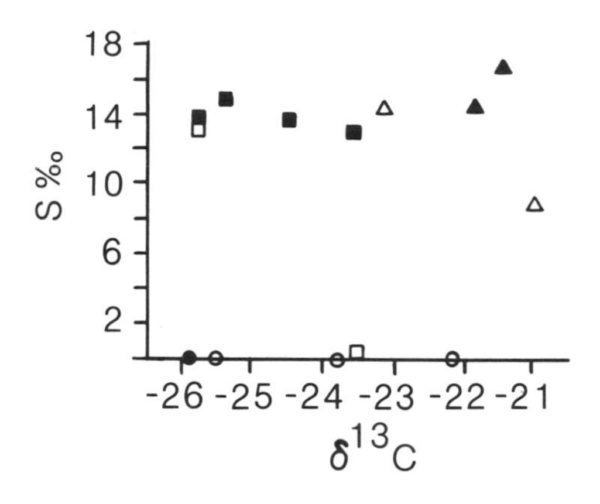

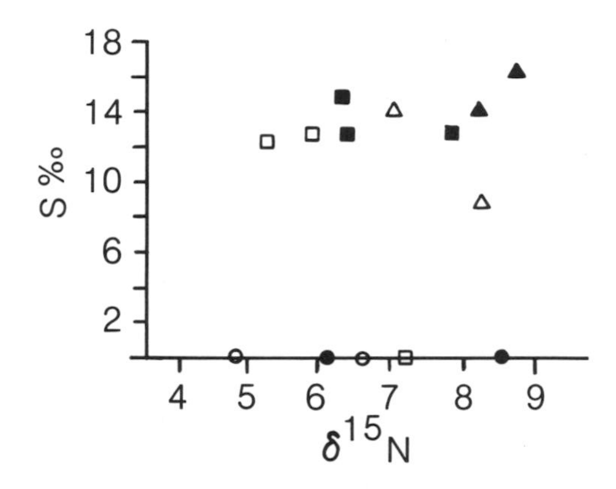

Figure 5. Particulate carbon isotope ratios (top) and particulate nitrogen isotope ratios (bottom) relative to salinity for surface samples (open symbols) and deep samples (closed symbols) at Lower Marlboro(O), Marsh Point (□) and St. Leonard Creek (Δ).

an inverse linear relationship between nitrate concentrations and salinity, indicating that freshwater inflow is the primary source for nitrate [2]. During the summer benthic nitrate fluxes are small when water column concentrations are low, but high and directed into the sediments in winter when water column concentrations are high [2]. Estimates of ammonium turnover rates and phytoplankton rates of uptake led to the conclusion that, unlike nitrate, most ammonium is produced locally. Boynton et al. [2] concluded that nutrient fluxes across the sediment-water interface represent an important nutrient source in summer, and serve as a nutrient sink in winter when demand is low.

The primary process of nitrogen generation from Patuxent sediments in April is denitrification (up to 275 umoles/m^2/hr near Marsh Point in 1984) [37]. In summer, however, ammonium regeneration to the water column is the dominant benthic process (up to 207 umoles/m^2/hr), providing the primary source of nitrogen for plankton productivity [2,37]. During the summer months when the water is stratified, anoxic conditions in the sediments inhibit nitrification, an aerobic process, which eliminates the source of nitrate for denitrification [37,38]. In early summer, as the bottom waters become anoxic, ammonium is no longer converted to nitrate and N_2, permitting large fluxes of ammonium across the sediment-water interface. Thus, by July, the isotopic composition of particulates would be determined mainly by the ammonium transported from depth. The nitrogen isotope data for early November, after the river has destratified and the deeper waters are again oxygenated, show a decrease from the summer high values (Fig. 4) suggesting that denitrification is again a major process. Data on particulate matter from the Scheldt estuary (Belgium) also have been interpreted to show a seasonal trend, although the isotopic differences were more scattered [5].

Patuxent colloidal materials were enriched in ^{15}N in the surface waters and generally were more abundant relative to deeper samples. Colloidal material in surface waters is derived from living and fresh cells [15]. These extracellular products are enriched in the heavier isotope relative to the whole cell, suggesting that, to conserve an isotopic mass balance, planktonic organisims are preferentially releasing material containing the heavier isotope. This suggestion may be justified because intracellular proteins are up to 3 per mil heavier than whole cells [32]. During the sinking of particulates, labile material is consumed, leaving a residual depleted in light isotopes [10]. Additional colloidal material released at depth from the residual particules would become increasingly heavy relative to surface material. During the summer there is less difference between the colloidal and particulate nitrogen isotope values in the deep waters (1.1 per mil) relative to those of surface waters (2.9 per mil).

The mechanisms and fractionation factors have been established for only a few processes involving nitrogen isotopes. In soil, for example, microbial nitrogen isotope fractionation in favor of the light isotope is most significant in the order of nitrification, denitrification and ammonium assimilation [39]. The fractionation of nitrogen isotopes during bacterial utilization of organic nitrogen substrates is thought to occur during the process of transamination in which amino acids entering the bacterial cell are deaminated to form ammonia and organic acids [32]. The ammonia and acids are rearranged to produce glutamic acid, which is then used to form the required protein amino acids. In laboratory experiments cultures utilizing alanine, serine, and threonine were depleted in ^{15}N up to 12.9 per mil relative to the initial composition of the substrate [32]. Bacteria grown on aspartic and glutamic acids, which can be utilized directly, were enriched in ^{15}N relative to the substrate [32]. This experiment suggests that preferential utilization of specific amino acids could produce isotopically heavy ammonium, which, when utilized by phytoplankton would result in ^{15}N enriched particulates.

Isotopically heavy ammonium in summer has been reported in the Scheldt estuary where the nitrogen isotope ratios can vary from 10 per mil upstream to 29 per mil downstream. In this system, the process responsible for the fractionation was nitrification, which left a residual of increasingly heavy ammonium [5]. In the Patuxent, the high summer enrichments of ^{15}N also may be a product of nitrification. During the winter and early spring, residual benthic ammonium would become heavier due to nitrification. After the transition to anoxic benthic conditions, the resulting flux of ammonium would be strongly enriched in ^{15}N.

The amount of isotopic fractionation in organisms varies depending on the growth rate, growth phase, light intensity, nitrogen source, and species [33], which complicates the interpretation of estuarine data. Furthermore, the present data set cannot be interpreted completely without data for nitrate, ammonium and dissolved material, in addition to data from intermediate points between surface and deep samples. In particular, intermediate points might provide information on the amount of deep material derived from sediment resuspension, as well as the amount of degradation occurring at the picnocline.

The Patuxent in the summer is one of the most productive estuaries for which data are available [3]. This high rate of productivity (up to 43 mg/m^3 chlorophyll *a*) is reflected in high DCAA and DFAA concentrations in the water column. A factor which promotes the high productivity is efficient and rapid recycling of nutrients between primary producers and heterotrophs [1,2].

The increase in ^{15}N contents in summer particulates can be correlated with a reduction of nitrification in Patuxent sediments during the summer. The decrease in nitrification results in a large benthic flux of ammonium to the water column. The carbon isotope ratios for particulates also show seasonal differences, which may be produced by benthic remineralization of nutrients. The results of this study indicate that different *in situ* processes produce distinct isotope ratios which may be useful for tracing nitrogen and carbon cycles in aquatic systems.

ACKNOWLEDGMENTS

We thank A. J. Bottenfield and R. H. Coupe for assistance in data reduction and computer graphics. W.R. Boynton, Chesapeake Biological Laboratory, and G.R. Helz, Department of Chemistry, the University of Maryland, kindly provided samples (April) and boat time (July). The manuscript has benefited from helpful reviews by T. Saino, E. T. Sundquist and M. J. Baedecker.

REFERENCES

1. Nixon, S. W. "Remineralization and Nutrient Cycling in Coastal Marine Ecosystems," in Estuaries and Nutrients. Neilson, B. L., and Cronin, L. E., eds. (The Humana Press, Clifton, N.J.), (1981), pp. 111-138.
2. Boynton, W. R., W. M. Kemp, and C. G. Osborne, "Nutrient Fluxes Across the Sediment-Water Interface in the Turbid Zone of a Coastal Plain Estuary," in Estuarine Perspectives, Kennedy, V. S., ed. (Academic Press, N.Y.), (1980), pp. 93-109.
3. Boynton, W. R., W. M. Kemp, and C. W. Keefe. "A Comparative Analysis of Nutrients and Other Factors Influencing Estuarine Phytoplankton Production, in Estuarine Comparisons, Kennedy, V. S., ed. (Academic Press, N.Y.), (1982), pp. 69-90.
4. Ryther, J. H., and W. M. Dunstan. "Nitrogen, Phosphorus, and Eutrophication in the Coastal Marine Environment," Science. 171:1008-1013 (1971).
5. Mariotti, A., C. Lancelot, and G. Billen. "Natural Isotopic Composition of Nitrogen as a Tracer of Origin for Suspended Organic Matter in the Scheldt Estuary," Geochim. Cosmochim. Acta. 48:549-555 (1984).
6. Crisp, P. T., S. Brenner, M. I. Venkatesan, E. Ruth, and I. Kaplan. "Organic Chemical Characterization of Sediment-Trap Particulates from San Nicolas, Santa Barbara, Santa Monica, and San Pedro Basins, California," Geochim. Cosmochim. Acta. 43:1791-1801 (1979).
7. Peters, K. E., R. E. Sweeney, and I. R. Kaplan. "Correlation of Carbon and Nitrogen Stable Isotope Ratios in Sedimentary Organic Matter," Limnol. Oceanogr. 23:598-604 (1978).
8. Minagawa, M., and E. Wada. "Stepwise Enrichment of ^{15}N Along Food Chains: Futher Evidence and the Relation Between ^{15}N and Animal Age," Geochim. Cosmochim. Acta. 48:1135-1140 (1984).
9. Wada, E., and A. Hattori. "Natural Abundance of ^{15}N in Particulate Organic Matter in the North Pacific Ocean," Geochim. Cosmochim. Acta. 40:249-251 (1976).
10. Saino, T., and A. Hattori. "^{15}N Natural Abundance in Oceanic Suspended Particulate Matter," Nature. 283:752-754 (1980).
11. Macko, S. A., M. L. F. Estep, P. E. Hare, and T. C. Hoering. "Stable Nitrogen and Carbon Isotopic Composition of Individual Amino Acids Isolated from Cultured Micro-Organisms," in Annual Report of the Director Geophysical Laboratory, Carnegie Institution, 1982-1983, pp. 404-410 (1983).
12. Pang, P. C., and J. O. Nriagu. "Isotopic Variations of the Nitrogen in Lake Superior," Geochim. Cosmochim. Acta. 41:811-814 (1977).

13. Macko, S. A. "Source of Organic Nitrogen in Mid-Atlantic Coastal Bays and Continental Shelf Sediments of the United States: Isotopic Evidence," in Annual Report of the Director Geophysical Laboratory, Carnegie Institution, 1982-1983, pp. 390-394 (1983).
14. Yoshida, N., A. Hattori, T. Saino, S. Matsuo, and E. Wada. "$^{15}N/^{14}N$ Ratio of Dissolved N_2O in the Eastern Tropical Pacific Ocean," Nature. 307:444 (1984).
15. Sigleo, A. C., P. E. Hare, and G. Helz. "The Amino Acid Composition of Estuarine Colloidal Material," Estuar. Coast. Shelf Sci. 17:87-96 (1983).
16. Hare, P. E. "Subnanomole-Range Amino Acid Analysis," in Methods in Enzymology. 47: Enzyme Structure, Part E., Hirs, C. H. W., and Timasheff, S. N., eds. (Academic Press, N.Y.), (1977), pp. 3-183-18 (1977).
17. Jones, B. N., S. Paabo, and S. Stein. "Amino Acid Analysis and Enzymatic Sequence Determination of Peptides by an Improved o-Phthaldialdehyde Precolumn Labeling Procedure," Jour. Liquid Chromatogr. 4(4):565-586 (1981).
18. Macko, S. A., M. F. Estep, and T. C. Hoering. "Nitrogen Isotope Fractionation by Blue-Green Algae Cultured on Molecular Nitrogen and Nitrate," in Annual Report of the Director Geophysical Laboratory, Carnegie Institution, 1981, pp. 413-417 (1982).
19. Menzel, D. W., and R. F. Vaccaro. "The Measurement of Dissolved and Particulate Carbon in Seawater," Limnol. Oceanogr. 9:138-142 (1964).
20. Stanier, R. Y., E. A. Adelberg, and J. L. Ingraham. The Microbial World, 4th Ed (Prentice Hall, 1976) 871 pp.
21. Riley, J. P., and D. A. Seger. "The Seasonal Variations of the Free and Combined Dissolved Amino Acids in the Irish Sea," Jour. of the Mar. Biol. Assoc. of the U.K. 50:713-720 (1970).
22. Jorgensen, N. O. G. "Heterotrophic Assimilation and Occurrence of Dissolved Free Amino Acids in a Shallow Estuary," Mar. Ecol. Prog. Ser. 8:145-159 (1982).
23. Ittekot, V. "Variations of Dissolved Organic Matter During a Plankton Bloom: Qualitative Aspects, Based on Sugar and Amino Acid Analyses," Mar. Chem. 11:143-158 (1982).
24. Gagosian, R. B., and C. Lee. "Process Controlling the Distribution of Biogenic Organic Compounds in Seawater," in Marine Organic Chemistry, Duursma, E. K., and Dawson, A., eds. (Elsevier), (1981), pp. 91-123.
25. Hellebust, J. A. "Excretion of Some Organic Compounds by Marine Photoplankton," Limnol. Oceanogr. 10:192-206 (1965).
26. Fogg, G. E. "The Extracellular Products of Algae," Oceanogr. Mar. Biol. Ann. Rev. 4:195-212 (1966).
27. Storch, T. A., and G. W. Saunders. "Phytoplankton Extracellular Release and its Relation to the Seasonal Cycle of Dissolved Organic Carbon in a Eutrophic Lake," Limnol. Oceanogr. 23:112-119 (1978).

28. Lampert, W. "Release of Dissolved Organic Carbon by Grazing Zooplankton," Limnol. Oceanogr. 23:831-834 (1978).
29. Crawford, C. C., J. E. Hobbie, and K. L. Webb. "The Utilization of Dissolved Free Amino Acids by Estuarine Microorganisms," Ecology. 55:551-563 (1974).
30. Keller, M. D., T. H. Mague, M. Badenhausen, and H. E. Glover. "Seasonal Variations in the Production and Consumption of Amino Acids by Coastal Microplankton," Estuar., Coast. Shelf Sci. 15:301-315 (1982).
31. Bell, W. H. "Bacterial Utilization of Algal Extracellular Products. 3. The Specificity of Algal-Bacterial Interaction," Limnol. Oceanogr. 28:1131-1143 (1983).
32. Macko, S. A., and M. L. F. Estep. "Microbial Alteration of Stable Nitrogen and Carbon Isotopic Compositions of Organic Matter," in Annual Report of the Director Geophysical Laboratory, Carnegie Institution, 1982-1983, pp. 394-398 (1983).
33. Wada, E., and A. Hattori. "Nitrogen Isotope Effects in the Assimilation of Inorganic Nitrogenous Compounds by Marine Diatoms," Geomicrobiol. Jour. 1:85-101 (1978).
34. Tan, F. C., and P. M. Strain. "Sources, Sinks and Distribution of Organic Carbon in the St. Lawrence Estuary, Canada," Geochim. Cosmochim. Acta. 47:125-132 (1982).
35. Hedges, J. I., H. J. Turin, and J. R. Ertel. "Sources and Distributions of Sedimentary Organic Matter in the Columbia River Drainage Basin, Washington and Oregon," Limnol. Oceanogr. 29:35-46 (1984).
36. Spiker, E. C. "The Behavior of ^{14}C and ^{13}C in Estuarine Water: Effects of In Situ CO_2 Production and Atmospheric Exchange," Radiocarbon. 22:647-654 (1980).
37. Twilley, R. R., W. M. Kemp, W. R. Boynton, and J. C. Stevenson. "Significance of Sediment Denitrification to the Nitrogen Budgets of Two Tributaries of the Chesapeake Bay," (abstract) Trans. Am. Geophys. Union, EOS. 65:914 (1984).
38. Jenkins, M. C., and W. M. Kemp. "The Coupling of Nitrification and Denitrification in Two Estuarine Sediments," Limnol. Oceanogr. 29:609-619 (1984).
39. Delwiche, C. C., and P. L. Steyne. Nitrogen isotope fractionation in soils and microbial reactions. Environ. Sci. Tech. 4:929-935 (1970).

CHAPTER 4

CLUES TO THE STRUCTURE OF MARINE ORGANIC MATERIAL FROM THE STUDY OF PHYSICAL PROPERTIES OF SURFACE FILMS*

William R. Barger
Chemistry Division, Surface Chemistry Branch
Naval Research Laboratory
Washington, D. C. 20375

Jay C. Means
Department of Chemistry, University of Maryland
Chesapeake Biological Laboratory
Solomons, Maryland 20688

ABSTRACT

Naturally-occurring surface-active organic films reduce the surface tension of water samples. Films adsorbed on the surface of water collected from 16 Atlantic and 8 Chesapeake Bay stations were studied in detail. Film pressure vs. area characteristics were determined. A modified van der Waals type equation that describes the data when the films are modeled as two-dimensional gases suggests an effective size range for the molecules that make up the films. The model enables the number of moles of film-forming material in each sample and the molecular weight of this material to be estimated. Amounts on the surface of 450 ml water samples ranged from 14×10^{-10} to 119×10^{-10} moles. Molecular weights ranged from 1400 to 4900. Coefficients of compressibility were also determined. An average value of 0.054 ± 0.009 cm/dyne was found. Chemical analyses of surface microlayer films have often found fatty acids, triglycerides, or other monolayer-forming compounds. However, when our physical data for natural films are compared to data for films of a series of pure compounds, the general results indicate that natural films are not composed primarily of free fatty acids, alcohols, or hydrocarbons. More oxygenated molecules of higher molecular weight are indicated.

INTRODUCTION

For more than thirty years it has been recognized that visible marks on the surface of natural water bodies can be caused by naturally-occurring surface-active organic compounds [1,2]. In recent years there has been a renewed interest in this phenomenon from geochemists and geophysicists using remote sensing instruments to make environmental observations from

---

* Contribution No. 1605, Center for Environmental and Estuarine Studies of the University of Maryland.

MARINE AND ESTUARINE GEOCHEMISTRY, Sigleo, A. C., and A. Hattori (Editors)

space and from aboard high-flying aircraft. Occasional unexplained images are recorded by remote-sensing instruments in regions where naturally-occurring surface films have been previously reported.

To determine the causes of these surface marks, a number of investigators have collected sea surface samples for chemical analysis [3,4,5,6]. There are, of course, countless compounds that might be present, so each investigator has sought the compounds that he thought likely to be present, or those which he had the capability to analyze. A review of general data on the composition of the sea-surface microlayer is contained in the chapter by Liss [7], and more specific details on surface films are given by Hunter and Liss [8]. To complicate matters, there is no really satisfactory method for collecting a sample for analysis from the sea surface, although many different methods have been used [9,10]. Since sampling methods may influence the results, there is disagreement among investigators about the importance of different classes of compounds.

Because of the considerations mentioned above, we were interested in applying the techniques used by surface chemists studying monomolecular films of pure compounds on water to films formed on water samples collected from the sea surface. Earlier work on natural surface films described briefly the physical properties of the films [11]. The purpose of this paper is to describe the apparent chemical properties of surface-active material found on natural bodies of water on the basis of physical measurements on the surface films. At best this method is inexact, but it can provide some information contributing to the understanding of properties and composition of natural surface films.

For a series of films formed on water samples collected from the field, calculations were made of the number of moles of surface-active material present on the surface, the surface area occupied per molecule of film material, and the approximate molecular weight of this material. The coefficient of compressibility for all films was also determined.

Sampling Locations

Samples were collected on two separate cruises, one during August, 1981, from 8 stations surrounding Hart and Miller Islands in Chesapeake Bay and another cruise during July, 1982, in the Atlantic Ocean from Baltimore to Providence, Rhode Island, and back to the York River in Chesapeake Bay (Fig. 1).

Sample Collection and Analysis Techniques

Three methods were used to collect 500 ml water samples for surface film studies. Water from the air/water interface was collected with a screen, and bulk water was collected by Niskin bottle casts or from a bow-penetration pumping system aboard the USNS Hayes. Samples were placed in 1-liter glass-stoppered bottles for transport to the laboratory. This volume was chosen so that nearly the entire sample could be transferred to the 450 ml capacity film balance. Samples that could not be run immediately were poisoned with 30 ul of a solution of 1 gm of sodium azide per 30 ml of HPLC water and stored at 5 ^{o}C.

Surface water from Chesapeake Bay was captured between the wires of a 73.5 cm x 57.2 cm horizontal aluminum screen. Several dips were necessary to recover 500 ml from the surface. From the area of the screen and the volume of water collected, the thickness of the recovered layer was found to be 292 $\pm$ 22 um.

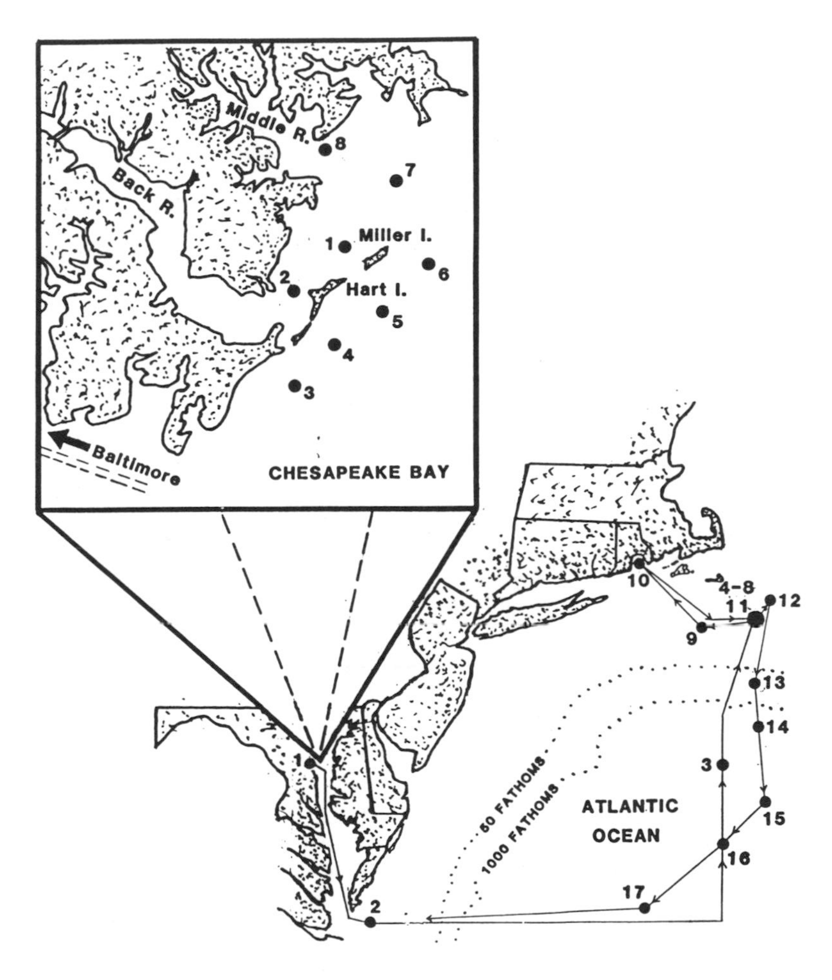

Figure 1. Field sampling locations in Chesapeake Bay and the Atlantic Ocean. Samples 4-8 and 11 are from Phelps Bank.

Even if all the water on the screen was recovered from only the 300 um layer near the air/water interface, a condition only achieved in perfectly calm conditions, the sample would still contain a rather large proportion of bulk water compared to interfacially adsorbed material. Nevertheless, enrichments at the surface are reflected in samples collected by this method. The merits of this collection method compared to others have been reviewed in detail by Huhnerfuss [10]. It has been the most commonly used method for collecting water from the air/water interface. A smaller screen, 51.5 x 57.2 cm, was fabricated for use aboard the USNS Hayes in a 60 cm x 60 cm x 15 cm polyethylene trough into which water was pumped and allowed to stand for five minutes before sampling.

Aboard the Hayes, a continuous supply of water was pumped to the wet laboratory at a rate of 15 liters/min, and periodically 500 ml samples of this water were collected. All samples were taken while the ship was underway or just prior to stopping, to avoid the possibility of contamination from the ship. Clean, 2.5-cm diameter polyethylene tubing was used to line the system just prior to each cruise. The tubing passed from just outside the bow, at a depth of 7 meters, to a magnetically-driven, teflon-lined pump and then to the laboratory. The pump was turned on at the beginning of the cruise and kept on until the end to keep the system clean. Subsurface water was also collected in Chesapeake Bay and the Atlantic with polyvinyl chloride sampling bottles.

High-molecular-weight colloidal material from 5 water samples of 40 liters each was concentrated into 1-liter samples by ultrafiltration after prefiltration through Gelman type A-E, 47-mm, glass-fiber filters (precombusted at 400° overnight) which were backed with 47-mm, 0.4 um Nuclepore filters. The sample was then recycled through and Amicon H_1P_5 (5000 molecular weight cut off) hollow fiber ultrafiltration system [12] and reduced to 1 liter aboard ship. Theoretically, the soluble material that passed through the 0.4 um Nuclepore filter and that had a molecular weight greater than 5000 was enriched in the 1-liter samples 40-fold.

Sample names indicate the stations from which samples were collected and the type of sample. For example, CS-1 indicates a sample collected from Chesapeake Bay by a screen at station 1 near Miller Island, whereas AB-15c indicates an Atlantic bulk water sample from station 15.

Film Pressure vs. Area Measurement Techniques

A film balance consisting of a shallow rectangular tray to hold water, a sensor to measure changes in surface tension, and a bar for reducing the area of the water surface, was used to study physical properties of films formed on natural waters samples. Good reviews of such devices and the data that they can provide are given by Gaines [13] and by Gershfeld [14]. Our apparatus consisted of a rectangular solid glass tray with a 12 cm x 70 cm trough etched into it 3 mm deep. It was coated with paraffin to make it non-wettable. It held approximately 450 ml of water which rose above the 3mm walls but did not spill out of the tray because of the coating. A paraffin-coated glass bar was placed across the tray to confine the surface film. When the bar was driven forward with a motorized screw, the area of water surface over which film had spread decreased, and a change in surface tension was measured. Surface tension was measured with a 1 cm^2 platinum foil by the Wilhelmy plate technique [13]. The motor drive and the Statham UC2 strain gauge which acts as an electrobalance for the platinum Wilhelmy plate were interfaced to a small microcomputer.

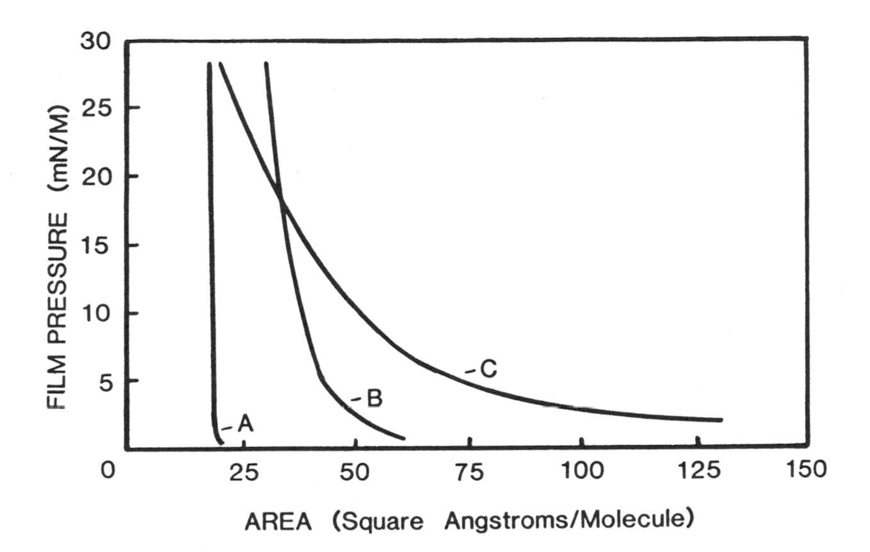

Figure 2. Typical film pressure vs. area per molecule curves for solid (A), liquid (B), and gas-like (C) monomolecular films on water. Curves for stearyl alcohol (A), oleyl alcohol (B), and polyethylene glycol (300) monolaurate are shown.

The reduction in surface tension which results from the compression of the surface film is interpreted as a film pressure. Film pressure, P, is defined as the difference between the surface tension of clean water, Sw, and water covered by a film, Sf.

$$P = Sw - Sf \tag{1}$$

The isotherm obtained upon compression of a surface film is called a force-area curve and is analogous to a pressure vs. volume isotherm for a gas. Two-dimensional equations of state for surface films analogous to the equations of state commonly used to describe gases can be written [13,15,16]. Arguments presented here are based on this analogy.

Equations of State

Figure 2 is an illustration of three typical film pressure vs. area diagrams. The three curves shown illustrate solid (A), liquid (B) and gaseous-type (C) monolayer films. The simplest equation of state to describe such films is the two-dimensional analog of the ideal gas law:

$$PA = kT \qquad (2)$$

where k is the Boltzmann constant and T the temperature in degrees Kelvin. When A is the area in square Angstroms per molecule and P is in millinewtons per meter, kT = 411.6 at 25° C. Langmuir [17] used a modified version of Equation 2 to account for the finite area, Ao, occupied by the adsorbed hydrophilic head groups of surface-active molecules:

$$P(A - Ao) = kT. \qquad (3)$$

This is also known as the Volmer equation of state [15]. The Ao here is analogous to the volume correction in the van der Waals equation for gases. Equation 3 gives a much better fit to the data for monolayer films than Equation 2 but still is not very satisfactory since the PA product, even when corrected by Ao, seldom gives the accepted value for k.

A further refinement was used by Schofield and Rideal [18]:

$$P(A - Ao) = XkT \qquad (4)$$

where X is related to the forces of interaction in the monolayer.

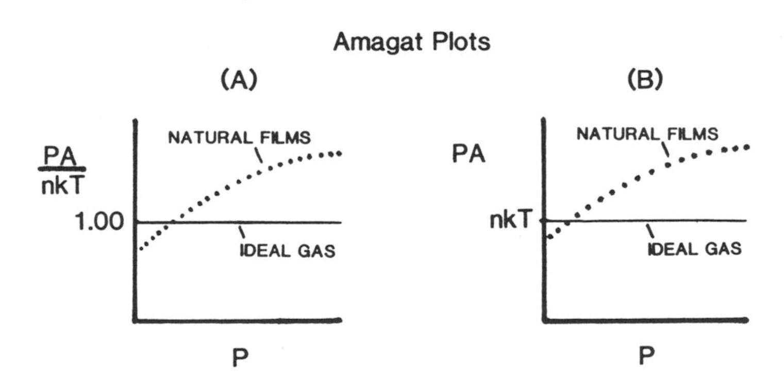

Figure 3. Amagat plots for natural surface films compared to the ideal gaseous monolayer.

Real gases deviate from the ideal gas law, but it is still the primary equation used for routine chemical calculations. PV/nRT may be plotted vs. P to display deviations from ideality in an Amagat diagram [15]. Since PV = nRT, a graph of this type for the ideal gas would give a horizontal line intercepting the PV/nRT axis at 1.00. The analogous approach can be taken with data for monolayer films where PA/nkT is plotted against P. Here, P is film pressure and A is area occupied by the film. An intercept at 1.00 would be expected as illustrated in Figure 3a.

We shall make a slight modification and plot PA vs P such as shown in Figure 3b. This simply has the effect of changing the intercept for the ideal film to nkT from 1.00. This form allows data for natural films to be examined. The primary difficulty in examining natural films is that almost nothing is known about them except the variation in film pressure with the area confining the film. The number of molecules, molecular weight, and mass of the sample are unknown.

For the natural films examined, curves such as depicted in Figure 3 are obtained, and a very good fit for the values of PA vs P can be made with a least-squares best fit parabola through the data. The data are well represented by the following quadratic equation of state.

$$PA = C_0 + C_1P + C_2P^2 \qquad (5)$$

where C_0, C_1, and C_2 are constants, and A is the actual area (not area per molecule). The constants determined by fitting this equation to the field-collected samples are shown in the Results section in Table III. The film pressure vs. area diagram can be easily reconstructed from the equation obtained by dividing through by P:

$$A = {C_0}/_P + C_1 + C_2P \qquad (6)$$

And the relationship of Equation 5 to Equation 4 is seen more clearly by a slight rearrangement of Equation 5:

$$P(A - C_1) = C_0 + C_2P^2 \qquad (7)$$

When area instead of area per molecule is used in Equation 4, XkT is replaced by XnkT. Therefore, C_1 can be interpreted as the finite area occupied by the molecules in the film, and C_0 can be taken as XnkT in the limiting case when P approaches zero.

$$C_0 = XnkT \qquad (8)$$

Since C_0 is easily determined from film pressure vs. area measurements, if X were known, then the number of molecules in an unknown film could be determined.

Determination of the Interaction Constant, X

A reasonably good estimate for the value of X in Equation 8 can be made. A series of known compounds were studied in the film balance, and data for others were computed from film pressure vs. area curves in the literature to compare them to natural films. Equation 5 with A in units of area per molecule was applied to this series of known compounds and fitted the data very well -- even for the solid type films. Constants for the known compounds are listed in Table I. The value of kT is also listed in corresponding units.

Table I. Equation Constants for Pure Compounds

Type	Compound	C_0	C_1	C_2	Ref.
SOLID	Stearyl Alcohol	0.82	19.0	-0.044	TW
	Cholesterol	0.80	37.3	-0.054	TW
	Cholesterol	1.08	40.5	-0.064	[13]
	Stearic Acid	2.60	20.7	-0.073	TW
	Tristearin	6.40	59.1	+0.020	[13]
LIQUID	Triolein	9.61	114.	-1.57	[13]
	Chlorophyll-A	11.1	117.	-1.73	[13]
	Vitamin A	12.1	35.7	-0.33	[13]
	Sorbitan Monooleate	14.7	70.5	-1.29	TW
	Oleyl Alcohol	23.5	36.0	-0.29	[13]
	Oleyl Alcohol	23.9	38.9	-0.35	TW
	Vitamin A	45.2	31.0	-0.57	TW
	Sucrose Monostearate	53.8	52.8	-0.53	TW
GAS	Sucrose Monomyristate	134.	41.8	-0.63	TW
	PEG(200)Monolaurate	185.	25.1	-0.52	TW
	PEG(300)Monolaurate	176.	43.3	-1.05	TW
	POE(6)Dodecanol	207	111.	-3.13	[19]
	POE(8)Dodecanol	399	136.	-4.49	[19]
Ideal Gas:	$P(A - Ao) = kT =$	411.6	Ao	0.00	at 25°C

Units: C_0 : ergs x 10^{-16}/molecule
C_1 : square Angstroms / molecule.
C_2 : cm^3 x 10^{-16}/dyne molecule.

TW = This work -- Run in same apparatus as natural films.

To compare the curves for known compounds in Table I to curves for the natural films, the quantity of known film required to give the same area in the film balance as a typical unknown film, CS-1, at a film pressure of 20 mN/M was determined, and the curve for that amount of known material was plotted in Figure 4. The amount of each compound required to normalize the curves in this way is shown in Table II.

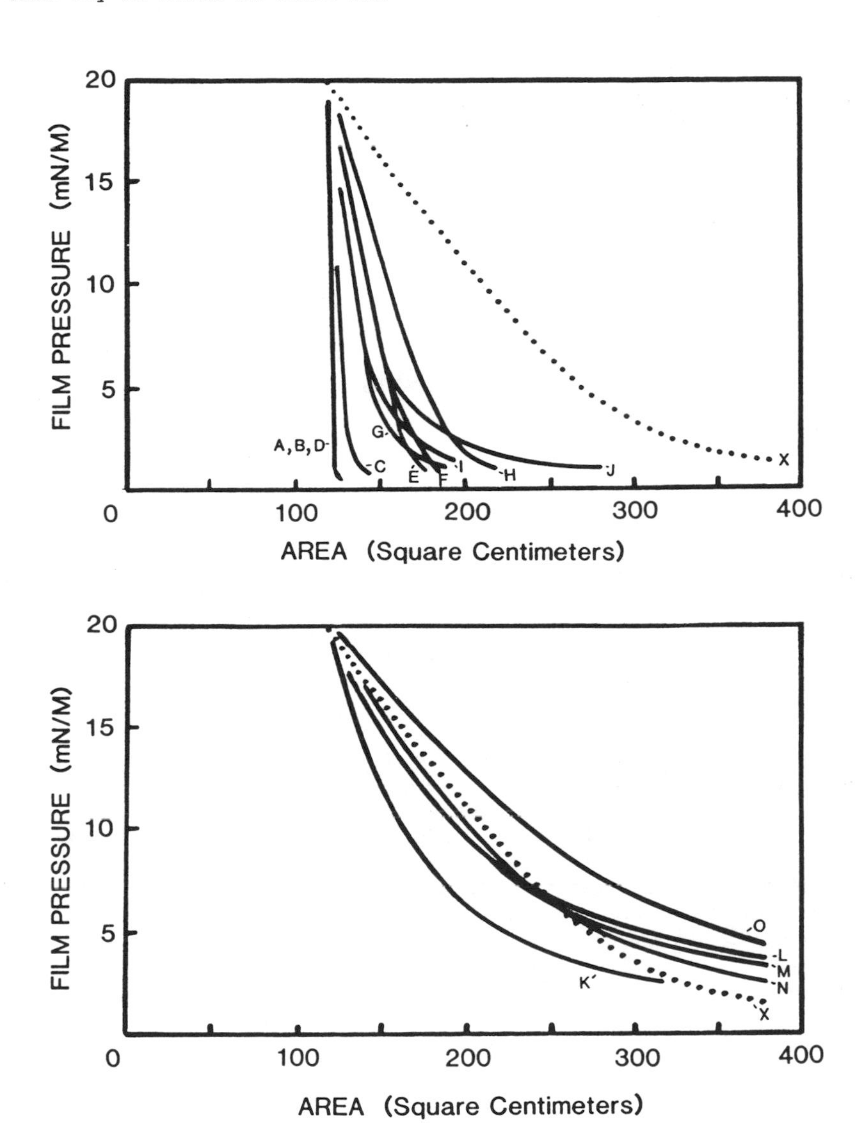

Figure 4. Film pressure vs. area in square centimeters for known quantities of pure compounds of the solid and liquid type (top) and of the gaseous type (bottom) compared to an unknown quantity of natural sample CS-1 (X). Pure compounds are identified in Table II.

Table II. Mass of Pure Compounds Graphed in Figure 4

Compound		Micrograms
A	Stearyl Alcohol	29.6
B	Cholesterol	21.1
C	Stearic Acid	29.1
D	Tristearin	29.6
E	Triolein	21.2
F	Chlorophyll-A	21.4
G	Vitamin A	19.1
H	Sorbitan Monooleate	20.3
I	Oleyl Alcohol	16.1
J	Sucrose Monostearate	26.9
K	Sucrose Monomyristate	30.6
L	PEG(200)Monolaurate	32.6
M	PEG(300)Monolaurate	31.5
N	POE(6)Dodecanol	15.2
O	POE(8)Dodecanol	16.1
	Mean:	24.0 ± 6.2 ug

This mass produces a surface film that occupies the same area as sample CS-1 at a film pressure of 20 mN/M. (119.5 cm^2)

The curve for Chesapeake Bay Sample CS-1 coincides with a curve based on the mean values of the constants for all CS samples. It should be realized that none of the plots for solid or liquid-type films in Figure 4 can ever be made to coincide with the plot for CS-1 by any choice of the amount of material placed on the surface. More material just displaces the curve to the right, effectively multiplying all of the X-axis values by a constant. The natural sample, CS-1, most closely resembles the gaseous-type films shown in Figure 4. Curves for POE(6)Dodecanol, PEG(200)Monolaurate, and PEG(300)Monolaurate are closest in form to the curve for CS-1. The C_0 values for these compounds in Table I are 207, 176, and 185 respectively, vs. 411.6 in the same units for the kT value of the two-dimensional ideal gas. Since these values are all about 0.5 kT, we used 0.5 for the value of X in Equation 8. Therefore,

$$C_0 = 0.5 \text{ nkT}. \qquad (9)$$

RESULTS

A measurable film formed on the surface of every water sample that was collected. Surface films on all the samples were remarkably similar in physical characteristics.

Table III. Constants for Pressure vs. Area Plots and Number of Moles and Molecular Size Estimates from the Gas Model

Sample	C_0	C_1	C_2	M x 10^{10} (moles)	Ao (A^2/molec)
Chesapeake Bay Screen-Collected Water					
CS-1	180.6	277.9	-8.37	145.8	316
CS-2	96.8	172.1	-4.59	78.0	366
CS-3	127.2	410.5	-10.54	102.6	664
CS-4	90.3	154.9	-4.06	72.8	353
CS-5	96.4	151.0	-4.98	77.8	322
CS-6	99.8	244.3	-6.66	80.6	504
CS-7	163.4	428.8	-13.68	131.8	540
CS-8	325.2	414.8	-13.51	262.4	262
Chesapeake Bay Bulk Water					
CB-1	37.2	131.4	-4.66	30.0	727
CB-2	52.7	144.0	-3.82	42.6	562
CB-3	60.3	193.8	-5.59	48.6	661
CB-4	69.7	77.9	-2.29	56.2	230
CB-5	62.2	100.2	-2.33	50.2	331
CB-6	37.3	121.9	-3.37	30.0	673
CB-7	104.8	103.6	-2.74	84.6	204
CB-8	79.5	208.7	-4.59	64.2	540
Atlantic Screen-Collected Water					
AS-2	551.2	323.6	-8.84	444.8	121
AS-3	257.0	293.7	-7.93	207.4	235
AS-4	96.0	129.5	-3.94	77.4	278
AS-5	71.6	111.4	-3.19	57.8	322
AS-6	96.1	91.5	-2.37	77.6	196
AS-7	60.7	82.3	-2.28	49.0	278
AS-8	27.5	73.0	-2.29	22.2	546
AS-11	45.0	83.6	-2.13	36.4	382
AS-12	27.1	76.4	-2.20	21.8	580
Atlantic Bulk Water Samples					
AB-1	71.2	83.1	-2.26	57.6	240
AB-2	34.6	77.9	-1.97	28.0	464
AB-4	12.1	43.6	-1.20	9.8	741
AB-5	34.9	85.0	-1.73	14.1	501
AB-6	26.4	73.6	-1.46	21.2	574
AB-7	14.6	69.6	-1.47	11.8	982
AB-8	21.5	110.8	-2.44	17.4	1061
AB-9	9.9	44.7	-1.03	8.0	926
AB-10	16.3	51.5	-1.02	13.2	652
AB-11	18.5	53.4	-1.13	15.0	592
AB-12a	20.9	62.5	-1.51	16.8	614
AB-13	8.9	73.7	-1.89	7.2	1694
AB-14	16.8	54.9	-1.47	13.6	673
AB-15a	22.3	71.8	-1.52	18.0	664
AB-16	10.0	37.2	-1.22	8.0	764
AB-17	15.1	40.4	-1.08	12.2	550

Table III Continued. Constants for Pressure vs. Area Plots and Number of Moles and Molecular Size Estimates

Sample	C_0	C_1	C_2	$M \times 10^{10}$ (moles)	Ao (A^2/molec)
----------------	Atlantic Deep-Water Samples	----------------			
AB-12b	17.8	48.0	-1.09	14.4	556
AB-12c	11.3	35.0	-0.69	9.2	635
AB-15b	25.1	79.5	-1.73	20.2	650
AB-15c	57.1	61.8	-1.52	46.0	223
AB-15d	8.9	24.9	-0.71	6.4	636
AB-15e	6.1	36.9	-1.35	5.0	1242
----------------	Atlantic Concentrated Samples	----------------			
AC-2	109.3	57.5	-1.98	88.2	108
AC-4	31.5	73.0	-2.10	25.4	476
AC-9	95.4	44.3	-0.96	77.0	96
AC-11	61.5	82.0	-1.82	49.6	275
AC-15	30.6	48.5	-1.84	24.6	326

Units: C_0: (mN/M) x cm^2 ; C_1: cm^2; C_2: $(Mcm^2)/mN$
Depths of AD 12b and 12c: 20 M; 50 M
Depths of AD 15b-15e: 30, 100, 500 and 2850 M

Calculations of Amounts and Molecular Areas

With Equation 9, the number of molecules on the surface of each sample in Table III was calculated. From the finite area in square centimeters occupied by the molecules of the film (C_1), the area per molecule was also determined. Amounts on the surface determined by this method ranged from 5 x 10^{-10} to 444 x 10^{-10} moles. Apparent areas per molecule ranged from 96 to 1694 square Angstroms. A summary of the average values for each of the different classes of field samples is given in Table IV.

In general, more material is indicated in Chesapeake Bay than in the Atlantic, and screen samples are enriched relative to subsurface samples in both the Chesapeake Bay and the Atlantic as expected. According to Table IV, the area per molecule of compounds in films collected using screens tends to be smaller than the area per molecule of compounds in films formed on bulk water samples. Compounds from Atlantic bulk water, which was collected from a greater depth, appear to be larger than compounds collected from surface or depth in Chesapeake Bay and also larger than compounds in corresponding screen samples from the Atlantic. The small relative size of the compounds from Atlantic concentrated water may be due to losses of the very high molecular weight compounds during the filtration, since these samples were filtered through glass fiber filters and through 0.4 um Nuclepore filters prior to concentration by ultrafiltration. Numbers in Table IV represent average properties of the complex mixture of which the films are composed.

Table IV. Calculated Amounts, Areas, and Molecular Weights from the Two-Dimensional Gas Model and Pure Compound Data

	Sample Type	Moles x 10^{10}	A^2/molecule	MW
8	Ches. Bay Screen	119±64	420±140	2300± 900
8	Ches. Bay Bulk Water	51±18	490±210	2700±1000
7	Atlantic Screen *	49±24	370±140	1800± 700
15	Atlantic Bulk Water **	14± 6	760±310	4900±2200
6	Atlantic Deep-Water	17±15	660±330	3700±1200
5	Atlantic Concentrated	53±29	260±160	1400± 900

* Samples AS-2 and AS-3, collected using a different screen method were not included to obtain the number of moles.
** Sample AB-1 from Baltimore Harbor is not included.

Molecular Weights

Some compounds have film properties that are essentially the same on a mass for mass basis. This can be seen from Figure 4 and the data in Table II. Curves A, B and D for 29 micrograms of stearyl alcohol, stearic acid and tristearin are quite similar. it can be seen from Table I that tristearin occupies about three times the area per molecule as stearic acid, but Figure 4 indicates that both have about the same area per gram. Similar behavior is exhibited by the oleyl alcohol-triolein pair. Since the properties of the monomers are additive in the polymer, surface films of polymers are usually described in the literature in terms of area per monomer unit rather than area per molecule.

Because approximately 24 ug of the known compounds in Table II were required to match the area of natural sample CS-1 at 20 mN/M in every case regardless of film type, the assumption can be made that CS-1 also had a mass of 24 ug. If this assumption is made, the molecular weight of CS-1 can be calculated. The 24 ug of CS-1 would be equivalent to 145.8 x 10^{-10} moles (from Table III), and therefore, its molecular weight would be 1646.

Since the area covered by a film of a pure substance at a constant value of film pressure is directly proportional to the mass on the surface, it is possible to extend this computation mathematically to all the natural films listed in Table III. First the area, A_{20}, at P = 20 mN/M can be calculated from Equation 6 using the constants from Table III. The proportional number of micrograms of the average pure compound from Table II is then 24 x (A_{20}/119.54). The resulting number of micrograms is taken as equivalent to the number of moles listed in Table III, and a simple division yields the molecular weight estimate. For all the samples listed in Table III, the results were averaged to give the mean value for each class of sample. Results are listed in Table IV.

Average molecular weights determined by this method ranged from 1400 ± 900 to 4900 ± 2200. Atlantic water that had been filtered and then processed through the hollow fiber concentrator had the lowest estimated molecular weight. Since we were dealing, by definition, with surface-active material, sample material may have been lost in the processing steps. The other samples were not processed in any way--just poisoned and refrigerated. There is a suggestion that material recovered from the sea surface has a slightly lower average molecular weight. Bulk samples from the Atlantic, collected at a depth of 7 meters were higher in molecular weight than bulk samples from Chesapeake Bay collected from approximately 1/3 meter. Estimated molecular weights listed in Table IV are consistent with the molecular weight range for estuarine colloidal material reported by Means et. al. [20].

Coefficients of Compressibility

In the preceding sections the calculation of the number of moles, area per molecule, and molecular weight were necessarily based on the use of data from model compounds to obtain reasonable constants for use in the appropriate equations of state. Another characteristic property of natural surface films, the coefficient of compressibility, can be determined from film balance measurements without assertions of similarities with models. The coefficient of compressibility defined by the equation

$$\beta = -\frac{1}{Vo}\left(\frac{dV}{dP}\right)_T \qquad (10)$$

where P, V, and T have their usual meaning, and Vo is for the standard state, is a thermodynamic property of real gases that is often tabulated. The two-dimensional analog for monomolecular surface films is

$$\beta = -\frac{1}{Ao}\left(\frac{dA}{dP}\right)_T \qquad (11)$$

where P is film pressure, A is area, and Ao is the area at 25° C. Area values are usually given in area per molecule.

It is possible to determine the coefficient of compressibility for natural films without knowing the quantity of material present on the surface or its molecular weight or size. Because of the gas-like behavior of the natural films, i.e., the fact that they approach the area axis asymptotically, the film pressure vs. area data can be well fitted to an equation of the form

$$P = M \text{ Log } (A) + B. \qquad (12)$$

where P is the film pressure, A is the area in square centimeters, and M and B are constants. Film pressure vs. Log(A) data for all samples gave an excellent straight line fit to Equation 12 by the method of least squares. The constants M and B and the correlation coefficients are shown in Table V.

In plots of P vs. Log (A), the slope of the line is M and the intercept is B. The size of the B constant is determined by the amount of material present. As in the P vs. A curves, more material displaces the curve to the right, and, in this case, alters only the B constant.

Differentiation of Equation 12 eliminates B.

$$P = \frac{M \text{ Ln}(A)}{2.303} + B. \qquad (13)$$

$$\frac{dP}{dA} = \frac{M}{2.303\ A}. \qquad (14)$$

The compressibility is simply the reciprocal of Equation 14 multiplied by 1/A:

$$\beta = -\left(\frac{dA}{dP}\right)\frac{1}{A} = \frac{-2.303}{M} \qquad (15)$$

This value is independent of the quantity of material in the surface film. Compressibilities for films on all water samples are listed in Table V.

Table VI summarizes the coefficient of compressibility data for each class of sample. Values for several types of known compounds are listed for comparison. A more extensive table of compressibility values for soluble surface-active agents is given in Meader and Criddle [21]. Since the compressibility coefficients are numerical indicators of the shapes of the film pressure vs. area curves, again we see the similarity of the physical characteristics of natural films to the highly-oxygenated model surface-active compounds and the remarkable similarity of properties of films collected from many locations.

Table V. Constants for Film Pressure vs. Log(Area) Plots and Coefficients of Compressibility

Sample	M	B	corr. coeff.	Xo	E	$-\left(\frac{dA}{dp}\right)\frac{1}{A}$
---------- Chesapeake Bay Screen-Collected Water ---------						
CS-1	-34.4	90.6	-0.977	431	2.7	0.0669
CS-2	-47.4	111.9	-0.999	229	1.3	0.0486
CS-3	-44.8	123.7	-0.996	576	2.6	0.0656
CS-4	-46.5	109.1	-0.999	221	1.9	0.0495
CS-5	-35.4	81.7	-0.997	203	1.5	0.0650
CS-6	-39.1	100.6	-0.991	377	2.3	0.0589
CS-7	-35.4	98.1	-0.994	589	4.1	0.0650
CS-8	-33.2	95.5	-0.990	749	2.9	0.0694
--------------- Chesapeake Bay Bulk Water ---------------						
CB-1	-41.0	90.2	-0.995	158		0.0562
CB-2	-53.7	120.0	-0.998	172		0.0429
CB-3	-50.6	119.0	-0.999	225		0.0455
CB-4	-39.6	81.6	-0.994	114		0.0581
CB-5	-54.3	115.1	-0.999	132		0.0424
CB-6	-38.3	84.7	-0.978	163		0.0601
CB-7	-47.0	101.3	-0.998	143		0.0490
CB-8	-56.4	136.4	-0.991	262		0.0408
------------- Atlantic Screen-Collected Water -------------						
AS-3	-43.7	114.6	-0.997			0.0527
AS-4	-38.9	87.9	-0.999			0.0592
AS-5	-43.2	94.2	-0.997			0.0533
AS-6	-43.9	93.6	-0.998			0.0524
AS-7	-40.8	84.2	-0.998			0.0565
AS-8	-41.2	80.9	-0.999			0.0559
AS-11	-41.1	85.2	-0.989			0.0560
AS-12	-47.0	92.5	-0.999			0.0490
------------- Atlantic Bulk Water (7 meters) -------------						
AB-1	-45.1	92.9	-0.999			0.0511
AB-2	-44.7	90.1	-0.995			0.0515
AB-4	-41.9	74.0	-0.987			0.0549
AB-5	-51.2	106.1	-0.989			0.0450
AB-6	-57.1	112.7	-0.993			0.0403
AB-7	-56.5	109.3	-0.994			0.0408
AB-8	-43.2	95.8	-0.967			0.0532
AB-9	-59.2	101.2	-0.999			0.0389
AB-10	-64.2	114.9	-0.999			0.0359
AB-11	-58.3	106.2	-0.999			0.0395
AB-12a	-55.3	103.4	-0.998			0.0417
AB-13	-48.1	93.7	-0.989			0.0479
AB-14	-48.8	89.0	-0.997			0.0472
AB-15a	-51.9	102.6	-0.984			0.0444
AB-16	-41.9	69.1	-0.999			0.0549
AB-17	-49.4	83.7	-0.999			0.0466

Table V Continued. Constants for Film Pressure vs. Log(Area) Plots and Coefficients of Compressibility

Sample	M	B	corr. coeff.	Depth (Meters)	$-\left(\frac{dA}{dP}\right)\frac{1}{A}$
Atlantic Deep-Water Samples					
AB-12b	-57.3	101.3	-0.998	20	0.0402
AB-12c	-68.9	111.7	-0.977	50	0.0334
AB-15b	-46.1	94.2	-0.965	30	0.0500
AB-15c	-48.1	92.4	-0.997	100	0.0479
AB-15d	-39.5	59.2	-0.993	500	0.0583
AB-15e	-36.3	59.8	-0.994	2850	0.0635
Atlantic Concentrated Samples					
AC-2	-29.3	59.5	-0.992		0.0785
AC-4	-45.9	89.9	-0.998		0.0502
AC-9	-33.1	64.4	-0.978		0.0697
AC-11	-55.5	113.1	-0.997		0.0415
AC-15	-28.6	52.7	-0.992		0.0805

Units of Compressibility Coefficient: cm/dyne
Units of M and B: mN/M. Xo = Area Intercept at P = 0
E = Enrichment ratio, Screen/Bulk

Table VI. Summary of Coefficient of Compressibility Data

	Sample Type	$-\left(\frac{dA}{dP}\right)\frac{1}{A}$ $\frac{cm}{dyne}$
8	Chesapeake Bay Screen Samples	0.061 ± 0.008
9	Atlantic Screen Samples	0.055 ± 0.003
5	Atlantic Concentrated Samples	0.064 ± 0.017
8	Chesapeake Bay Bulk Water Samples	0.049 ± 0.008
16	Atlantic Bulk Water Samples	0.046 ± 0.006
6	Atlantic Deep Samples	0.049 ± 0.011
	Mean of all of the above groupings:	0.054
	Nonionic Surfactant: POE(6)dodecanol	0.056
	Chlorophyll	0.021
	Vitamin A	0.019
	Ionic Surfactant: Na dodecyl benzene sulfonate	0.017
	Proteins: Egg Albumin and Hemoglobin	0.009

DISCUSSION

Compounds with film behavior similar to that of the natural films all have large, highly oxygenated hydrophilic head groups associated with the water. The solid-like film materials all have relatively few oxygens in the hydrophilic end and long hydrophobic tails that can be closely packed in a compressed monolayer (Table I). Liquid-like films may have relatively large hydrophobic tails, but usually these tails are branched or bent chains that prevent close packing in the monomolecular films. The gas-like films have a relatively large, highly-oxygenated hydrophilic head group compared to the size of the hydrophobic tail. Figure 6 illustrates the structure of two model compounds exhibiting physical behavior similar to that of natural films. The compounds are drawn roughly in the orientation that they might assume at the air/water interface when the oxygen groups are held in the water by hydrogen bonding. Many materials with the same hydrophil-lipophil balance could produce similar film pressure vs. area curves. The sucrose head group of sucrose monomyristate is probably closer to the true structure of marine surface-active material than the head groups of the synthetic surfactants are, since carbohydrates constitute a significant fraction of marine organic material [7,8,12,20,22]. Sucrose monolaurate with a 12- instead of 14-carbon tail should have a curve that coincides more closely with natural sample CS-1.

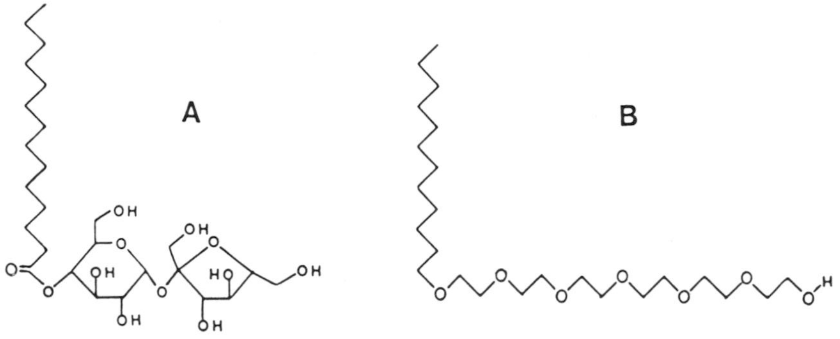

Figure 6. Structural formulas of sucrose monomyristate (A) and polyoxyethylene (6) dodecanol (B). These compounds and natural films have similarly-shaped film pressure vs. area isotherms.

The presently known average composition of organic material in seawater is summarized in the chapter by Hunter and Liss [8]. Carbohydrates make up a large relative proportion of this material compared to fatty acids, and may be important components of the surface films. To investigate the possibility that films with properties silmilar to natural films might be produced by a simple mixture of a pure carbohydrate with a known film-forming compound, studies of soluble starch, bovine plasma albumin, and a mixture of the two were conducted using a film balance. Concentrations of soluble starch as high as 10 mg/L in 450 ml samples produced no measurable film pressure on compression of the surface. Bovine albumin (0.5 mg/L) produced a film pressure vs. area curve very different in shape from curves due to natural material. When up to 20% soluble starch was mixed with the albumin (maintaining albumin at 0.5 mg/L) there was no change in the shape of the film pressure vs. area curve.

Recently D'Arrigo, et. al. [23,24], using a water-soluble extract from Hawaiian forest soil, reported on the composition and surface properties of the surfactant mixture surrounding natural microbubbles. The presence of a glycopeptide-lipid-oligosaccharide complex held together by both hydrogen bonding and nonpolar interactions was indicated. On initial compression the oligosaccharide material was selectively desorbed from the surface film of this material, resulting in a highly insoluble monolayer containing glycopeptide-acyl lipid complexes. The presence of such material in our samples would be consistent with the observed film properties.

CONCLUSIONS

The general results here agree with the observation of van Vleet and Williams [25] that natural films are not composed primarily of free fatty acids, alcohols, or hydrocarbons, although these compounds have all been found in surface film samples. More oxygenated molecules of higher molecular weight are indicated. Simple mixtures of biochemical films do not adequately describe the observed behavior of natural films. The A glycopeptide-lipid-oligosaccharide complex, such as described by D'Arrigo [23,24] is more consistent with our observations of physical characteristics of natural surface films. The same hydrophil-lipophil balance observed in the model compounds must occur as an average property of marine surface-active material. None of the model compounds have molecular weights of several thousand, so very large polymer-like materials with similar film characteristics are suggested. Nanomolar amounts of material which occupied a rather large surface area per molecule (approximately 400-800 square Angstroms) and had a high molecular weight (approximately 2000 to 5000) were indicated in our field samples by the calculation techniques discussed in this paper. This picture of marine organic material is consistent with evidence obtained by other methods.

ACKNOWLEDGEMENT

The work described in this chapter is from a dissertation to be submitted to the Graduate School, University of

Maryland, by W. R. Barger in partial fulfillment of the requirements for the Ph. D. degree in chemistry.

REFERENCES

1. Ewing, G. C. "Slicks, Surface Films and Internal Waves," J. Marine Res. 9:161-187 (1950).
2. LaFond, E. C. and K. G. LaFond, "Perspectives of Slicks, Streaks, and Internal Waves," Bull. Jap. Soc. Fish. Oceanogr. 49-57 (1969).
3. Garrett, W. D. "Collection of Slick-Forming Material from the Sea Surface," Limnol. Oceanogr. 10:602-604 (1965).
4. Garrett, W. D. "The Organic Chemical Composition of the Ocean Surface," Deep-Sea Res. 14:221-227 (1967).
5. Williams, P. M. "Sea Surface Chemistry: Organic Carbon and Inorganic Nitrogen and Phosphorous in Surface Films and Subsurface Waters," Deep-Sea Res. 14:791-800 (1967).
6. Jarvis, N. L., W. D. Garrett, M. A. Scheiman and C. O. Timmons," Surface Chemical Characterization of Surface-Active Material in Seawater," Limnol. Oceanogr. 12:88-96 (1967).
7. Liss, P. S., "Chemistry of the Sea Surface Microlayer," in Chemical Oceanography, Second Edition, Vol. 2, J. P. Riley and G. Skirrow, Eds. (New York: Academic Press, 1975), pp. 193-243.
8. Hunter, K. A., and P. S. Liss, "Organic Sea Surface Films," in Marine Organic Chemistry, E. K. Duursma and R. Dawson, Eds. (Amsterdam: Elsevier Scientific Publishing Co., 1981), pp. 259-298.
9. Garrett, W. D. and R. A. Duce, "Surface Microlayer Samplers," in Air-Sea Interaction Instruments and Methods, F. Dobson, L. Hasse and R. Davis, Eds. (New York: Plenum Press, 1980), pp. 471-490.
10. Huhnerfuss, H., "On the Problem of Sea Surface Film Sampling: a Comparison of 21 Microlayer-, 2 Multilayer, and 4 Selected Subsurface-Samplers-Part 1," Meerestechnik 12 (5):137-142 (1981). and "....-Part 2," Meerestechnik 12 (6):170-173 (1981).
11. Barger, W. R., W. H. Daniel and W. D. Garrett, "Surface Chemical Properties of Banded Sea Slicks," Deep-Sea Res. 21: 83-89 (1974).
12. Means, J. C., and R. Wijayaratne, "Role of Natural Colloids in the Transport of Hydrophobic Pollutants," Science 215: 968-970 (1982).
13. Gaines, G. L. Insoluble Monolayers at Liquid-Gas Interfaces (New York: Interscience Publishers, 1966).
14. Gershfeld, N. L., "Film Balance and the Evaluation of Intermolecular Energies in Monolayers," in Techniques of Surface and Colloid Chemistry and Physics, Vol. 1, R. J. Good, R. R. Stromberg and R. L. Patrick, Eds. (New York: Marcel Dekker, Inc., 1972), pp. 1-39.
15. Ross, S. and E. S. Chen, "Adsorption and Thermodynamics at

the Liquid-Liquid Interface," in Chemistry and Physics of Interfaces, D. E. Gushee, Ed. (Washington, D.C.: American Chemical Society Publications, 1965), pp.44-56.
16. Cadenhead, D. A. "Monomolecular Films at the Air-Water Interface, Some Practical Applications," in Chemistry and D. C.: American Chemical Society Publications, 1971), pp. 28-34.
17. Langmuir, I., "Oil Lenses on Water and the Nature of Monomolecular Expanded Films," J. Chem. Phys. 1:756-776 (1933).
18. Schofield, R. K. and E. K. Rideal, "The Kinetic Theory of Surface Films. Part I. -- The Surfaces of Solutions," Proc. Roy. Soc. (London) A109:57-77 (1925).
19. Lange, H. "Surface Films," in Nonionic Surfactants, M. J. Schick, Ed. (New York: Marcel Dekker, Inc., 1967), pp. 443-477.
20. Means, J. C., R. D. Wijayaratne and W. R. Boynton, "Fate and Transport of Selected Pesticides in Estuarine Environments," Canadian Journal of Fisheries and Aquatic Sciences 40 (Suppl. 2):337-345 (1983).
21. Meader, Jr., A. L. and D. W. Criddle, "Force-Area Curves of Surface Films of Soluble Surface-Active Agents," J. Colloid Sci. 8:170-178 (1953).
22. Sigleo, A. C., T. C. Hoering and G. R. Helz, "Composition of Estuarine Colloidal Material: Organic Components," Geochim. Acta 46:1619-1626 (1982).
23. D'Arrigo, J. S., C. Saiz-Jimenez and N. S. Reimer, "Geochemical Properties and Biochemical Composition of the Surfactant Mixture Surrounding Natural Microbubbles in Aqueous Media," J. Colloid Interface Sci 100:96-105 (1984).
24. D'Arrigo, J. S. "Surface Properties of Microbubble-Surfactant Monolayers at the Air/Water Interface," J. Colloid Interface Sci. 100:106-111 (1984).
25. van Vleet, E. S. and P. M. Williams, "Surface Potential and Film Pressure Measurements in Seawater Systems," Limnol. Oceanogr. 28 (3):401-414 (1983).

CHAPTER 5

SEDIMENT RESPONSE TO SEASONAL VARIATIONS IN ORGANIC MATTER INPUT

Norman Silverberg, Harry M. Edenborn and Nelson Belzile
Department d'Oceanographie
Universite du Quebec
Rimouski, Quebec
Canada G5L 3A1

INTRODUCTION

When modeling the biogeochemical reactions and exchange processes which occur near the sediment-water interface during early diagenesis, it is most convenient to assume that temporal variations are slight or can be averaged out over the time scales of the processes under study. This permits simplification of the non-linear diagenetic equations by assuming steady-state conditions, and allows one to estimate quantitative values for parameters such as the rates of dissolution, precipitation, diffusion, sedimentation, and bioturbation.

It has been observed previously that strong seasonal variations in the rates of bioturbation and release of dissolved manganese from coastal sediments can occur (1). Shifts in the depth profiles of some dissolved porewater components have also been reported (2). These differences are most noticeable between summer and winter and are probably related to the strong changes in water temperature and storm activity which affect both the water column and bottom sediments in shallow regions. In the deep sea, the physical environment remains stable over long periods of time and sedimentation rates are very slow, making the assumption of steady-state conditions reasonable. Recent sediment trap studies indicate that, even in the deep sea, the flux of particulate matter through the water column varies with time and is related to seasonal variations in primary production in surface waters (3, 4, 5, 6).

Our study area, the 300-500 m deep Laurentian Trough (Fig. 1), shares some of the characteristics of the deep sea: the bottom water temperature is almost constant at about $4^{o}C$, salinities vary little from $35^{o}/oo$, the residence time of the bottom water is 2-3 years, and bottom currents are sluggish. The sediment also displays a well-developed redox zonation with depth (indicated by the successive depletion of oxygen, nitrate, Mn- and Fe-oxides and hydroxides, and sulphate). Given the physical stability of the bottom environment, we have felt at ease using the steady-state assumption (7,8). We have detected occasional deviations from steady-state conditions. These include the presence of a significant dissolved-Fe gradient at the sediment-water interface (implying export of Fe from the sediment), the persistence of nitrate well below the sediment-water interface, and variations in thickness of the oxidized layer at the sediment surface.

MARINE AND ESTUARINE GEOCHEMISTRY, Sigleo, A. C., and A. Hattori (Editors)

LAURENTIAN TROUGH

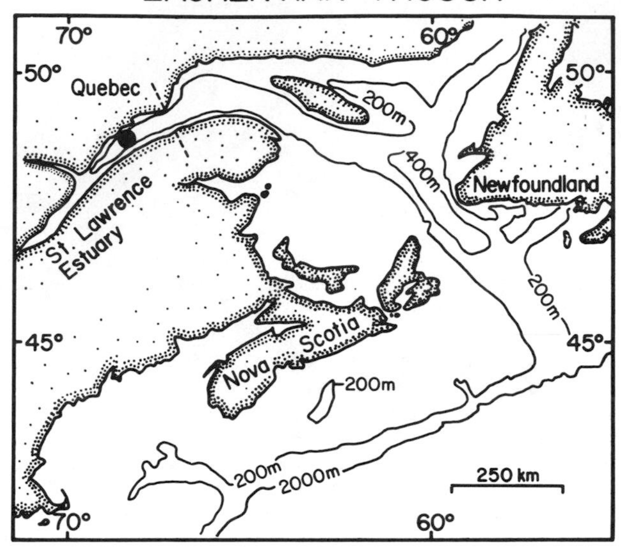

Figure 1: Map showing the extent of the Laurentian Trough and the location of station 23.

In this note we report on observations of significant variations in the rate of sedimentation of total solids and organic carbon, as well as in the quality of the organic matter reaching the sediment, and we examine the response of the sediment to these variations in input.

INPUT TO THE SEDIMENTS FROM THE WATER COLUMN

As part of a study designed to provide valid estimates of modern sedimentation rates in the Laurentian Trough, we have been using a free-drifting sediment trap, modeled after that of Staresinic (9). The four cylinders of the trap have a total collecting surface of about $0.5m^2$ and could collect several hundred milligrams of particulate matter during deployment times lasting between 9 and 24 hours. The samples were recovered from the base of each cylinder directly into 2 L containers. A sample from one of the cylinders was kept at in situ temperature for microbiological experiments and microscopic examination. On each cruise the material was consistently dominated by various forms of fecal matter, with variations over the months reflecting changes in the zooplankton population. After most of the material had settled in the other three samples, some water was decanted and the remainder centrifuged to collect all particulate matter. The material was freeze-dried and the dry weight recorded. Although there was some evidence that live copepods may have invaded the traps, these were routinely removed with the supernatant water during sample recovery. Poisoning of the traps was not deemed to be necessary during these short-period measurements. The sediment trap material was ground to a uniform powder, and subsamples were analysed for organic carbon and nitrogen using a Perkin Elmer Model 240 CHN analyser.

Repeated measurements of the sedimentation rate were made between 1981 and 1984 at a reference location (station 23, Fig. 1) in the middle of the estuary portion of the Trough. Traps were placed at approximately 150 m depth, in the zone of minimum suspended particulate matter (Fig. 2). At this depth, the trap avoids sampling the euphotic zone, in which there is a continuous production and destruction of particles. It also avoids sampling the bottom 50-100 m of the water column, in which there is contamination from resuspended sediment. We thus believe that the fluxes measured represent accurate estimates of the true net flux of large particles to the sediment-water interface from the water column. In 1984, several particle-flux measurements were made each month during the sampling period. Some variations between nightime, daytime, and 24hr trap deployments were observed, but these were not statistically significant compared to the monthly variations. Therefore only the mean values are presented in this note.

The sediment trap data for all four years are presented in Figure 3. It is apparent that there is a gradual decrease in the input of total solids to the sediment between the spring and late fall. The fluxes correspond to a decrease in sedimentation rate from about 4.5 mm/yr to about 1 mm/yr. Over the same period of time, the carbon content of the input increased gradually. These measurements most likely reflect the input of riverborne detritus, which is most abundant during spring run-off, and an increase in primary production within the estuary during the summer. The C/N ratio decreased during the same period, which also indicates a progressive decrease in the influence of terrigenous organic matter in favour of marine organic matter production. Overall, the input of organic matter to the sediment surface varies by a factor of 3-4 over the ice-free season. There are some differences in the absolute values from year to year, and there are also indications of periodic increases in carbon flux during the early summer and late fall periods (perhaps associated with phytoplankton blooms). Nevertheless, the data set collected over four years indicates that the observed seasonal trends in both the quantity and quality of organic matter input to the underlying sediment are real.

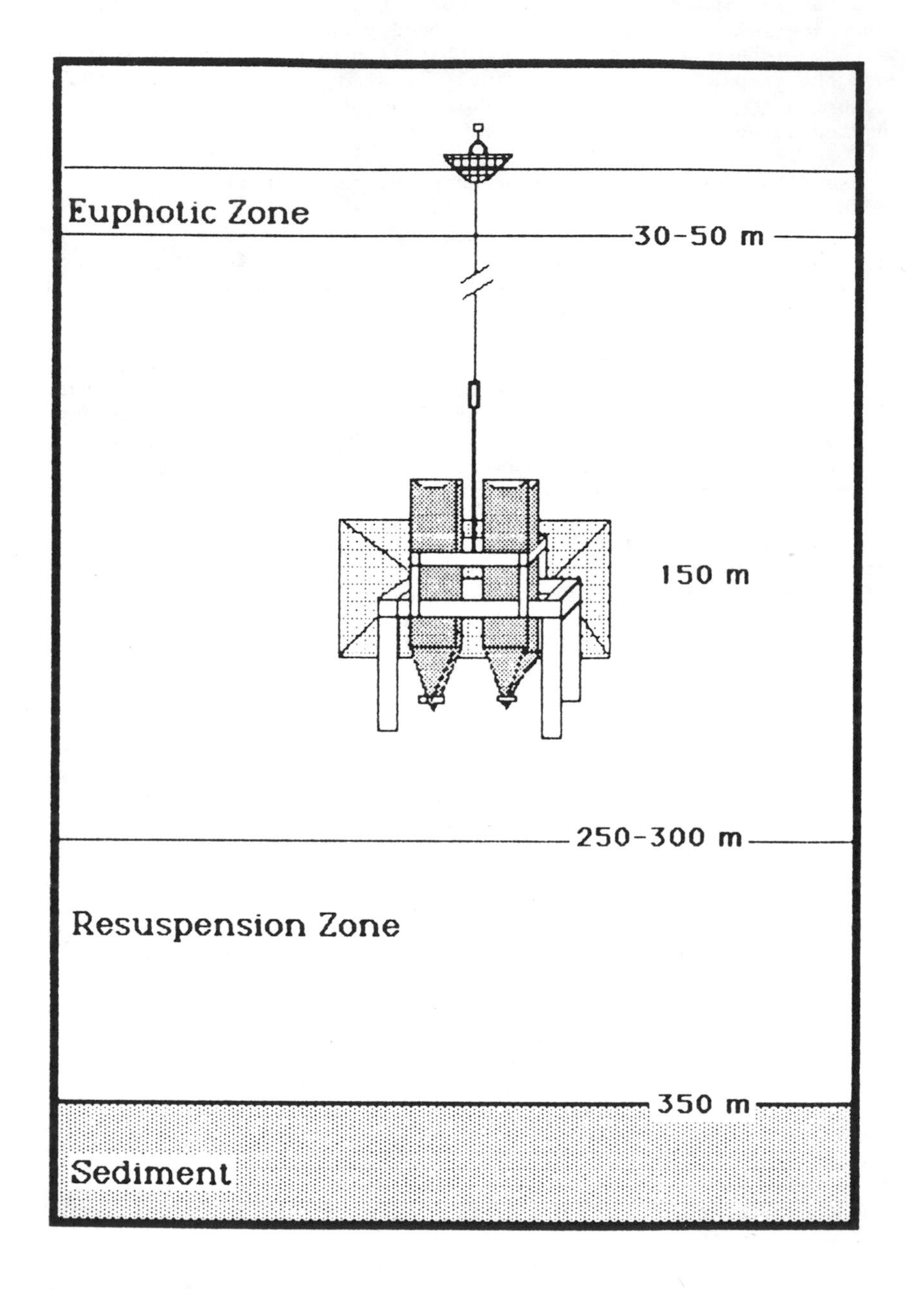

Figure 2: Placement of the free-drifting sediment trap in the water column.

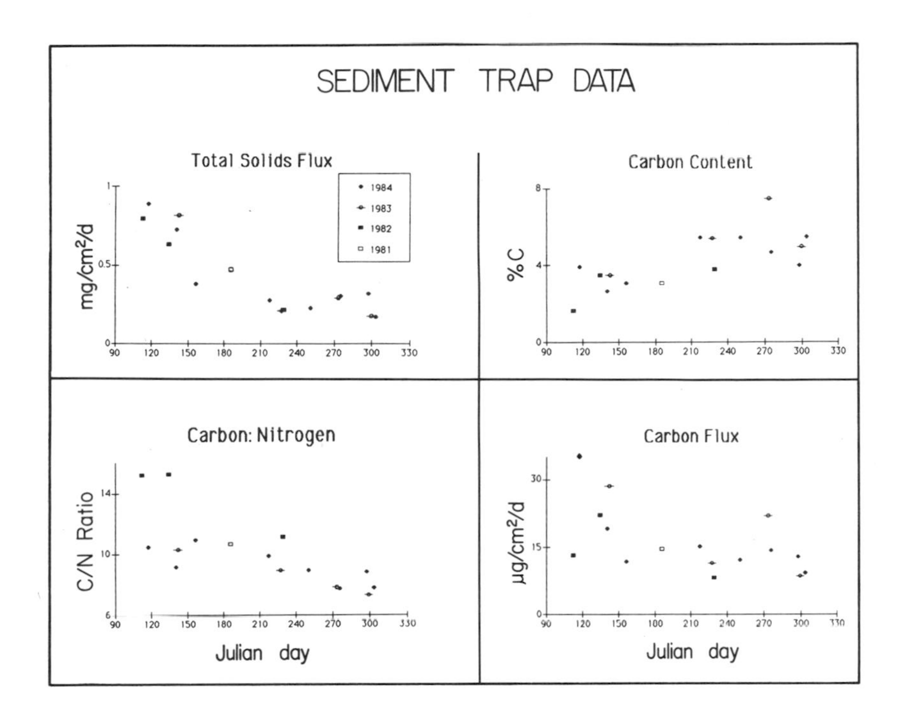

Figure 3: Characteristics of sedimenting material at station 23 between 1981 and 1984.

RESPONSE OF THE UNDERLYING SEDIMENT

Undisturbed bottom sediments were recovered during each sediment trap deployment using an Ocean Instruments, Mark II, 0.1 m^2 box corer. Fine subsampling at 2-10 mm depth intervals was carried out aboard ship using a specially-designed, nitrogen-filled glove box (10). Subsamples were then analysed in the laboratory for porosity, organic carbon and nitrogen, and, towards the end of the sampling season, for microbiological sulphate reduction rates and various porewater chemical constituents. The organic carbon values in the 0-2 mm surface layer were always close to 2% C by weight, regardless of the carbon content of the material measured in the traps, and the C/N ratios were always higher than that of the trap material. Respirometer studies, using cultures of surface sediment bacteria, also indicated that the sediment organic matter was about one order of magnitude less available for bacterial use than that of the corresponding trap material (unpublished data). Considerable variation was seen in the shape of the carbon profile in the upper 3 cm of the box cores between sampling periods. Below this depth the profiles were more regular and generally followed similar linear slopes from box to box.

The results indicated that the freshly deposited solids were being inmixed by bioturbation and rapidly diluted with older, more degraded, sediment. The carbon loss at station 23 is about 30% from the surface to 40 cm depth (11). The rate of carbon loss over the top 10 cm of box cores obtained in 1983-84 ranged between -0.01 to -0.04% C/cm, based upon regression analyses. Using these carbon gradients, and estimates for the effective bioturbational diffusion coefficient of 10^{-2} to 10^{-3} cm^2/day (calculated from excess Th-234 activities in the sediment (12,13), the downward flux of carbon into the sediment was calculated. The values obtained by this procedure amounted to only 10% of the incoming flux of fresh organic carbon as measured in the sediment traps. This implied that about 90% of the freshly deposited organic matter was being degraded at the sediment-water interface, without ever being buried within the sediment itself. If this were true, then seasonal variations in organic matter flux would have little influence on diagenetic processes just below the interface, and steady-state would be a valid assumption for the subsurface conditions.

In order to test this hypothesis, and to determine if the 0-2 mm sampling interval provided enough resolution to detect seasonal flux changes on the order of micrograms/cm^2/day, we also obtained samples using a Pamatmat-style multiple corer. This apparatus recovers short cores with a perfectly intact sediment-water interface. Following recovery, the cores were immediately frozen using liquid nitrogen. In the laboratory, the surficial sediment was very finely subsampled by carefully scraping off successive layers of sediment using a scalpel. Each subsample was then rinsed with distilled water to remove sea salts, centrifuged, freeze-dried, homogenized, and analysed for organic carbon. The results of this experiment are presented in Figure 4. Carbon levels were never as high as those measured in the corresponding trap samples, and again, all of the surface values were close to 2% C. What was revealed, however, were carbon gradients 30-40 times greater than those obtained from the box-core samples. These are conservative estimates since we assumed that each scraping was 0.5 mm thick, when they were in fact often considerably thinner. Such rapid carbon losses must be limited to a very shallow depth within the sediment, otherwise all of the carbon would disappear in the top two or three centimeters, but they are representive of the crucial interface gradient.

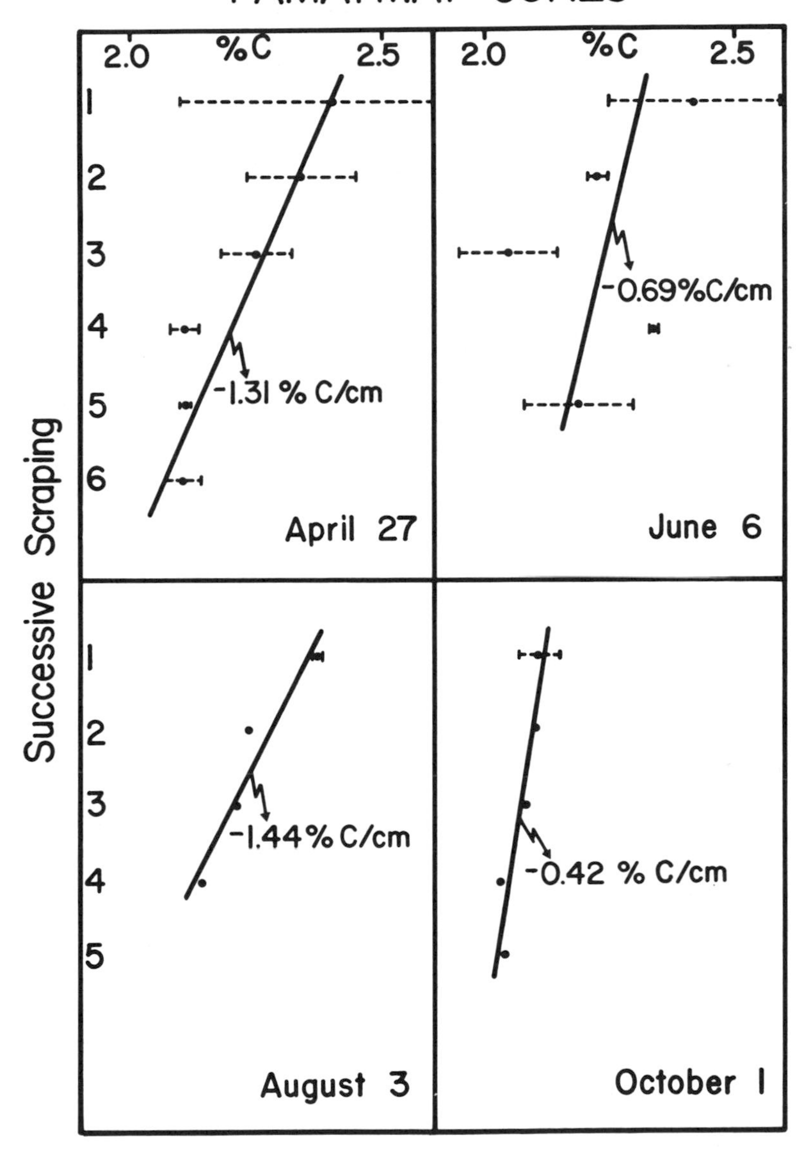

Figure 4: Best-fit regression lines for carbon gradients at the surface of Pamatmat cores taken in different months. Error bars equal SD about the mean.

When we then calculated the downward flux of carbon across the sediment-water interface using the Pamatmat core gradients, we obtained values of the same order of magnitude as that of the downward flux of carbon through the water column to the sediment-water interface (Table 1). From these calculations, we can infer that variations in the quantity and quality of organic matter input from the water column are probably influential to some depth within the underlying sediments.

Table 1

DOWNWARD FLUXES OF CARBON ACROSS THE SEDIMENT-WATER INTERFACE AND FROM THE WATER COLUMN TO THE SEDIMENT

Date	Carbon gradient	Bioturbation flux (range)*	Flux from water column
1984	% C/cm	$\mu gC/cm^2/day$	$\mu gC/cm^2/day$
Apr. 27	-1.31	5.2 - 52.5	43.9
June 6	-0.69	2.2 - 22.3	11.9
Aug. 3	-1.44	5.8 - 57.7	15.2
Oct. 1	-0.42	1.7 - 17.0	14.3

*$J = -Db\ \partial C/\partial z$ (uses $Db = 10^{-2} - 10^{-3} cm^2/d$; assumes water content content = 85%)

Another experiment we performed lends support to this interpretation. The lack of a carbon-rich layer at the sediment surface might also be interpreted to be the result of rapid decomposition of organic matter by sediment bacteria. To test this possibility, we measured the glucose mineralization potential of bacteria in the surface sediment using the method of Litchfield et al. (14). Surface sediment samples were incubated at in situ temperature in sterile 60 mL serum bottles with 0.11 nmol. of D-[UL-C^{14}] glucose (Amersham; specific activity-250 mCi/mmol) for 6 hours. During that time, $^{14}CO_2$ produced by the bacterial breakdown of the labeled glucose was trapped on a phenethylamine-soaked filter paper held in a plastic bucket suspended from the gas-tight rubber cap of each serum bottle. After the incubation time had elapsed, 1 mL of 2N H_2SO_4 was added to the reaction mixture by needle and syringe, stopping the bacterial activity and releasing the $^{14}CO_2$ from solution. The filter paper wicks were added to 10 mL of Aquasol scintillation cocktail (New England Nuclear

Corp.) and were radioassayed in a Beckman LS5801 liquid scintillation spectrometer. Quench corrections were made using the external standard (H-number) method, and the final data were expressed as dpm $^{14}CO_2$ evolved per gram dry weight of sediment per hour.

If we compare the observed bacterial glucose mineralization rate and the actual carbon budget for station 23 for August 3, 1984, approximately 6 x 10^{-6} μg of glucose carbon per gram-dry weight of surface sediment were mineralized to CO_2 per day. This is equivalent to 1.7 x 10^{-3} μgC/cm^2/d. The actual carbon input to the sediment surface on this date, based on a flux of 280 μg/cm^2/d of solids containing 5.4% C, amounted to approximately 15 μgC/cm^2/d. If we make the liberal assumption that the naturally-occurring particulate organic carbon is decomposed at the same rate, then four orders of magnitude more carbon reaches the sediment than would be broken down to carbon dioxide by in situ bacterial activity. Much of the organic matter deposited on the sediment surface would consist of compounds more biochemically-complex than glucose, and would be broken down even more slowly. Therefore, bacterial decomposition of organic matter at the sediment surface does not appear to be rapid enough to account for the depletion of a significant amount of organic matter before it can be mixed via bioturbation into the underlying sediment.

Since the processes of organic matter degradation are responsible for many of the other chemical and biochemical changes which occur during early diagenesis, the reality of steady-state conditions should be examined more closely for these sediments. If changes in the input from the water column are in fact "seen" by the underlying sediment one might anticipate some changes in the depths at which the different zones of redox or biochemical reactions occur. Figure 5 shows some data for porewater Mn and Fe, as well as rates of bacterial sulphate reduction obtained towards the close of the 1984 sampling season. There are some noticeable shifts in the depth at which dissolution of manganese becomes important. As expected, the depth at which dissolved Fe increased in the porewaters was deeper than that for manganese, but for the same dates the Fe behaviour was more variable. The sulphate reduction rate profile also showed some shifting from month to month, but the integrated reduction rates over 35 cm (783, 965, 846, and 893 pmol SO_4/cm^2/hr for June, August, September and October respectively) are not significantly different statistically, and reflect the generally low rates observed in Trough sediments. These data suggest that bioturbational inmixing may not attain the depth where sulphate reduction becomes important.

Changes in the water column fluxes were quite small during the last few months of 1984. The degree of shifting observed in the porewater profiles suggests that events taking place in the euphotic zone may influence early diagenetic processes in the underlying sediment. We believe that greater variations in porewater profiles probably accompany major changes in the flux of

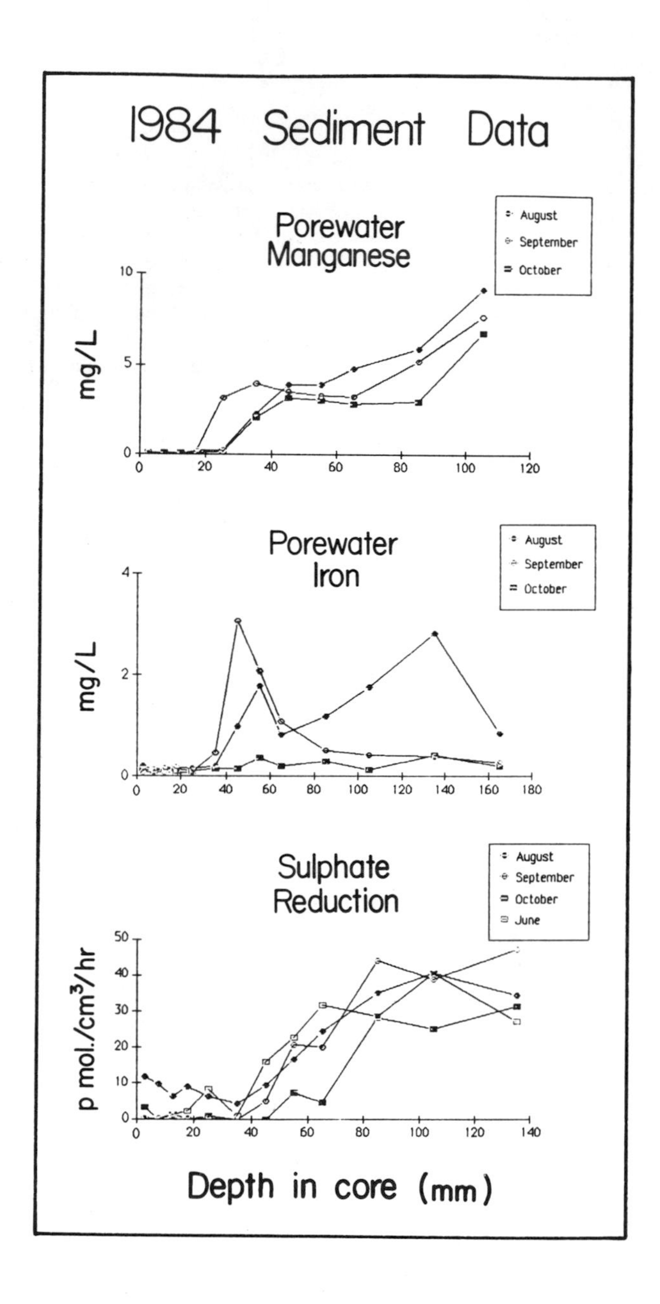

Figure 5: Profiles of manganese, iron and the rate of sulphate reduction in sediment cores from different months in 1984.

total solids, carbon, and nitrogen, which we observe consistently between April and July.

SUMMARY

Significant variations in the sedimentation rate, flux of organic matter, and in the quality of the organic matter reaching the sediment surface have been observed in a physically stable, deep coastal environment. Bioturbational mixing is sufficiently active to ensure that freshly arriving organic matter is incorporated below the sediment surface before it can be significantly degraded. Seasonal variations in the overlying water column may induce corresponding temporal variations in diagenetic parameters in the bottom sediments. We are continuing our study of the St. Lawrence region to further define the time-scales over which steady-state assumptions may be validly applied. Caution is advised when extending diagenetic rate constant determinations, based upon measurements obtained on a single date, to long-term models.

REFERENCES

1. Aller, R.C. & L.K. Benninger. "Spatial and temporal patterns of dissolved ammonium, manganese, and silica fluxes from bottom sediments of Long Island Sound, U.S.A.," *J. Mar. Res.* 39: 295-314 (1981).

2. Hines, M.E., W.B. Lyons, P.B. Armstrong, W.H. Oren, M.J. Spencer, H.E. Gaudette & G.E. Jones. "Seasonal metal remobilization in the sediments of Great Bay, New Hampshire," *Mar. Chem.* 15: 173-187 (1984).

3. Billet, D.S.M., R.S. Lampitt, A.L. Rice & R.F.C. Mantoura. "Seasonal sedimentation of phytoplankton to the deep-sea benthos," *Nature* 302: 520-522 (1983).

4. Bishop, J.K. & J. Marra. "Variations in primary production and particulate carbon flux through the base of the euphotic zone at the site of the Sediment Trap Intercomparison Experiment (Panama Basin)," *J. Mar. Res.* 42: 189-206 (1984)

5. Deuser, W.G. & E.H. Ross. "Seasonal change in the flux of organic carbon to the deep Sargasso Sea," *Nature* 283: 364-365 (1980).

6. Deuser, W.G., P.G. Brewer, T.D. Jickells & R.F. Commeau. "Biological control of the removal of abiogenic particles from the surface ocean," *Science* 219: 388-391 (1983).

7. Sundby, B., N. Silverberg & R. Chesselet. "Pathways of manganese in an open estuarine system," *Geochim. Cosmochim. Acta* 45: 293-307 (1981).

8. Sundby, B. & N. Silverberg. "Manganese fluxes in the benthic boundary layer," *Limnol. Oceanogr.* (in press).

9. Staresinic, N., G.T. Rowe, D. Shaughnessy & A.J. Williams. "Measurement of the vertical flux of particulate matter with a free-drifting sediment trap," *Limnol. Oceanogr.* 23: 559-563 (1978).

10. Edenborn, H.M., A. Mucci, N. Belzile, J. Lebel, N. Silverberg & B. Sundby. "A glove box for the fine-scale subsampling of sediment boxcores," (Submitted to *Deep-Sea Res.*).

11. Sundby, B., G. Bouchard, J. Lebel & N. Silverberg. "Rates of organic matter oxidation and carbon transport in early diagenesis of marine sediments," *Advances in Organic Geochemistry 1981*. J. Wiley & Sons, 350-354 (1983).

12. Aller, R.C., L.K. Benninger & J.K. Cochran. "Tracking particle-associated processes in nearshore environments by use of $^{234}Th/^{238}U$ disequilibrium," *Earth Planet. Sci. Lett.* 47: 161-175 (1980).

13. Silverberg, N., H.V. Nguyen, G. Delibrias, M. Koide, B. Sundby, Y. Yokoyama & R. Chesselet. "Radionuclide profiles, sedimentation rates and bioturbation in modern sediments of the Laurentian Trough," Joint Oceanographic Assembly, Halifax, Aug. 2-13 (1982)

14. Litchfield, C.D., M.A. Devanas, J. Zindulis, C.E. Carty, J.P. Nakas & E.L. Martin. "Application of the ^{14}C organic mineralization technique to marine sediments," in *Methodology for Biomass Determinations and Microbial Activities in Sediments, ASTM* (Philadelphia, PA: American Society for Testing and Materials, 1979), pp. 128-147.

Chapter 6

THERMAL DEGRADATION PRODUCTS OF NON-VOLATILE ORGANIC MATTER AS INDICATORS OF ANTHROPOGENIC INPUTS TO ESTUARINE AND COASTAL SEDIMENTS

A.G. Requejo, John Brown, Paul D. Boehm

Battelle New England Marine Research Laboratory
397 Washington Street
Duxbury, MA 02332

ABSTRACT

Anthropogenic inputs to surficial sediments in Boston Harbor, Massachusetts Bay and Cape Cod Bay were evaluated using stepwise pyrolysis of sediments followed by capillary gas chromatographic and gas chromatographic/mass spectrometric analyses of the volatile degradation products. A positive correlation was found between the ratio styrene/C_2-benzenes in sediment pyrolysates and the distribution of several classes of trace organic pollutants (PCB, PAH and the fecal sterol coprostanol) determined by conventional extraction methods. This correlation suggests an anthropogenic origin for these pyrolysates. A consideration of various substances which might yield styrene as a pyrolysate indicates that thermal degradation of synthetic polymers present in the sediments is the most likely mechanism of origin. The highest correlations with the styrene/C_2-benzenes pyrolysis ratio involved the sewage tracer coprostanol, suggesting that wastewater effluent discharges may be a source of styrene-generating substances to the sediments. The results demonstrate the potential of pyrolysis as an analytical tool in marine pollution studies.

INTRODUCTION

Analytical pyrolysis techniques such as pyrolysis gas chromatography (PGC) and pyrolysis gas chromatography/mass spectrometry (PGC/MS) have been applied extensively in organic geochemical studies of fossil fuels and ancient sediments (reviewed by Larter and Douglas (1)). By comparison, applications of these techniques in marine pollution studies have been limited. Whelan et al., (2) described a thermal distillation-PGC technique which, through

MARINE AND ESTUARINE GEOCHEMISTRY, Sigleo, A. C., and A. Hattori (Editors)

controlled heating, is capable of sequentially determining the content and composition of both volatile organic matter (termed P_1) andthermally degraded non-volatile organic matter (termed P_2) in sediments and particulate matter. Marine sediments contaminated by a fuel oil spill which were characterized by this technique exhibited elevated P_1/P_2 ratios and P_1 contents which correlated closely with total hydrocarbon concentrations obtained using conventional extraction methods (2). In a study of organic particulates in Boston Harbor, the P_1 and P_2 composition of sediment-trap particulates and sediments were found to differ from those of sewage sludge and more closely resembled those of phytoplankton, suggesting rapid reworking or mineralization of the sewage material (3). Other pollution studies employing pyrolysis techniques have examined PCB contamination in sediments (4) and the areal distribution of drilling fluid discharges in sediments (5).

The results of these initial studies indicate that PGC and PGC/MS can be a useful tool by which to characterize the distribution of pollutants in the marine environment. Their application in the majority of cases has relied primarily on gas chromatographic pattern "matching" of the P_1 and P_2 (or their equivalent) produced by various natural and anthropogenic organic substances with those of environmental samples, in order to infer the presence and relative abundance of these substances. This approach is based on the ability to discern a pattern characteristic of a given source which, in a sediment matrix containing organic matter from several sources, can be difficult. In addition, the overall pattern produced by different substances may not vary markedly. An alternative approach would be to examine the occurrence of specific pyrolysates whose presence and/or relative abundance might be indicative of organic inputs from different sources. This "marker" approach, analogous to the biomarker approach widely employed in organic geochemical studies, makes the tacit assumption that pyrolysates of sediment and particulate matter exhibit subtle but detectable qualitative and/or quantitative differences depending on organic matter origins, which can be used to differentiate between various natural and anthropogenic inputs. Although such differences have not been exploited in the various pollution studies which have employed pyrolysis, source distinctions using a similar approach have been reported in several marine geochemical studies (6-8).

We report here the results of a study in which surface sediments collected from Boston Harbor, Massachusetts Bay and Cape Cod Bay on the Atlantic Coast, U.S.A., were characterized using PGC and PGC/MS. The areal distributions of selected pyrolysates were contrasted with those of three classes of anthropogenic organic compounds - polychlorinated biphenyls (PCB), polycyclic aromatic hydrocarbons (PAH) and the fecal sterol coprostanol - determined by conventional solvent-extraction and liquid and gas chromatographic techniques using subsamples of the same sediments. The pollutant biogeochemistry of this region has been studied in detail by Boehm et al., (9). The objective of this study was to examine the relationship between sediment pyrolysate composition and the distribution of adsorbed trace organic compounds which have been previously documented as markers of anthropogenic organic inputs, in order to determine which pyrolysates might be suitable as pollution indicators.

STUDY AREA

Massachusetts and Cape Cod Bays are delineated by Cape Ann to the north and Cape Cod to the south (Figure 1) and are characterized by moderate rainfall, an absence of inputs from large rivers and seawater of uniformly high salinity (30-32 o/oo). Most of the region is approximately 80 m deep, although maximum depth in the Stellwagen Basin, located in the center of Massachusetts Bay, is over 100 m. The bottom sediments have been grouped according to location and sediment type (10). Nearshore, adjacent to the rocky coast from Cape Ann south to the Cape Cod Canal, the bottom is a patchwork of gravel, sand, mud and bedrock, while along the coast of Cape Cod the generally smooth bottom consists of well-sorted sand mixed with gravel. Finer grained sediments of clay, silt and sand are found in the deep basin and in the channels entering the bay.

Boston Harbor occupies an area of approximately 120 km^2 in the westernmost portion of Massachusetts Bay (Figure 1). Harbor waters have a salinity of about 25 o/oo and a residence time of approximately two tidal cycles, with an apparent net anticyclonic circulation. The harbor opening to Massachusetts Bay is partially blocked by Stellwagen Bank, an important offshore submarine feature which rises to within 20 m of the surface. Sediments in Boston Harbor are primarily silts and clays containing high levels of organic carbon and nitrogen. The principal source of anthropogenic contaminants to the offshore region is Boston Harbor, with its extensive industrialization, sewage discharges and heavy ship traffic. Two municipal waste treatment plants situated within the harbor on Deer and Nut Islands (see inset in Figure 1) discharge an average of 120 and 45 x 10^9 gallons of chlorinated primary sewage per year, respectively. Solids associated with these discharges can account for as much as 10% of the sediment deposited in the harbor (11). In addition, five major sewage systems on the northern side of Massachusetts Bay release approximately 25 x 10^9 gallons of sludge and effluent per year.

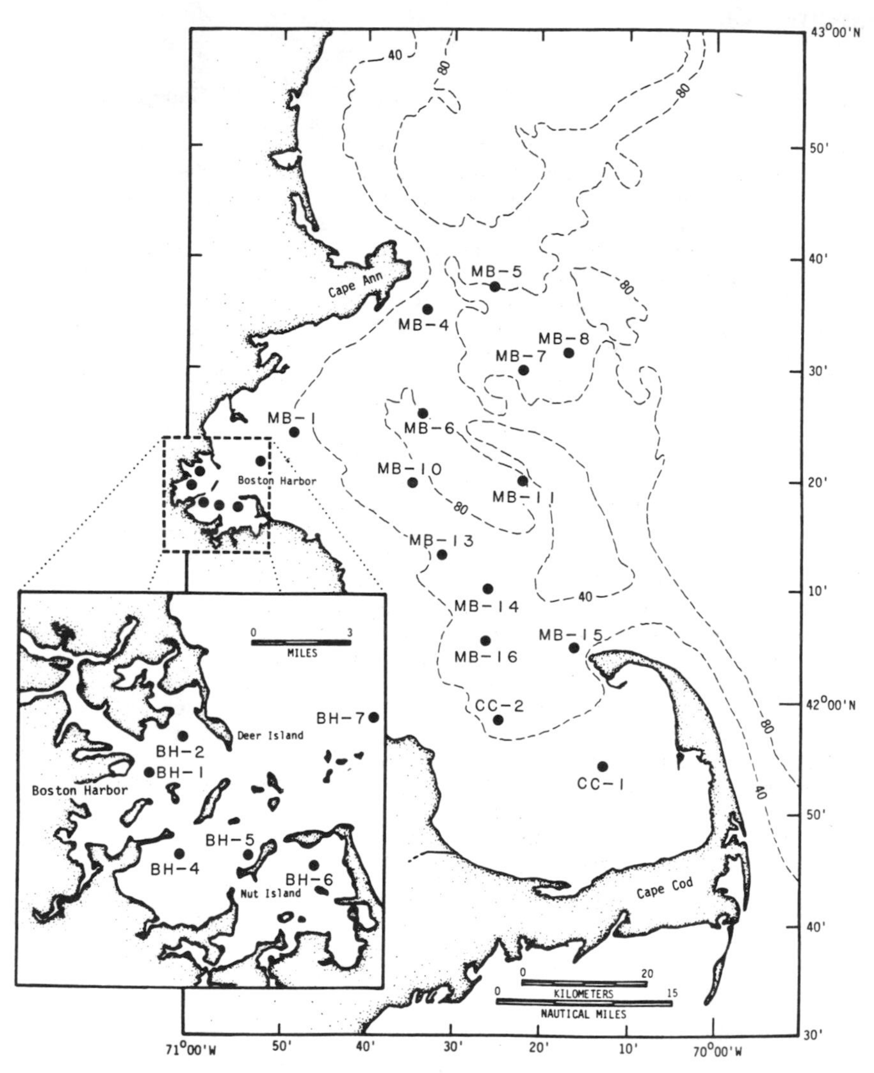

Figure 1. **Location of sediment sampling stations in Boston Harbor (BH), Massachusetts Bay (MB), and Cape Cod Bay (CC).**

MATERIALS AND METHODS

Sediment samples were collected in Boston Harbor and Massachusetts and Cape Cod Bays during June 1983 using a 0.1 m^2 modified Van Veen grab sampler. The sampling station grid is shown in Figure 1. The upper 2 cm from each grab were subsampled and transferred to clean, solvent-rinsed glass jars, which were refrigerated on board ship and frozen at -20°C after shipment to the laboratory.

Sediments for PGC and PGC/MS were acidified with an excess of 6N HCl (24 h) to remove carbonates. Each sample was then filtered through a precombusted Whatman GF/C glass fiber filter, rinsed thoroughly with organic-free deionized water, air dried and ground using a mortar and pestle.

The pyrolysis unit (Chemical Data Systems Model 121 Pyroprobe) is housed in a heated GC interface which is connected directly to the capillary inlet of a Hewlett/Packard 5840A gas chromatograph by a 1/8" stainless steel welded needle assembly. The interface is continuously swept with a stream of helium, which then flows into the capillary inlet. Separate mass flow controllers are used to regulate the flow of helium through the interface and the capillary inlet. A schematic diagram of the PGC apparatus is shown in Figure 2.

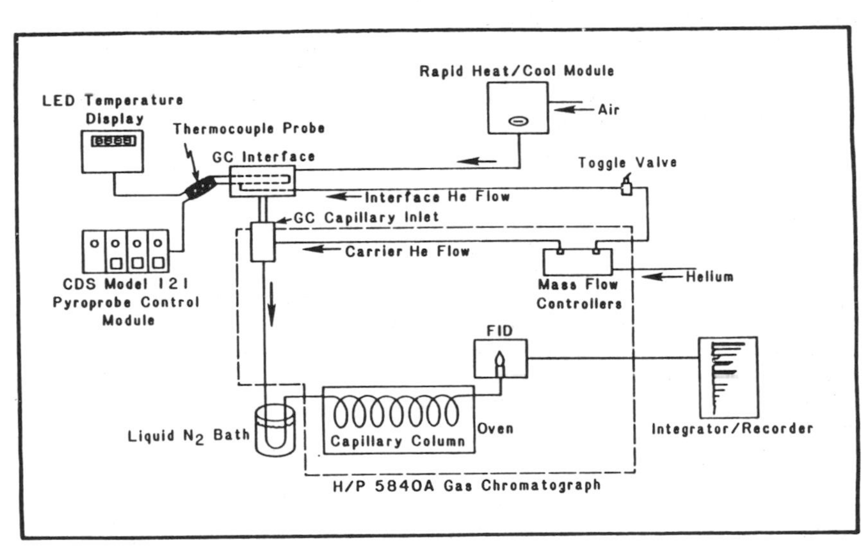

Figure 2. Schematic diagram of pyrolysis gas chromatography (PGC) apparatus.

Acidified sediments (5-35 mg) were weighed into 2 mm x 26 mm quartz tubes using plugs of prewashed quartz wool to hold the sample in place. In order to eliminate interferences from low temperature components, the quartz tube containing the sample was initially placed within the platinum coil of the probe and heated to 270°C in the GC interface. Helium was swept over the sample and vented into the ambient air for several minutes to remove any sorbed volatiles. The efficiency of this procedure was evaluated by injecting hexane solutions of a volatile hydrocarbon standard mixture (n-heptane to n-nonadecane) onto a sample and monitoring for the presence of standard components in sample pyrolysates. Pyrolysis gas chromatograms of sediment fortified in this manner were identical to those of unfortified samples, indicating that all added volatiles were removed prior to pyrolysis. The possibility of partial sample pyrolysis during the course of this procedure was examined by collecting and characterizing the volatile composition of several samples (detailed methods described below). In no case were any of the major pyrolysates identified in this study detected among the volatilization products. Volatilization yields were generally significantly lower than pyrolysis yields, with few resolved peaks evident.

Samples were pyrolyzed sequentially using set temperatures of 450°, 550° and 650°C, a temperature ramp of 20°C/msec and an interval of 10 sec. This stepwise technique was similar to programmed pyrolysis techniques used in previous studies (2,3,6-8). By monitoring the thermocouple display during pyrolysis, we were able to establish that the set temperatures employed corresponded to actual sample temperatures of 338 ± 4°, 403 ± 7° and 469 ± 12°C, respectively (n=4 to 7). This observation agrees with that of Sigleo et al., (7), who reported that for a Pyroprobe unit actual temperatures are approximately 125° to 150°C cooler than set temperatures. The products generated are carried by the helium stream into the GC capillary inlet, where they are introduced onto a fused silica capillary column. The inlet was operated in the split mode, using the combined flows through the GC interface and inlet to give a split ratio of approximately 50:1. A cold trap was used to collect and concentrate all pyrolysates prior to gas chromatography. Preliminary experiments using standards indicated that, under the flow conditions employed, 4 min trapping time was sufficient to recover n-triacontane. Most pyrolysis products however, were found to elute from the GC column well before this alkane.

For all gas chromatographic analyses, a 30 m x 0.25 mm DB-5 bonded phase fused silica capillary column (J&W Scientific, Inc.) was connected to the GC capillary inlet by a 1 m x 0.25 mm deactivated fused silica capillary guard column using a zero dead volume capillary column butt connector (Supelco, Inc.) and programmed from 40° to 290°C at 6°C/min after a 4 min isothermal period.

To identify unknown pyrolysates selected samples were analyzed by PGC/MS using a Finnigan/MAT Model 4530 GC/MS equipped with an INCOS 2300 data system. Identifications were made by matching the mass spectrum of unknowns with those contained in a computerized spectra library which forms part of the data system. Where possible, these identifications were confirmed by direct comparison of GC retention times and mass spectra using authentic standards. The same methods and GC conditions outlined above were also employed for the PGC/MS analyses.

Pyrolysis temperatures were calibrated using a variation of the model molecular thermometer technique of Levy and Walker (12), which employs the styrene/isoprene copolymer Kraton 1107 as a standardization material. Calibration of pyrolysis conditions was achieved by thermal desorption of a standard mixture containing saturated, unsaturated and aromatic hydrocarbons. The standard was injected daily onto a plug of quartz wool contained inside a quartz tube housed within the platinum pyrolysis coil.

Analytical methods for extraction and isolation of PCB, PAH, and coprostanol from sediments are described in Boehm (13) and Boehm et al., (9). Total PCB (sum of Aroclors 1242, 1254 and 1260) and total PAH (sum of selected 2-ring to 5-ring compounds) concentrations were determined by capillary GC/ECD (electron capture detection) and GC/MS, respectively, using internal standard techniques. Concentrations of coprostanol and other selected sterols (cholesterol, 5α-cholestanol, β-sitosterol, stigmasterol) were determined by capillary GC/FID (flame ionization detection) relative to an external standard.

Acidified sediments were analyzed for total organic carbon (TOC) content by high temperature combustion using a Hewlett Packard CHN analyzer (9).

RESULTS AND DISCUSSION

The major pyrolysates consisted of a series of early-eluting peaks, which could not be adequately resolved under the chromatographic conditions employed (Figure 3). Several of these components were identified by PGC/MS from their molecular ions as acetone, acetic acid, methyl acetate and methyl furans. Because of the difficulties encountered in their resolution, and thus their unambiguous identification, no attempts were made to examine the distribution of these compounds. Instead, our characterizations focused on later-eluting pyrolysates which could be more efficiently resolved and reproducibly detected. Compounds identified within this category include pyrrole, toluene, styrene and other C_2-benzenes such as xylenes and ethyl benzenes phenol, m- and p-cresol and a series of 1-alkene/n-alkane pairs ranging in carbon numbers from C_8 to C_{17} (Figure 3). All of these compounds have been previously reported as thermal degradation products of sediments and particulate matter (2, 3, 6).

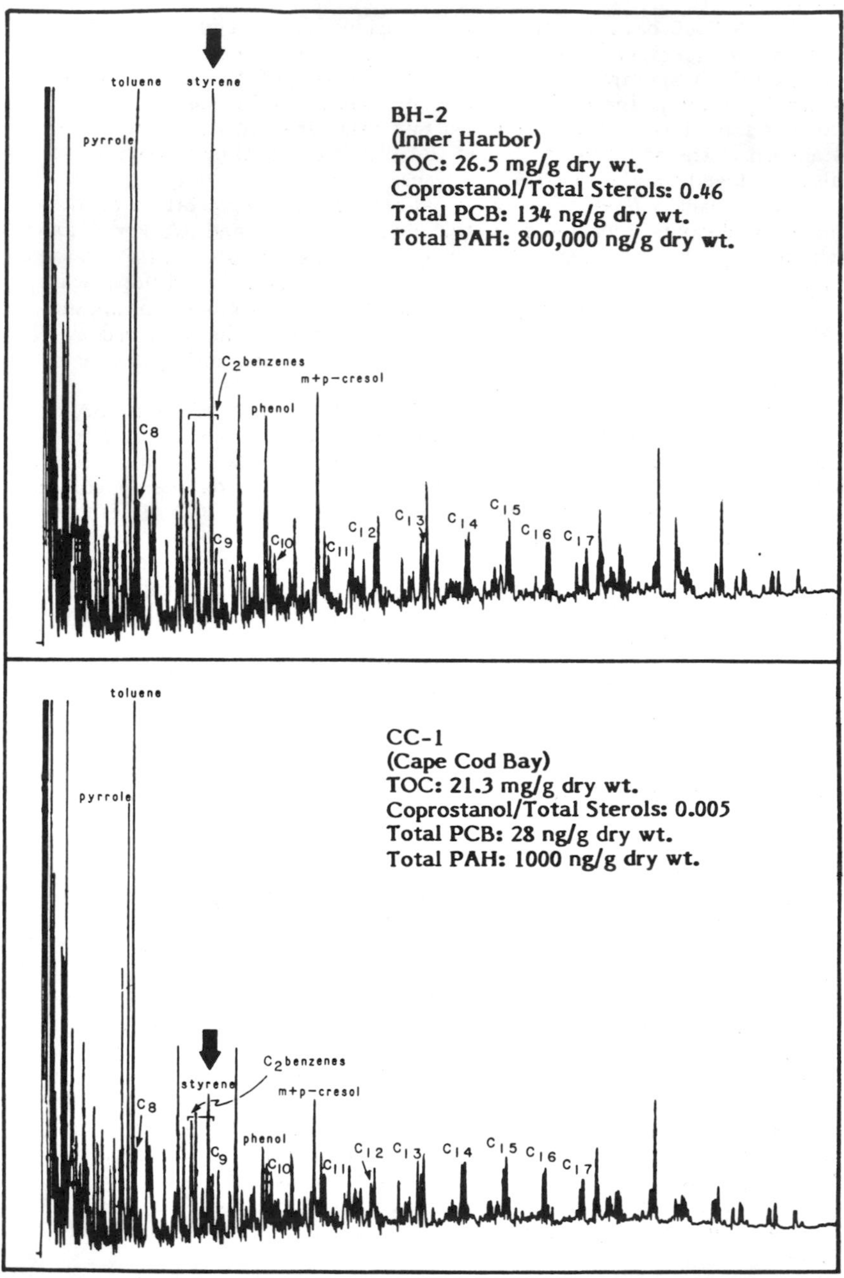

Figure 3. Typical PGC chromatograms of surface sediments from sites BH-2 and CC-1. Identities of selected components are given above the corresponding peak. TOC and trace organic pollutant data for each sample are also given. For experimental conditions, see text.

The overall PGC chromatographic patterns exhibited by most samples were strikingly similar. However, upon closer examination, one pyrolysate - styrene -was found to exhibit large variations in both its occurrence and relative abundance (exemplified by the bold arrow in Figure 3). To quantify this variation a pyrolysis ratio was formulated based on the chromatographic peak areas corresponding to:

$$\text{Styrene/C}_2\text{-benzenes} = \frac{\text{styrene}}{\text{o-xylene + m+p-xylene + ethyl benzene}} \quad (1)$$

The xylenes and ethyl benzene were selected because, together with styrene, they comprise the total C_2-benzenes detected and should exhibit a similar PGC response. Values for styrene/C_2-benzenes in the sediments analyzed, together with their TOC content and the concentrations of PCB, PAH and coprostanol relative to organic carbon, are listed in Table I. The highest styrene/C_2-benzenes values (1.27 to 2.50) were found in the sediments from Boston Harbor. With the exception of two inshore sites (MB-1 and MB-4), the samples from Massachusetts and Cape Cod Bays had values of less than 1.00. The distribution of organic pollutants exhibits the same general trend, with elevated values in the harbor sediments which for each parameter decrease to uniformly lower levels in the offshore regions (Table I).

The areal distribution of styrene/C_2-benzenes in surface sediments of the region is depicted in Figure 4. The highest value was found at site BH-2, which is located at close proximity to the Deer Island waste treatment plant (inset in Figure 4). Somewhat elevated values are also evident in Massachusetts Bay along the northern shoreline and in a region off the tip of Cape Cod, to the south and east of the harbor. Overall, however, the geographic trend is toward decreasing values with increasing distance from the harbor. All values less than 0.50 were found either at the outermost Massachusetts Bay sites or in central Cape Cod Bay (Figure 4). This distribution suggests that the occurrence of styrene is not likely associated with organic contributions from marine sources, which would be greatest at the outermost sites, but is instead associated with some input common to the harbor region. Given the magnitude of wastewater and industrial discharges to the harbor and the documented contamination of harbor sediments, it would seem likely that anthropogenic sources directly or indirectly influence the observed distribution.

A detailed comparison of the organic pollutant data with styrene/C_2-benzenes values supports the apparent anthropogenic origin for styrene. Figure 5 shows the results of linear regression analyses between styrene/C_2-benzenes and four anthropogenic geochemical ratios - coprostanol/total sterols, coprostanol/TOC, PCB/TOC and PAH/TOC. A significant positive correlation is evident in each case. The strongest correlations are obtained versus the two parametric ratios involving the fecal sterol coprostanol. Coprostanol has been used previously as a geochemical tracer of sewage discharge

Table I. PGC Styrene/C_2-Benzenes Ratio, Total Organic Carbon (TOC), and Organic Pollutant Geochemical Ratios for Boston Harbor-Massachusetts Bay-Cape Cod Bay Sediments. Station Locations are Shown in Figure 1.

Station	Styrene/ C_2-Benzenes	TOC (mg/g Dry Wt.)	Coprostanol/ Total Sterols	Coprostanol/ TOC X 10^6	PCB/ TOC X 10^6	PAH/ TOC X 10^6
			Boston Harbor			
BH-1	1.94	8.8	0.31	372	8.90	222
BH-2	2.50	26.5	0.46	672	5.07	40400
BH-4	1.77	43.1	NC	NC	NC	NC
BH-5	1.72	24.9	0.20	244	4.93	329
BH-6	1.27	13.6	0.23	121	4.90	140
BH-7	1.62	3.8	0.16	157	3.66	121
			Massachusetts Bay			
MB-1	1.00	8.9	0.03	15.7	4.99	1370
MB-4	1.14	13.3	0.02	6.8	1.85	94.7
MB-5	0.38	13.5	0.02	3.7	0.43	72.6
MB-6	0.78	24.0	0.02	12.9	3.43	156
MB-7	0.41	20.3	0.03	1.5	1.37	76.8
MB-8	0.24	10.6	0.03	13.2	0.94	56.6
MB-10	0.74	17.5	0.02	5.7	1.19	80.0
MB-11	0.40	23.7	0.09	16.5	1.14	81.0
MB-13	0.96	6.2	0.02	16.1	1.02	71.0
MB-14	0.82	8.8	0.02	21.6	1.18	78.4
MB-15	0.83	8.6	NC	NC	NC	NC
MB-16	0.78	10.2	0.03	5.9	0.70	9.8
			Cape Cod Bay			
CC-1	0.50	21.3	0.01	2.8	1.31	56.3
CC-2	0.70	15.0	0.03	12.7	1.64	84.7

NC = Not Calculated.

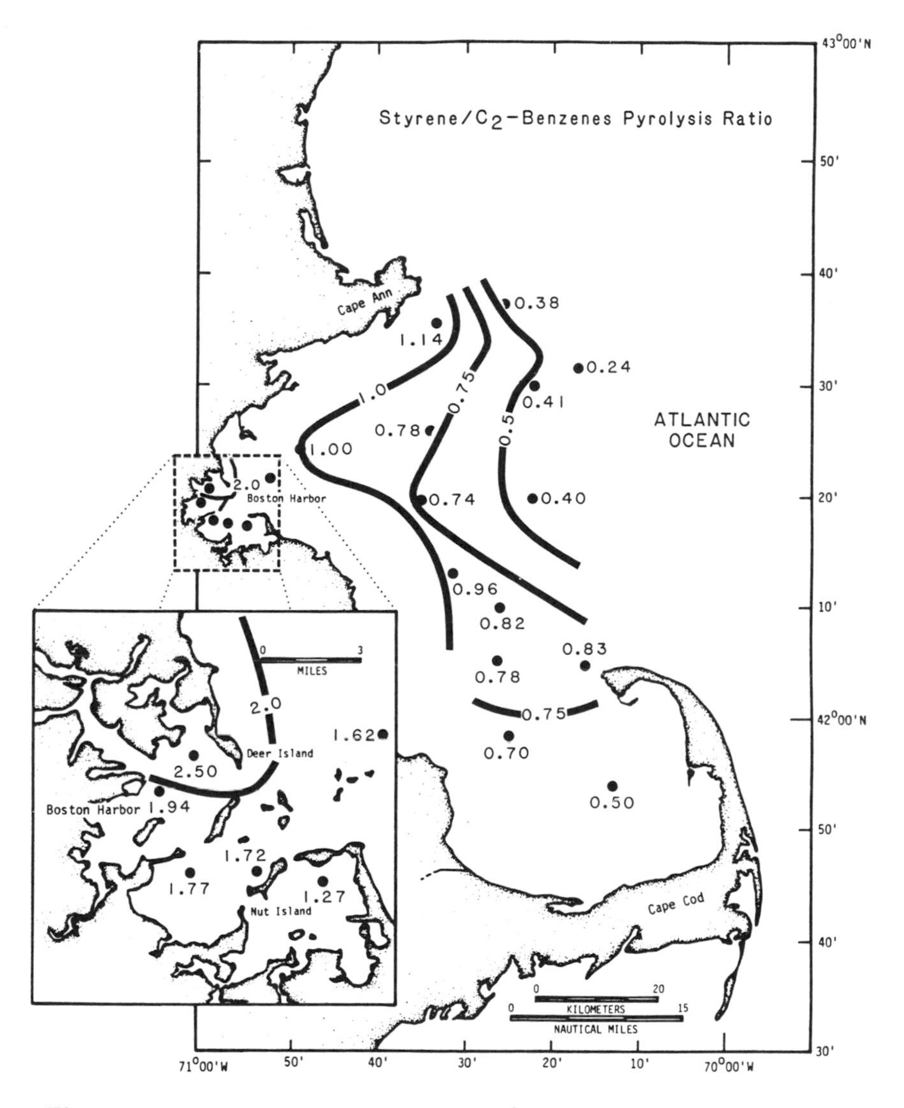

Figure 4. Areal distribution of styrene/C_2-benzenes pyrolysates ratio.

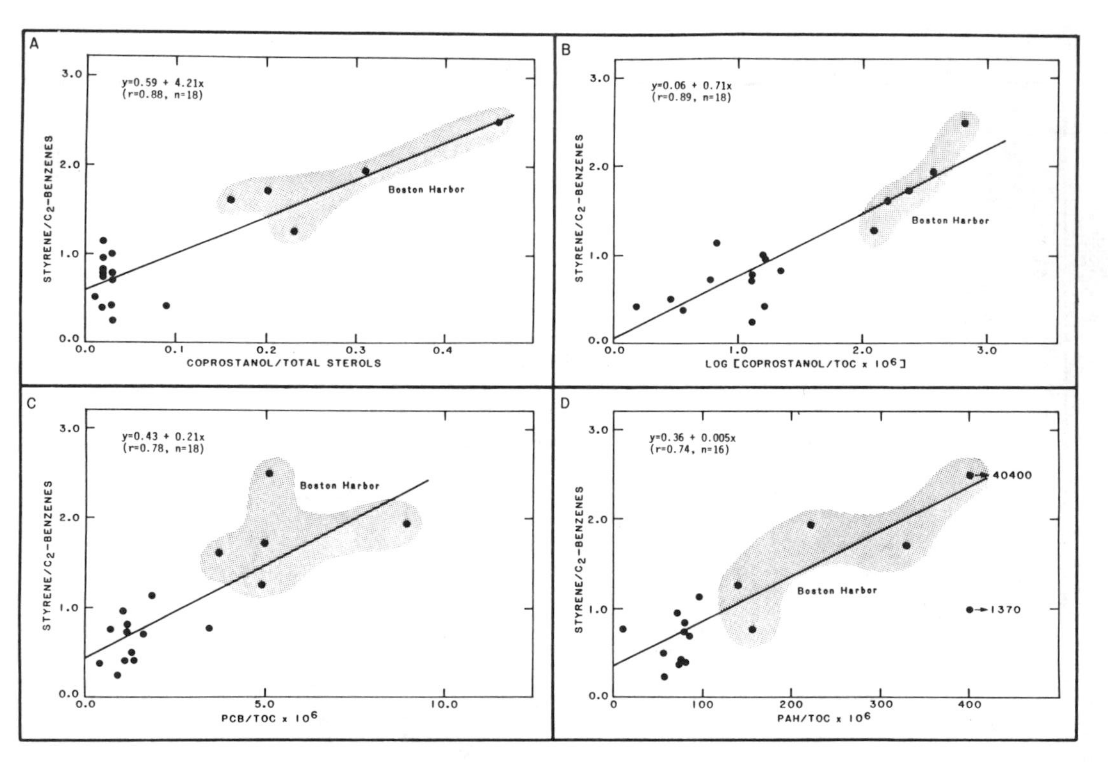

Figure 5. Linear regression analysis for styrene/C_2-benzenes pyrolysis ratio and (A) coprostanol/total sterols, (B) log coprostanol/TOC x 10^6 , (C) PCB/TOC x 10^6, (D) PAH/TOC x 10^6. Note that in (D) the two offscale points are not included in the regression calculation. The shaded data points correspond to samples from the Boston Harbor region.

(14-16). Sewage sludge has a coprostanol/total sterol value between 0.5 and 0.8 (9); thus the value of 0.46 found at site BH-2 (which also exhibits the highest styrene/C_2-benzenes value) indicates a substantial sewage input, consistent with its proximity to the Deer Island outfall. The strong correlation between these sewage indicators and styrene/C_2-benzenes suggests that this pyrolysis parameter may be related to a very specific anthropogenic input i.e., sewage-derived solids which become incorporated in the sediments, and thus might serve as a sewage tracer in future environmental studies. The fact that lesser correlations are obtained versus the ratios PCB/TOC ($r = 0.78$) and PAH/TOC ($r = 0.74$) also indirectly supports the inferred sewage-specific source. Although both of these compound classes are present in significant quantities in sewage, each can also originate through alternate sources or transport pathways (e.g. industrial inputs and atmospheric deposition in the case of PCB, petroleum inputs associated with shipping traffic or accidental spills and generation through combustion processes in the case of PAH). All four geochemical ratios can be determined with comparable precision ($\pm$ 10-20%), and differences in degree of linear correlation are not likely to be attributed to variable analytical precision.

We examined the possibility that the pyrolytic generation of styrene might be an artifact of the sample preparation method. Sediments not acidified prior to PGC were analyzed and the results compared with those of the same samples after acidification. Deleting the acid treatment resulted in moderate or no increases in the styrene/C_2-benzenes values (0 to + 27%). Although some increases were slightly greater than the $\pm$ 17% expected solely from analytical variability ($n = 5$), they clearly indicate that formation of styrene is not related to acidification. Similarly, comparison of solvent-extracted versus unextracted samples resulted in variations ranging from -20% to +10%, indicating that styrene/C_2-benzenes values are also not affected by solvent extractable material. It would appear therefore, that the variations in styrene result from differences in the composition of the macromolecular organic matter present in the sediments. The alternate possibility that styrene variations are due to varying contents of smectic clays in harbor versus offshore sediments, which might result in catalytically-enhanced styrene generation (17), is not favored because the clay content of samples exhibiting large styrene/C_2-benzenes differences did not vary sufficiently to account for such an effect and also because the abundance of other pyrolysates such as toluene and some of the lower molecular weight 1-alkene/n-alkane pairs which would also be subject to the same catalytic effect do not exhibit a comparable enhancement.

Styrene has also been reported as a pyrolysate of particulate matter and sediments collected in the Rhine River delta (6). Whelan et al., (3) characterized styrene as a pyrolysate common to "everything" and thus of little diagnostic value as an indicator of biogenic organic matter sources. Although styrene has been detected in pyrolysates of compositionally diverse substances (e.g., phytoplankton (7), peat (18) and coal (19)), it has not been reported as a major pyrolysate in biopolymers or geopolymers (1-3,6-8). Several synthetic polymers, on the other hand, have been shown to yield significant quantities of styrene upon pyrolysis. In addition to polystyrene plastics, Wuepper (20) reported that other common copolymers such as those in the carbon black filled rubbers used in friction pads (styrene-butadiene) and paint resins (styrene-ethyl acrylate) yield styrene as a major pyrolysate. The Kraton 1107 standardization material would be another example of a synthetic polymer which produces significant quantities of styrene upon pyrolysis. Given the areal distribution and the correlation with organic pollutants described above, we consider the most likely source of styrene in these sediments to be synthetic polymers originating from anthropogenic discharges. A biogenic origin is not favored by the previously cited compositional studies, which have shown styrene to be a minor pyrolysate of various marine and terrestrial biopolymers and geopolymers (a marine planktonic contribution should not be ruled out entirely, however, as the very few species analyzed have shown high benzyl contents and variable compositions (3,7)).

The styrene/C_2-benzenes ratio might be widely applicable as a general anthropogenic, rather than strictly a sewage tracer, since synthetic polymers may also originate from alternate sources, such as industrial effluents. The degree of specificity might vary from region to region; for example, in the Boston Harbor -Massachusetts Bay region, the major source of anthropogenic inputs is largely wastewater effluent discharges originating in the harbor and at other point sources, thus the strong correlation between sewage tracers and styrene/C_2-benzenes. Highly industrialized regions where associated discharges would be more significant or coastal areas receiving drainage from urban rivers might exhibit slightly different styrene/C_2-benzenes distributions. The exact case awaits the determinations of this pyrolysis ratio in other estuarine or coastal regions. The results of the present study indicate that this and similar pyrolysis ratios can be a valuable tool in future marine pollution studies. The principal advantages of pyrolysis techniques relative to the more conventional solvent extraction techniques are the small sample size requirements, minimal sample preparation and rapid analysis times.

Acknowledgements

We thank Don Cobb and the crew of the R/V Mya for collection of the sediment samples and Bill Steinhauer for coordinating much of the trace organic analyses. The study was sponsored in part by NOAA/National Marine Fisheries Service (Contract No. NA-83-GA-C-00022).

REFERENCES

1. Larter, S.R. and A.G. Douglas. "Pyrolysis Methods in Organic Geochemistry: An Overview." J. Anal. Appl. Pyrolysis 4:1-19 (1982).

2. Whelan, J.K., J.M. Hunt and A.Y. Huc. "Application of Thermal Distillation - Pyrolysis to Petroleum Source Rock Studies and Marine Pollution." J. Anal. Appl. Pyrolysis 2:79-96 (1980).

3. Whelan, J.K., M.G. Fitzgerald and M. Tarafa. "Analyses of Organic Particulates from Boston Harbor by Thermal Distillation - Pyrolysis." Environ. Sci. Tech. 17:292-298 (1983).

4. McMurtrey, K.D., N.J. Wildman and H. Tai. "Pyrolysis Gas Chromatography - Mass Spectrometry of Polychlorinated Biphenyls on Sediment." Bull. Environ. Contam. Toxicol. 31:734-737 (1983).

5. Kennicutt, M.C., W.L. Keeney - Kennicutt, B.J. Bresley and F. Fenner. "The Use of Pyrolysis and Barium Distributions to Assess the Areal Extent of Drilling Fluids in Surficial Marine Sediments." Environ. Geol. 4:239-249 (1983).

6. Van de Meent, D., J.W. de Leeuw and P.A. Schenk. "Chemical Characterization of Non-Volatile Organics in Suspended Matter and Sediments of the River Rhine Delta." J. Anal. Appl. Pyrolysis 2:249-263 (1980).

7. Sigleo, A.C., T.C. Hoering and G.R. Helz. "Composition of Estuarine Colloidal Material: Organic Components." Geochim. Cosmochim. Acta 46:1619-1626 (1982).

8. Wilson M.A., R.P. Philip, A.H. Gillam, T.D. Gilbert and K.R. Tate. "Comparison of the Structures of Humic Substances from Aquatic and Terrestrial Sources by Pyrolysis Gas Chromatography - Mass Spectrometry." Geochim. Cosmochim. Acta 47: 497-502 (1983).

9. Boehm, P.D., W. Steinhauer and J. Brown. "Organic Pollutant Biogeochemistry Studies-Northeast U.S. Marine Environment." Battelle Final Report to NOAA/National Marine Fisheries Service, Highlands, New Jersey, Contract No. NA-83-FA-C-00022 (1984).

10. Schlee, J. "Atlantic Continental Shelf and Slope of the United States - Sediment Texture of the Northeastern Part." U.S. Geol. Survey Prof. Paper 529-L (1973).

11. Fitzgerald, M.G. "Anthropogenic Influence of the Sedimentary Regime of an Urban Estuary - Boston Harbor." Ph.D. Dissertation, Woods Hole Oceanographic Institution, WHOI-80-38. (1980).

12. Levy, E.J. and J.Q. Walker. "The Model Molecular Thermometer: A Standardization Method for Pyrolysis Gas Chromatography". J. Chromatog. Sci. 22:49-55 (1984).

13. Boehm, P.D. "Estuarine - Continental Shelf and Benthic -Water Column Coupling of Organic Pollutants in the New York Bight Region." Can. J. Fish. Aqu. Sci. 40 (Suppl. 2): 262-276 (1983).

14. Goodfellow, R.M., J. Cardoso, G. Eglinton, J.P. Dawson and G.A. Best. "A Faecal Sterol Survey in the Clyde Estuary." Mar. Pollut. Bull. 8:272-276 (1977).

15. Hatcher, P.G. and P.A. McGillivary. "Sewage Contamination in the New York Bight. Coprostanol as an Indicator." Environ. Sci. Tech. 13:1225-1229 (1979).

16. Walker, R.M., C.K. Wun and W. Litsky. "Coprostanol as an Indicator of Fecal Pollution." CRC Critical Rev. Environ. Control. April: 91-112 (1982).

17. Davis, J.B. and J.P. Stanley. "Catalytic Effect of Smectite Clays in Hydrocarbon Generation Revealed by Pyrolysis-Gas Chromatography." J. Anal. Appl. Pyrolysis 4:227-240 (1982).

18. Roy, C., E. Chornet and C.H. Fuchsman. "The Pyrolysis of Peat. A Comprehensive Review of the Literature." J. Anal. Appl. Pyrolysis 6:262-332 (1984).

19. Hughes, B.M., J. Troost and R. Liotta. "Pyrolysis/$(GC)^2$/MS as a Coal Characterization Technique." in American Chemical Society, Division of Fuel Chemistry, Preprints of Papers Presented at Atlanta, Georgia, March 29-April 3, 1981. 26(2):107-120 (1981).

20. Wuepper, J.L. "Pyrolysis Gas Chromatographic - Mass Spectrometric Identification of Intractable Materials." Anal. Chem. 51:997-1000 (1979).

CHAPTER 7

PARTITIONING OF PCBs IN MARINE SEDIMENTS

Bruce J. Brownawell and John W. Farrington
Department of Chemistry
Woods Hole Oceanographic Institution
Woods Hole, Massachusetts 02543

ABSTRACT

Polychlorinated biphenyls (PCBs) are useful model compounds to study the physical-chemical processes which affect the biogeochemistry of hydrophobic organic compounds. In this study two box cores from New Bedford Harbor, Massachusetts, were analyzed for PCBs. Measurements are reported for total PCBs and several individual chlorobiphenyls for both the sediments and interstitial waters. Concentrations of total PCBs were highly elevated in the pore waters and reached a maximum of 20.1 μg/L at the Outer Harbor site. Results from these two cores combined with predictions from laboratory experiments indicate that most of the PCBs measured in interstitial waters are actuallly sorbed to organic colloids. The partitioning of chlorobiphenyls between water column particulates and filtrate shows a greater importance of dissolved compounds due to lower concentrations of organic colloids. A simple three-phase equilibrium model involving colloids, dissolved phase, and chlorobiphenyls sorbed to particulate organic matter is presented to explain the observed partitioning.

INTRODUCTION

Polychlorinated biphenyls (PCBs) have a wide range physical-chemical properties [1,2] which are representative of many hydrophobic organic compounds present in the marine environment. Because of their slow rates of chemical and biological degradation [3], PCBs provide excellent model compounds to study the physical-chemical processes important in the biogeochemistry of organic compounds in sediments. Contemporary sediments are repositories for PCBs and many other hydrophobic, recalcitrant pollutants released to the environment. Many depositional environments are biologically and physically active, and sediment-interstitial water partitioning processes will control the rates at which organic pollutants are released back to the water column or taken up by the biota. In the water column, the transport and fate of hydrophobic organic compounds will also be affected by associations with suspended particles and colloidal substances. The extent of these partitioning reactions influence rates of particulate removal, dispersion, volatilization, and biological uptake.

MARINE AND ESTUARINE GEOCHEMISTRY, Sigleo, A. C., and A. Hattori (Editors)

The sorption reactions of PCBs and other nonpolar organic compounds with freshwater sediments and soils have been well studied with laboratory experiments [4,5,6,7,8,9]. These and other studies have documented the importance of the organic fraction of sediments in controlling aqueous sorption of these compounds. Linear partition coefficients, K_p (L/Kg), can be predicted for a range of sediments by a single organic carbon normalized partition constant, K_{oc}:

$$K_p = f_{oc} K_{oc} \quad (1)$$

where f_{oc} is the fraction organic carbon of the sorbent. K_{oc} increases with decreasing solubility and increasing hydrophobicity of the compound. K_{oc} has been shown to be related to the octanol-water (K_{ow}) coefficient of the sorbate:

$$\log K_{oc} = a \log K_{ow} + b \quad (2)$$

where the slope (a) ranges from 0.72 to 1.0 in studies by different workers using different compound classes [4,6,7,8].

Organic colloids and dissolved humic substances have also been shown in the laboratory to have high sorption affinities for hydrophobic compounds [10,11,12]. Estuarine and marine colloids have been operationally defined as those substances which pass a 0.4 to 1 μm filter but are retained by a variety of ultrafiltration membranes which exclude most materials having dimensions greater than 1-4 nm. Colloidal substances include both submicron particles and macromolecular organic matter. The presense of colloidal organic matter has been shown to decrease the apparent sorption of 2,2',5,5'-tetrachlorobiphenyl and cholesterol onto riverine particles [13] and that of fatty acids with marine sediments [14]. Gschwend and Wu [9] have also demonstrated the effect of non-settling particles in decreasing partition coefficients of PCBs in laboratory sediment sorption experiments. Estuarine organic colloids appear to behave as sorbents for hydrophobic organic compounds in a manner similar to particulate organic matter [10,12]. Thus, there are experimental and field observation data establishing that associations of PCBs with colloidal substances will affect their rates and modes of transport in aquatic environments.

To study the partitioning of PCBs, we have investigated the distribution of PCBs in sediments from New Bedford Harbor, Massachusetts. Interstitial waters from these organic-rich, coastal sediments provide an environment with high concentrations of colloidal organic matter, and the observed distribution of individual chlorobiphenyls yields insights into the roles of sedimentary and colloidal organic matter in partitioning. Measured distribution coefficients or ratios, K'_d, are calculated as:

$$K'_d = \frac{S}{PW} \quad (L/Kg) \quad (3)$$

where S and PW are the sediment and pore water concentrations of individual chlorobiphenyls. The pore water concentration includes both dissolved and any colloid associated components. Comparison of measured K'_ds of PCBs in sediments provide a good test of predictive sorption models based on laboratory results.

In this chapter, we present results of PCB analysis from two box cores taken in New Bedford's inner and outer harbors. A more complete discussion of the biogeochemistry of PCBs in the outer harbor core is reported elsewhere [15], but we have incorporated examples of the partitioning data to better illustrate the role of organic colloids in sediment interstitial waters. To provide a useful comparison to the sediment investigations, we also report results from a water column partitioning study where two stations in New Bedford Harbor were occupied and sampled over a tidal cycle. Within the limitations of data sets available, the partitioning results from these studies are compared to a simple three-phase equilibrium partitioning model based on predictions from laboratory experiments.

SAMPLING SITES AND METHODS

The sampling sites of the sediment cores (Stations 67 and 84) and water samples (Stations 74 and 81) are shown in Figure 1. New Bedford Harbor is severely polluted by PCB contamination [16,17]. Concentrations of PCBs in surface sediments are highest in the upper part of the estuary, north of our Station 81, and generally decrease moving south through the outer harbor and out into Buzzards Bay.

Station 84 was sampled with a Soutar box corer, 0.04 m^2 x 1 m, off the RV Asterias on October 29, 1981. This site was in 3 meters of water and bottom water temperature was 13.1°C. Sediment samples of one to three cm depth were sectioned into glass quart jars shipboard. The upper three cm was heavily populated by a small (about one cm diameter) unidentified bivalve. Sediments were stored overnight at room temperature and pore water samples were extracted the following two days using a hydraulically powered, stainless-steel squeezer at 2000-2500 psi as discussed by Henrichs [18]. Interstitial water was filtered through two internal Reeve Angel glass fiber filters and an external Gelman Type AE glass fiber filter of a nominal pore size of 1.0 μm. 75 to 240 ml pore water samples were obtained for each sediment depth and were stored in glass with 25 ml of CH_2Cl_2 for later PCB analysis.

A large volume box core was obtained at Station 67 with a Sandia-Hessler type MK3 sediment corer on September 1, 1983. The temperature of the sediment was 20.5°C and inert atmosphere techniques were used in the sectioning of the core and subsequent extraction and filtration of pore water. These procedures are explained in detail elsewhere [19,15] and resulted in the isolation of about 1.5 liters of pore water per two cm depth

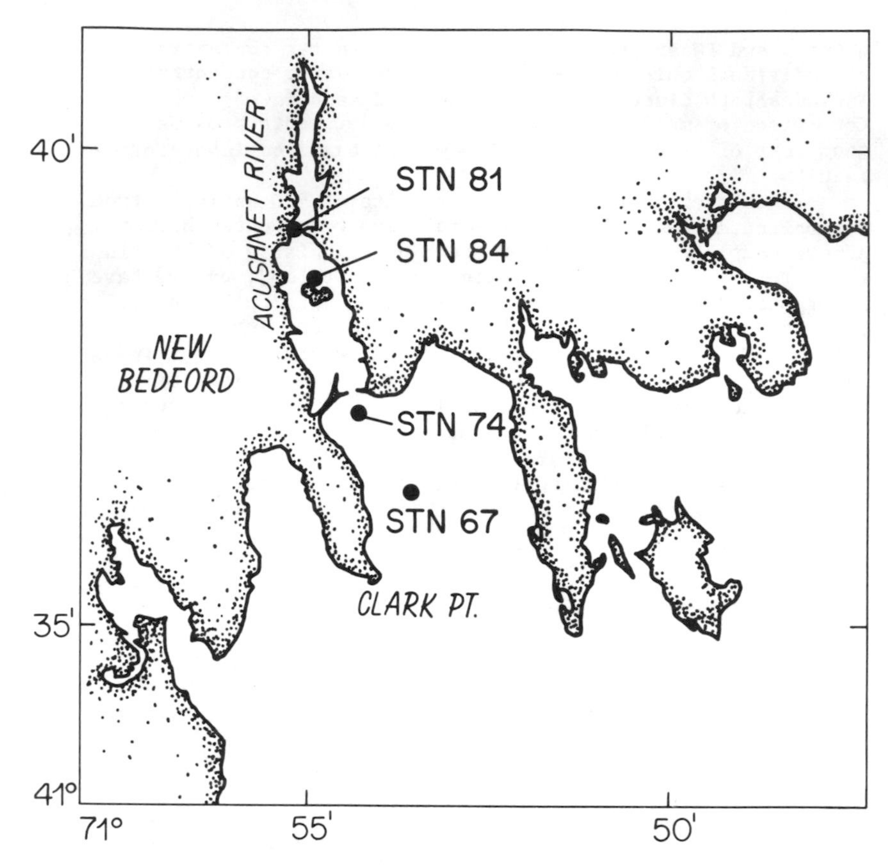

Figure 1. Map of New Bedford Harbor on the United States east coast.

horizon. Samples of pore water were stored and analyzed for PCBs, dissolved organic carbon (DOC), salinity, sulfate, and Fe and Mn. Frozen sediment samples were later analyzed for PCBs and CHN.

Water column samples were taken at Stations 81 and 74 throughout a complete tidal cycle on September 22, 1982. Samples were collected hourly in stoppered two liter flasks at both subsurface and near-bottom depths at both sites. CTD casts were also made at each site every hour. These samples, covering incoming and outgoing tides, were combined into 14-17 L composite samples. The eight combined samples were pressure-filtered through Gelman Type A/E glass fiber filters within ten hours of sampling, and filters were frozen until analysis. CH_2Cl_2 was added to the filtered samples and stirred vigorously in glass carboys.

Analysis of PCBs in all three sets of samples was similar with slight procedural modifications in chromatography and quantification of total PCBs. Filtered water and pore water samples were extracted three times with CH_2Cl_2 in separatory funnels. Particulate material, filtered from each water sample, and wet sediments were Soxhlet extracted for 24 hrs. with 1:1 hexane:acetone (water particulates were extracted for an additional day with fresh hexane:acetone). Extracts are combined when necessary, dryed over sodium sulfate, evaporated to near dryness, and chromatographed on silica gel columns. Chromatographic columns for sediment and interstitial water extracts employ a layer of activated copper to remove reduced and elemental sulfur. PCB fractions were eluted from the silica gel with either hexane or toluene in hexane and then analyzed by capillary column gas chromatography and electron capture detection on a HP5840 gas chromatograph equipped with a 30 meter DB5 fused silica column (J&W Scientific) using H_2 carrier gas. 2.0 μl syringe injections were made using the following conditions: 2 minutes at 40°C followed by rapid heating to 120°C (hold until 5 minutes; temperature programming at 2°C/min to 230°C and then 4°C/min to 270°C; with a makeup gas of 5% methane/95% argon at 30 ml/min.

The apparent distribution coefficients, K'_d (equation 3), of 51 chlorobiphenyl peaks are calculated as the ratio of the peak areas of sediment and pore water samples multiplied times appropriate sample factors. The chlorobiphenyl identification of these peaks is given in Table I and is based on identifications by other workers [20,21,22,23,24,25], retention times of over 70 pure chlorobiphenyl standards (from Analabs and UltraScientific), and GCMS spectra of Aroclor PCB standards. Table I also lists the reported values of log K_{ow} determined by reverse phase HPLC [2] for many of these compounds.

Concentrations of total PCBs have been estimated as mixtures of Aroclor 1242 and Aroclor 1254 which are industrial mixtures of PCBs consisting primarily of di- through tetrachloro-, and tetra- through hexachlorobiphenyls, respectively. Total PCB estimates of sediments and pore waters were calculated by capillary gas chromatography described above and in [15]. Estimations of total PCBs in water column samples were determined by comparison to Aroclor standards on a Perkin-Elmer 900 gas chromatograph run isothermally at 190°C on a two meter column packed with 1.5% OV-17/1.95% QF-1 on Chromosorb W HP (100/120 mesh). Chlorobiphenyls <u>29</u> and <u>143</u> were used as internal standards in Station 67 samples and in the sediment and particulate samples in the other studies. Average recoveries of these internal standards were 80-95% in the various studies. Results from Station 67 incorporate internal standards as quantifcation standards, while the other samples do not. Concentrations and K'_d calculations for the latter samples do not correct for recoveries and have higher estimated errors (±15% in concentrations and ±22% in K'_d) compared to those reported for Station 67 results [15].

Table I. PCB identification and associated values of log K_{ow} from Rapaport and Eisenreich [2].

Peak no.	#Cl	PCB Isomer no.[a]	Identification	Relative Retention Time	log K_{ow} [b]	(log K_{ow})[c]
1	2	4(d) +	2,2'	0.2776	4.89	4.89
	2	10	2,6		5.31	
2	2	7(d) +	2,4	0.3103	5.30	5.30
	2	9	2,5			
3	2	6	2,3'	0.3245	5.02	5.02
4	2	8(d) +	2,4'	0.3323	5.10	5.10
	2	5	2,3			
5	3	19	2,2',6	0.3594	5.48	5.48
6	3	18	2,2',5	0.3940	5.55	5.55
7	3 +	17(e) +	2,2',4	0.3964	5.76	
	2	15	4,4'		4.82	
8	3	24	2,3,6	0.4093	5.67	5.67
9	3	16 +	2,2',3	0.4215	5.31	5.53
		32	2,4',6		5.75	
10	3	26	2,3',5	0.4518	5.76	5.76
11	3	28(d) +	2,4,4'	0.4650	5.69	5.69
		31	2,4',5			
12	3	33 +	2',3,4	0.4800		
		21 +	2,3,4			
	4	53	2,2',5,6'			
13	3	22	2,3,4'	0.4916	5.42	5.42
14	4	UK (45)	2,2',3,6	0.4987		
15	4	52	2,2',5,5'	0.5226	6.09	6.09
16	4	49	2,2',4,5'	0.5288	6.22	6.22
17	4	47 +	2,2',4,4'	0.5331	6.29	6.29
		75	2,4,4',6			
18	4	44	2,2',3,5'	0.5536	5.81	5.81
19	3 +	37(e) +	3,4,4'	0.5581	4.94	
	4	42	2,2',3,4'			
20	4	41 +	2,2',3,4	0.5725	6.11	6.11
		64	2,3,4',6			
21	4	40	2,2',3,3'	0.5845	5.56	5.56
22	4	74 +	2,4,4',5	0.6112	6.67	6.67
		61	2,3,4,5			
23	4	70	2,3',4',5	0.6181	6.23	6.23
24	4 +	66(e) +	2,3',4,4'	0.6241	6.31	6.43
	5	95	2,2',3,5',6		6.55	
25	4	60	2,3,4,4'	0.6490	5.84	5.84
26	5	92 +	2,2',3,5,5'	0.6563	6.97	6.97
		84	2,2',3,3',6			
27	5	101	2,2',4,5,5'	0.6636	7.07	7.07
28	5	99	2,2',4,4',5	0.6714	7.21	7.21
29	5	83	2,2',3,3',5	0.6880		
30	5	97 +	2,2',3',4,5	0.6964	6.67	6.67
		86	2,2',3,4,5			
31	5	87	2,2',3,4,5'	0.7049	6.37	6.37
32	5	85	2,2',3,4,4'	0.7110	6.61	6.61
33	6 +	136 +	2,2',3,3',6,6'	0.7144	6.51	6.51
	4	UK				
34	5 +	110(d) +	2,3,3',4',6	0.7207		
	4	77	3,3',4,4'		5.62	
35	5	82	2,2',3,3',4	0.7370		
36	6	151	2,2',3,5,5',6	0.7415		
37	6	135 +	2,2',3,3',5,6'	0.7486	7.15	7.15
		144	2,2',3,4,5',6			

Table I continued.

Peak no.	#Cl	PCB Isomer no.[a]	Identification	Relative Retention Time	log K_{ow} [b]	(log K_{ow})[c]
38	6	149	2,2',3,4',5',6	0.7611	7.28	7.28
39	5	118	2,3',4,4',5	0.7643	7.12	7.12
40	6 +	153	2,2',4,4',5,5'	0.8028	7.75	7.75
	6	132	2,2',3,3',4,6'			
	5	105	2,3,3',4,4'			
41	6	141	2,2',3,4,5,5'	0.8217		
42	6	137	2,2',3,4,4',5	0.8326	7.71	7.71
43	6	138	2,2',3,4,4',5'	0.8450	7.44	7.44
44	6 +	129(d) +	2,2',3,3',4,5	0.8564	7.32	7.32
	7	UK				
45	6	128	2,2',3,3',4,4'	0.8867	6.96	6.96
46	7	174	2,2',3,3',4,5,6'	0.9092		
47	7	177	2,2',3,3',4',5,6	0.9174		
48	6	156	2,3,3',4,4',5	0.9269		
49	8 +	200 +	2,2',3,3',4.5',6,6'	0.9359		
	6	157	2,3,3',4,4',5'			
50	7	180	2,2',3,4,4',5,5'	0.9555		
51	7	170	2,2',3,3',4,4',5	1.000		

(a) From Ballschmiter and Zell (22). (b) PCB isomer K_{ow}s from (2). (c) Log K_{ow}s used in log K_d vs log K_{ow} relationships. Log K_{ow} for the peak is assumed to equal that of the major isomer when the isomers are not fully resolved, and in some cases an average K_{ow} of two isomers is used when their contributions are similar and log K_{ow}s are similar. (d) Reported to represent 70% or more of the mass of the peak (20,22,24,25) in Aroclor mixtures or as determined By GCMS. (e) Similar Contributions of isomers in peak (20,22,24,25 or GCMS).

RESULTS AND DISCUSSION

Sediment-Interstitial Water Studies

Station 84 sediments were organic rich and sulfate reducing. Organic carbon contents were determined at three depths, 0-1, 1-3, and 27-30 cm, and had corresponding per cent organic carbon contents of 6.83, 7.00 and 5.16 respectively. The profiles of total PCB concentration, expressed as the sum of Aroclor 1242 and 1254, in both the pore waters and sediments are illustrated in Figure 2. The pore water concentration of total PCBs in the 0-1 cm section is 1570 ng/L and can be compared to a overlying water "dissolved" PCB concentration of 170 ng/L. Interstitial water PCBs decrease rapidly below three cm and remain low with depth in the core. The blank for some of these samples cannot be neglected and would represent a pore water concentration of 50 ng/L for a 150 mL pore water sample. It is not certain to what extent contamination contributes to some of the deeper pore water results.

The sediment concentration of PCBs contrasts the pore water results in that the level of PCBs remains fairly constant

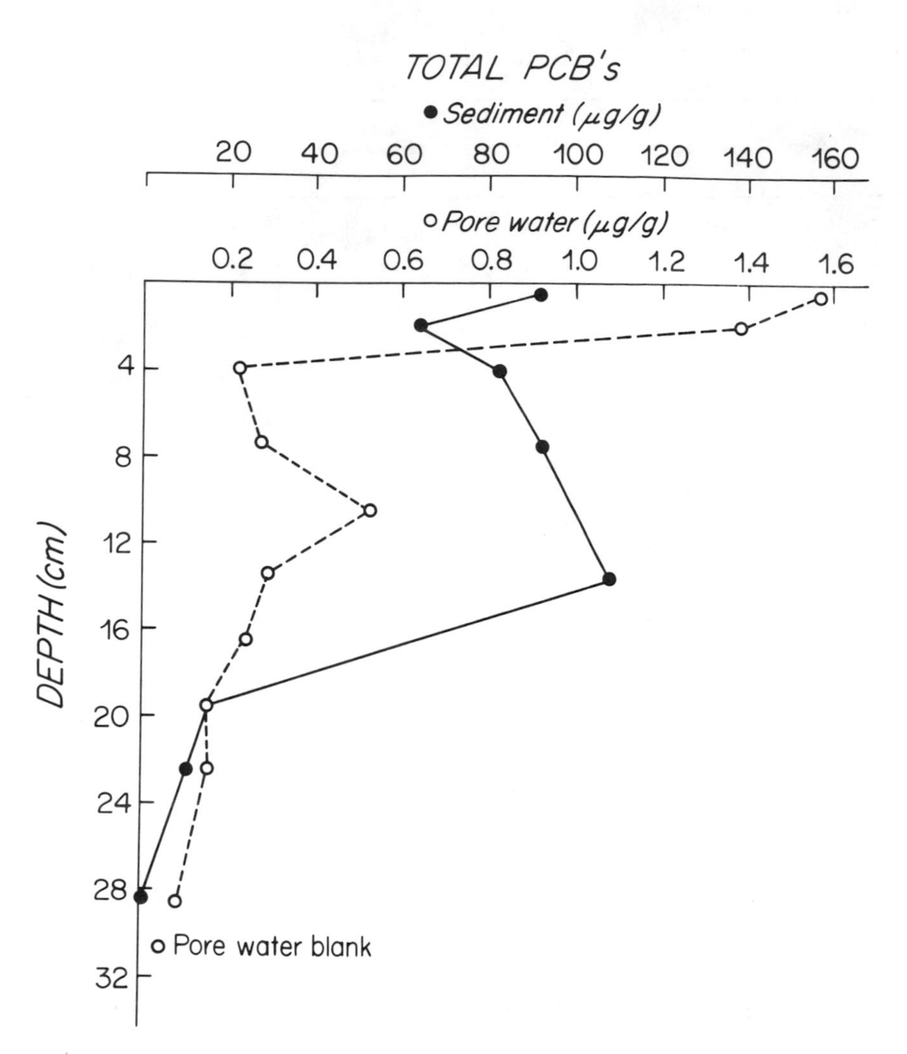

Figure 2. Total PCBs (Aroclors 1242 + 1254) in sediments and pore waters at Station 84.

(64 to 108 μg/g total PCB) over the upper 15 cm before decreasing to levels nearly an order of magnitude lower in the region between 18 and 24 cm. The concentration of PCBs in the sediments at 27-30 cm decreases another two orders of magnitude to 0.10 μg/g. It is interesting to point out that a sediment accumulation rate of 1.7 cm/yr was calculated for a core at a nearby site using ^{210}Pb geochronology [26]. That sediment accumulation rate was for sediment which had presumably accumulated after 1966 when the hurricane barrier seperating the inner and outer harbors was completed. Applying this rate to our core

gives a sediment accumulation of about 26 cm in the time since the construction of the hurricane barrier when the sedimentation rate in this area appears to have increased [26]. The marked drop in PCB concentration between 21-24 and 27-30 cm may either correspond to the changes in deposition corresponding to the construction of the hurricane barrier, or to an uncharacterized change in the source of PCB to the harbor.

The composition of individual chlorobiphenyls in the sediments is very similar to a mixture of Aroclor 1242 and Aroclor 1254. These two Aroclor mixtures were primarily used by the manufacturing plants in New Bedford along with Aroclor 1016 (Aroclors 1016 and 1242 are very similar in composition). This mixture of PCBs does not change appreciably with depth in the sediments which may indicate that the source of PCBs has remained fairly constant in composition over this time, and that there are not isomer specific diagenetic processes affecting the solid phase distribution of PCBs at this site.

The distribution of chlorobiphenyls in the interstitial waters are nearly identical to that in the sediments at the same depth. In other words, the apparent distribution coefficients, K'_d (equation 3), for all the individual compounds considered, generally vary by less than a factor of two. This constancy of K'_d over the wide range of K_{ow}s (Table I) exhibited by the different compounds is not expected if the observed distribution of PCBs represents equilibrium partitioning between the sediments and a dissolved phase in the pore water. In two-phase partitioning K'_d is equivelant to foc K_{oc} (equation 1). Combining this relationship with equation 2, a predicted slope (a) of 0.72 to 1.0, and log K_{ow} values given in Table I, it is seen that K'_d should increase by two or more orders of magnitude from the dichlorobiphenyls to the less soluble hexa- and heptachlorobiphenyls. This is not the case, thereby suggesting that PCBs in the pore waters are not dissolved but associated with a filterable colloidal phase.

The profiles of K'_d with depth in the sediment for three chlorobiphenyls are given in Figure 3. The range of profiles indicates the similarity of compositions in the pore waters and sediments noted above. The values of K'_d for 153 increase from 4.8×10^4 at 1-3 cm to a high of 6.5×10^5 at 12-15 cm and then decrease with depth in the sediment. It is difficult to ascribe an explaination for these increases in K'_d with depth without further characterization of the interstitial water. A decrease in colloid concentration with depth would result in increased apparent distribution coefficients. The sediment hydrology (i.e. possible groundwater advection), and effect benthonic biota on the observed pore water concentrations are not known. A quantitative comparison of K'_d values for this core with predictions from two- and three-phase partitioning models is presented in more detail later in this chapter.

The results from the core at Station 67 provide a study which much more clearly illustrate the role of organic colloids in the partitioning of PCBs in coastal marine sediments [15]. Profiles of total PCBs, TOC, DOC, and K'_ds of selected iso-

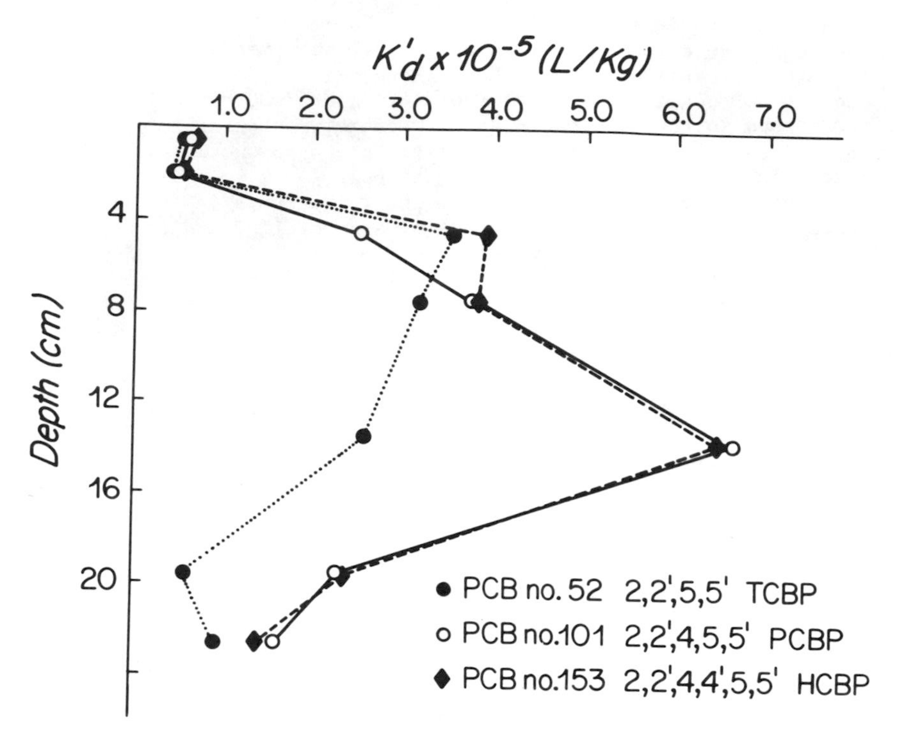

Figure 3. Depth profiles of K'_d for three chlorobiphenyls at Station 84.

mers are given in Table II. Pore water PCB concentrations are higher at this site, 1.31 to 20.1 μg/L, than at Station 84 although sediment concentrations are lower. The concentrations of PCBs in pore solution increase with depth over the upper 11 cm and remain high deeper in the core. Again, the composition of pore water PCBs is similar to the solid phase distribution. We have reported elsewhere [15] that the higher chlorinated, more hydrophobic chlorobiphenyls are slightly enriched in the pore waters and that K'_d actually tends to decrease with increasing K_{ow}. The profiles of K'_d for individual chlorobiphenyls decrease with depth over the upper 11 cm. These profiles of K'_d correlate well with the inverse of the DOC and pore water PCB profiles. It has been shown in other studies [27,28,29] that organic colloids can represent a high fraction of the DOC in interstitial waters of reducing sediments. Together, these observations suggest an important association of PCBs with organic colloids in the pore waters.

In the following section we further evaluate the partitioning results from these cores in terms of a three-phase partitioning model. Comparison of the data with model calculations based on predictions from laboratory experiments illus-

Table II. Profiles of TOC, DOC, Total PCBs, and K'_ds of Selected Isomers at Station 67.

Depth	TOC	DOC	Total PCBs		$K'_d \times 10^{-3}$ (L/Kg)		
cm	%Org C	mg/L	Sediments (μg/g)	Pore Waters (μg/L)	Chlorobiphenyl no.: 52	101	153
0-3	6.06	14.4	16.6	1.31	12.3	11.6	---
3-5	5.23	17.4	16.0	3.61	4.11	3.63	3.42
5-7	3.74	27.7	14.6	4.72	2.84	2.17	2.00
7-9	4.98	40.3	16.0	9.97	1.68	1.34	1.25
9-11	4.40	47.9	21.6	20.1	1.32	0.950	0.746
11-13	4.14	32.6	24.6	12.7	2.36	1.68	1.48
15-17	4.02	39.2	33.0	13.8	3.08	2.31	1.90
17-19	4.74	55.1	27.8	14.8	2.45	1.59	1.48
19-21	4.98	42.0	30.7	14.9	2.04	1.72	1.62
21-23	5.96	87.0	30.3	13.6	1.73	1.78	1.92
23-25	5.68	50.9	26.1	12.3	2.68	2.03	1.74
25-27	4.78	81.4	27.4	9.68	3.31	2.63	2.51
27-29	4.86	43.7	27.3	8.37	4.88	3.33	2.73
29-31	5.33	54.2	25.2	11.0	3.51	2.66	2.29
35-41	4.75	41.4	13.4	8.18	1.67	1.93	1.53

trates the quantitative importance of organic colloids and serves as a basis for evaluating the assumptions in the model.

Partitioning Model for PCBs in Sediments

The aqueous sorption of neutral, hydrophobic organic compounds with sediments and soils involves nonspecific, noncompetitive interactions with the organic fraction of the sorbent [4,5,6,8]. The driving force of sorption is largely entropic having relatively low enthalpic contributions [5,8,30]. The solvent partitioning model proposed by Chiou and co-workers [5] incorporates these and other observations, and appears to explain laboratory studies of sorption quite well. Among the assumptions implicit in this model are that sorption is: 1) independent of both solute and sorbent concentration, 2) reversible, and 3) that all of the sorptive "volume" of the organic matter is accessible to solutes for sorption. Assumptions one and two are somewhat controversial [9,31,32,33,34] but are supported by some laboratory experiments [9,35]. The discrepencies may be largely attributable to an experimental artifact in centrifugation [9,30,34]. The third assumption is supported by the constancy of K_{oc} over a large range of f_{oc} [4,6].

Partitioning of hydrophobic organic compounds with sediment organic matter or octanol depends primarily on solute solubility in water [4,6,7,8] and to a lesser extent on solute incompatibility with water-saturated organic phases [8,36]. The ability of octanol-water partitioning to mimic sorption by sediment organic matter is demonstrated by the good correlations of log K_{oc} and log K_{ow} (equation 2)[4,6,7,8] and is presumably due to similar polarities of the water-saturated organic phases.

Although not studied to the same extent, organic colloids appear to behave as sorbents in a manner similar to sediment organic matter [10,11,12]. Wijayaratne and Means [12] have shown a dependence of log K_{oc} and log K_{ow} of polycyclic aromatic hydrocarbons (PAH) with estuarine colloidal organic matter which corresponds well with one previously observed for freshwater sediment organic matter. It was suggested that K_{oc} of colloids ($K_{oc}c$) was several times that of sediments ($K_{oc}s$) but salinity differences in the two studies limit this comparison.

In natural waters dissolved PCBs will tend to approach equilibrium with both sediment and colloidal organic phases. At equilibrium, equations 1 and 3 can be combined to yield:

$$K'_d = \frac{f_{oc}s\,K_{oc}s\,D}{D + f_{oc}c\,K_{oc}c\,D} \qquad (4)$$

where D is the dissolved concentration and $f_{oc}s$ and $f_{oc}c$ are the fraction organic carbons of sediment and colloid phases respectively. $f_{oc}c$ is some fraction, α, of the total measured

DOC, and it is further assumed for this model that $K_{oc}s = K_{oc}c$. Equation 4 then becomes:

$$K'_d = \frac{f_{oc}s\ K_{oc}s}{1 + \alpha\ DOC\ K_{oc}s} \qquad (5)$$

Comparisons of our sediment-interstitial water K'_d data and model predictions of K'_d from two-phase (equation 1) and three-phase (equation 5) partitioning models are presented in Figures 4 and 5. The three solid lines are two-phase predictions of log K'_d vs log K_{ow} and are based on measured $f_{oc}s$ and experimentally determined log K_{oc} vs log K_{ow} relationships published by three workers [6,7,8]. Line 1 from Means et al. [6] is for a wide range of hydrophobic organic compounds including PAH and is nearly identical to a relationship reported by Karickhoff et al. [4]. Lines 2 [8] and 3 [7] are for chlorinated aromatic hydrocarbons and result in predictions of K'_d of PCBs which are significantly lower than line 1. Differences in the three correlations may be due to the different compound classes considered [7] or possibly to differences in experimental protocol.

Two other points need to be made in comparing these predictions to the data reported here. First, for the most part, these relationships are based on solutes with log $K_{ow} < 6.0$ while the PCBs considered here have log K_{ow}s ranging from 4.89 to 7.75. The validity of the extrapolations has not been adequately tested. Secondly, the effect of seawater ionic strength and major ion composition on sediment sorption has not been well studied. Seawater electrolytes may have effects on the activity of PCBs in both the aqueous and sediment organic matter phases. Salting-out of non-electrolytes has been well studied for smaller molecules [37,38] but good data for large chlorinated hydrocarbons are lacking. It can be assumed that increases in aqueous phase activity coefficients of PCBs will increase sorption (K_{oc}) by 0.1 to 0.4 log units, probably increasing with degree of chlorination. In the sediment organic matter phase, salinity increases could possibly change sorbate activity coefficients by affecting the conformation, charge density, or water content of the organic matter.

A model of three-phase partitioning (equation 5) is represented in Figure 4 by line 4. This curve assumes $f_{oc}c$ = 10 mg/L in the pore water at Station 84 at 1-3 cm. The major assumption is that both $K_{oc}s$ and $K_{oc}c$ can be predicted by the regression defining line 1 (log K_{oc} = 1.0 log K_{ow} - 0.32). This assumption is somewhat arbitrary for the reasons stated above, but fits the data better than the other lines and does illustrate the effect of colloidal organic matter on the observed distribution of PCBs. The three-phase model adequately describes the constancy of log K'_d with increasing log K_{ow} for this range of compounds but underpredicts the magnitude of K'_d by about 0.8 log units. Two-phase partitioning predicted by line 1 overpredicts log K'_d for log $K_{ow} > 6.0$, but under-

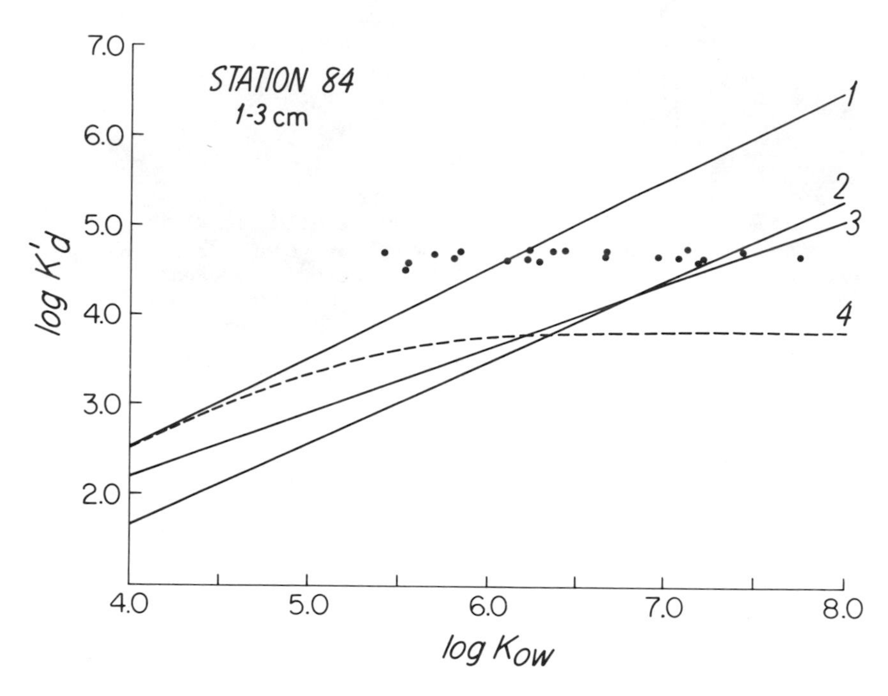

Figure 4. Log K'_d vs log K_{ow} of individual chlorobiphenyls from 1-3 cm section at Station 84. Solid lines 1.(6), 2.(8), and 3.(7) are predictions of two-phase partitioning of PCBs based on laboratory experiments. Line 4. represents the three-phase partitioning model (Equation 5.) predicted using assumptions given in the text.

predicts K'_d in lines 2 and 3. Prediction of K'_d by all four lines becomes worse as K'_d increases with depth in the core (Figure 3). These high K'd values cannot presently be explained. It is possible that processes such as diagenetic alteration of sediment organic matter and mineral precipitation in some aged sediments could "trap" sorbed PCBs and make them less accessible for aqueous desorption. There have been few desorption studies of sediments contaminated in the field.

Measured K'_ds of individual chlorobiphenyls at Station 67 are plotted for two representative depths in a similar manner (Figures 5a and b). The three-phase model curves incorporate measured values of DOC and assume $\alpha = 1$ in equation 5. These results demonstrate a suprisingly good fit of the data to the simple three-phase model, and clearly illustrate the effective lowering of K'_d from predictions based on two-phase partitioning. Results from 17-19 cm show the small decrease of log K'_d with increasing log K_{ow}, which is apparent in all sediment depths at this site [15]. This trend of K'_d being higher for the lower chlorinated biphenyls may be the result of the smaller

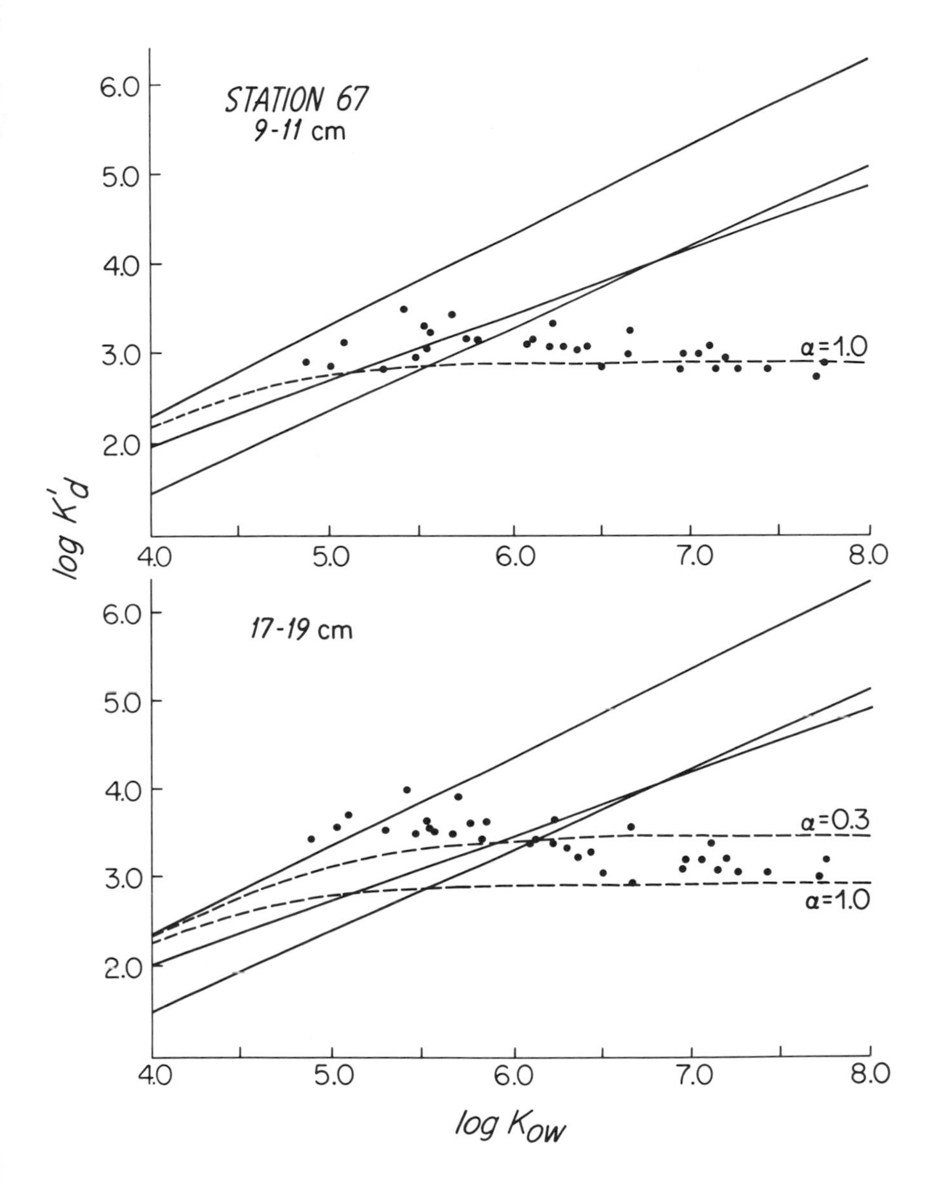

Figure 5. Log K'_d vs log K_{ow} for two representative sediment intervals at Station 67. Model lines are the same as in Fig. 4. Figure b. illustrates the effect of assuming a lower α (0.3) in Equation 5.

compounds being able to access more of the sorptive volume of the sediment organic matrix. However there is little evidence of steric hindrance to sorption in other sediment sorption studies, although sorption kinetics have been shown to be slower for larger more hydrophobic compounds [35]. There is evidence at Station 67 of microbial degradation of di- through tetra-chlorobiphenyls [15]. If this process occurs in pore waters at a rate comparable to sediment desorption, then measured K'_d may represent disequilibria and be higher than predicted.

Figure 5b shows the effect of assuming a lower colloid concentration (α = 0.3) on model calculations. The calculation of α from equation 5 and the observed K'_ds of 101 for all sediment depths yield values of 0.223 to 0.967 and average 0.573. These estimates of organic colloid concentrations in pore waters are consistent with those operationlly defined by ultrafiltration in other studies [27,28,29].

Water Column Partitioning Studies

The concentrations of total PCBs in the water column were five to ten times higher in the inner harbor samples at Station 81 than at Station 74 in the outer harbor (Table III). Large differences in concentration over the tidal cycle were not observed, nor was there a large gradient between surface and bottom water samples. CTD casts at both stations indicated that the water column was fairly well mixed at the time of sampling.

The concentrations of total PCBs are nearly evenly distributed between particulates and the filtrate or "dissolved" phases. However, the composition of PCBs in these two phases is very different. The more soluble chlorobiphenyls are enriched in the filtrate. This difference in composition is illustrated by the gas chromatograms of particulate and dissolved fractions of one sample (Figure 6). Not all of the peaks in

Table III. Tidal Cycle Concentrations of Total PCBs.

Sample	Particulates	Filtrate ng/L	Total
Incoming Tide			
Station 81, Surface	183	329	512
Station 81, Bottom	134	152	286
Station 74, Surface	28	52	80
Station 74, Bottom	22	28	50
Outgoing Tide			
Station 81, Surface	386	289	675
Station 81, Bottom	232	244	476
Station 74, Surface	27	36	63
Station 74, Bottom	30	41	71

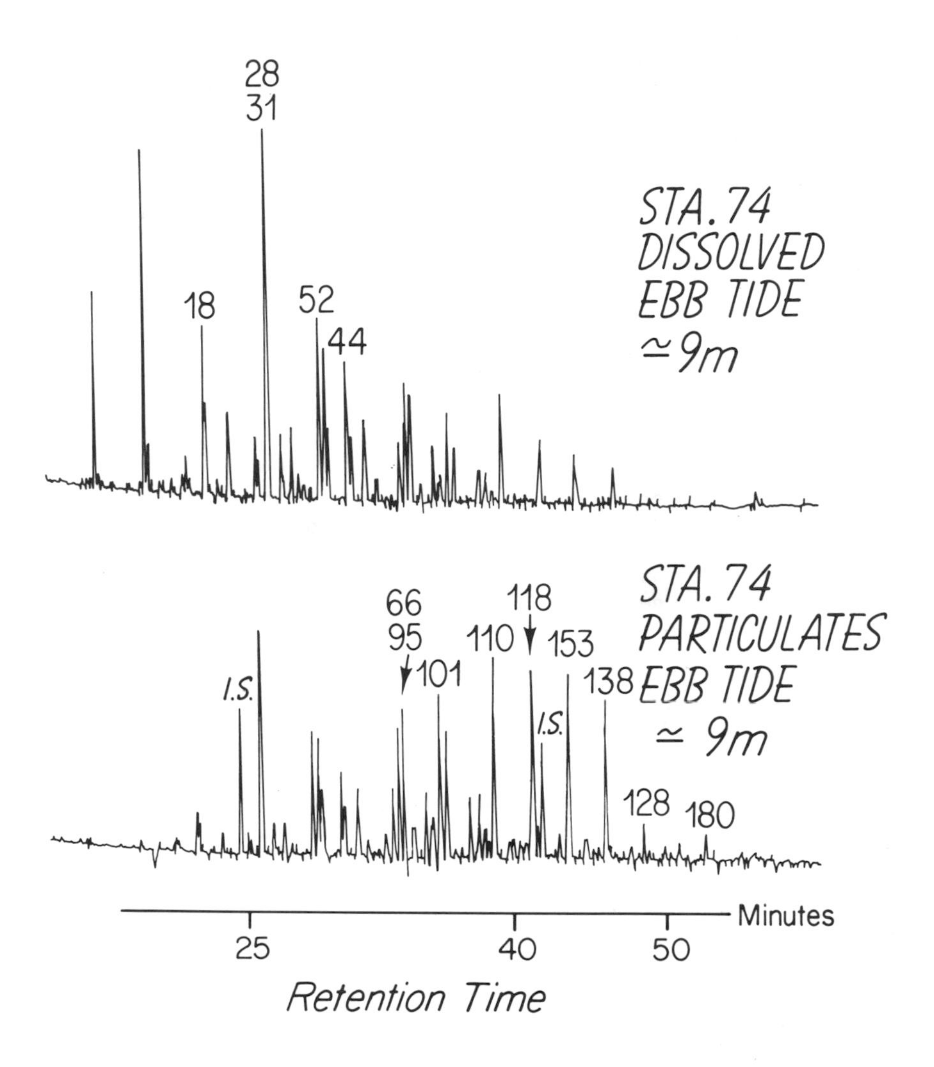

Figure 6. Capillary gas chromatograms of "dissolved" and particulate PCBs from New Bedford Harbor Station 74. Numbers are PCB numbers given in Table I. I.S.: internal standard.

the gas chromatograms are PCBs, some of which have been labeled with IUPAC chlorobiphenyl numbers (Table I). The trend of K'_d increasing with degree of chlorination or K_{ow} agrees with observed partitioning of PCBs in the Hudson River [39], Puget Sound [40], and Pacific Ocean [41], and is predicted by the models of two-phase partitioning described earlier.

Plots of log K'_d vs log K_{ow} for these eight water column samples provide a good comparison to the results from sediment-interstitial water studies. Because measurements of suspended-solids concentrations and $f_{oc}s$ were not made on the water samples, we have had to assume values in the range of those measured in an extensive survey of these properties in New Bedford Harbor [26], which covered tidal and seasonal cycles. Assumed values for suspended solids concentrations and $f_{oc}s$ are 3.0 mg/L and 0.20 for surface samples, and 5.0 mg/L and 0.15 for bottom samples. These assumptions are probably valid within a factor of three and do not affect calculated slopes of log K'_d vs log K_{ow} presented later.

Figure 7 shows the increase of log K'_d with log K_{ow} for the surface sample at Station 81 during the incoming tide. K'_d increases from 1.8×10^4 for chlorobiphenyl 4 to 1.6×10^6 for 137. Log K'_d parallels log K'_d predicted by line 1 for log K_{ow} $\leq$ 6.1 and then increases less rapidly for greater log $K_{ow}s$. The three-phase model depicted by line 4 assumes an $f_{oc}c$ of 0.25 mg/L, which is consistent with reported measurements made in coastal waters [42] and unpublished results from our own lab. The three-phase model calculations do show that reasonable assumptions of $f_{oc}s$ and $f_{oc}c$ can explain the observed partitioning of PCBs and that, under these conditions, the quantitative role of organic colloids on observed K'_ds is minimal until log $K_{ow}s$ exceed 6.0. This is not the case in pore

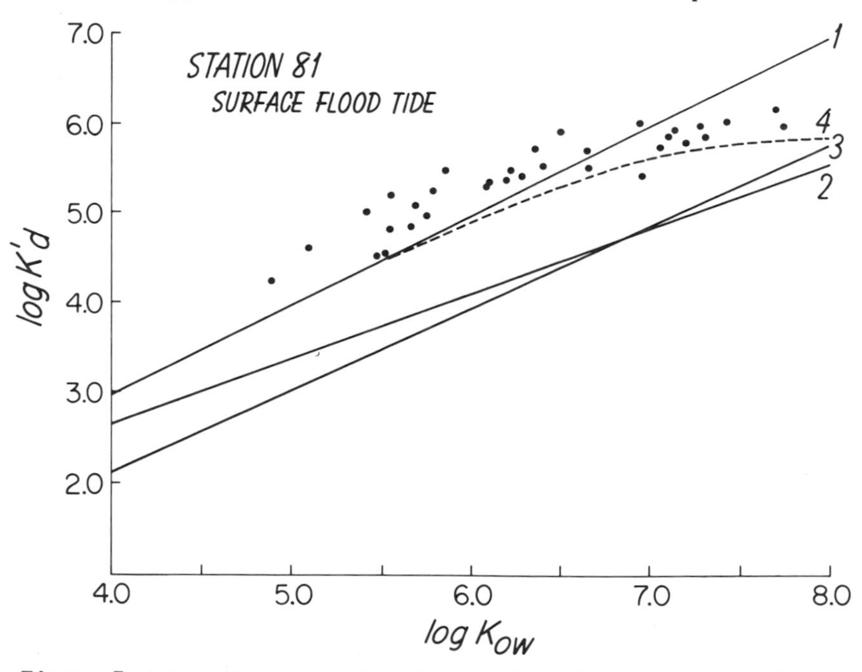

Figure 7. Log K'_d vs log K_{ow} for Station 74 water column sample. Lines are as given in preceeding figures with assumptions of $f_{oc}s$ and $f_{oc}c$ given in the text.

Table IV. Regression Parameters of $\log K_{OC} = a \log K_{OW} + b$ for Tidal Cycle Stations 81 and 74.

Sample	a	b	r
Incoming Tide			
Station 81, Surface	0.589	2.39	0.912
Station 81, Bottom	0.498	3.07	0.870
Station 74, Surface	0.514	2.87	0.874
Station 74, Bottom	0.556	2.23	0.855
Outgoing Tide			
Station 81, Surface	0.534	2.95	0.916
Station 81, Bottom	0.526	2.88	0.916
Station 74, Surface	0.562	2.53	0.886
Station 74, Bottom	0.558	2.52	0.914
Including only $\log K_{OW}$s $\leq$ 6.11			
Incoming Tide			
Station 81, Surface	0.879	0.75	0.853
Station 81, Bottom	0.579	2.58	0.690
Station 74, Surface	0.309	3.98	0.475
Station 74, Bottom	0.527	2.35	0.448
Outgoing Tide			
Station 81, Surface	0.860	1.12	0.863
Station 81, Bottom	0.640	2.21	0.740
Station 74, Surface	0.832	0.97	0.731
Station 74, Bottom	0.756	1.38	0.729
Predictions from Laboratory Experiments (6,7,8)			
Line 1. Means et al. (6)	1.00	-0.32	0.990
Line 2. Chiou et al. (8)	0.904	-0.78	0.994
Line 3. Schwarzenbach and Westall (7)	0.72	0.49	0.975

waters or in other organic rich waters where organic colloid concentrations can be many times higher.

The linear regression parameters, a and b, from equation 2 have been calculated. In Table IV, we first list these parameters derived from considering all data points, and compare them to those from the three model lines. It is seen that the slopes do not differ much between samples, 0.50 to 0.59, and are lower than the laboratory derived slopes of 0.72 to 1.0. The $\log K_{OC}$ intercepts are quite high and their absolute values are dependent on the assumptions of POC that we have made. However, we interpret both of these observations to be consistent with the expected curvilinearity of this relationship when considering chlorobiphenyls with high K_{OW}. The linear regressions have also been calculated by only considering data for isomers with log $K_{OW} \leq 6.11$ where two-phase partitioning is expected to be more nearly approximated. There are increased errors associated with the correlations derived from the smaller data sets. In the six samples which exhibit the least scatter in data, the slopes are increased closer to those predicted from two-phase partitioning. There are also corresponding decreases in b for those samples.

Departure from equilibrium partitioning can be expected to be greater in the water column than in sediment pore waters, where the latter can be considered a much less open and dynamic environment. The main source of PCBs to the water column in New Bedford Harbor is probably resuspension of contaminated sediments. A desorptive approach to equilibrium in the water column would result in higher log K'_d and the observed slope of log K'_d vs log K_{OW} would be incresed if desorption kinetics are slower for the more hydrophobic chlorobiphenyls [35]. Disequilibria caused by volatilization will also tend to increase observed partitioning and the effect on the slope would depend on the relative rates of vapor exchange and desorption kinetics of the individual isomers. Within the limitations of both the data here and models of seawater partitioning, it is not possible to distinguish whether equilibrium partitioning of PCBs is established in the water column. We suggest that the results do show that dissolved PCB phases are more important in observed water column distributions as is predicted by our simple three-phase partitioning model.

CONCLUSIONS

It is proposed that the observed partitioning of individual chlorobiphenyls between particulates and filtered water samples is controlled not only by solution-sediment sorption but also by partitioning between solution and organic colloids which can represent a high fraction of DOC in interstitial waters. We have presented the following evidence that most PCBs measured in the interstitial waters at these two sites are bound to organic colloids:

1) The compositions of PCBs in the pore water at both stations are very similar to those in the sediments. If sediment-solution partitioning were the only processes involved, the more soluble, lower chlorinated PCBs would have lower observed K'_ds and thus would be enriched in the pore solution. We have found this to be the case in the water column where colloids may be present in much lower concentrations. At Station 67, K'_d actually decreases with increasing degree of chlorine substititution. This trend is the opposite of what solubility considerations predict for a two phase system.

2) Pore water concentrations in surface sediments are elevated at both stations. At Station 67 the concentrations of all PCBs increase with depth and K'_ds decrease over the upper

11 cm. DOC concentrations increase over this same interval. We believe these trends are due to sorption of PCBs onto organic colloids whose concentrations are reflected by the DOC. The high values of K'_d at Station 84 and increase of K'_d with depth at this site are not easily explained.

3) A simple three-phase equilibrium partitioning model can describe the observed partitioning of PCBs at Station 67 and does predict the constancy of log K'_d vs log K_{ow} at Station 84.

The high concentrations of PCBs in pore waters provide a relatively large and potentially mobile pool of PCBs. Pore water PCBs may be taken up by benthic organisms in contact with the sediments or be transported back into the water column by diffusion or mixing of the sediments. There is little knowledge concerning the role of PCB-colloid associations in the bioavailability of PCBs or in the mobility of PCBs in sediments, which in light of these results are important considerations for understanding the biogeochemical cycling and long term fate of PCBs in coastal sediments.

A three-phase partitioning model for PCBs in the water column predicts a greater importance of dissolved PCBs due to much lower concentrations of organic colloids. The observed partitioning approaches predictions based purely on sediment-solution sorption when the effect of organic colloids on high K_{ow} compounds is considered. A more critical evaluation of organic matter-seawater partitioning and also of the kinetics of processes which may affect disequilibria, are necessary to further evaluate these sorption models.

ACKNOWLEDGEMENTS

We wish to thank Capt. A.D. Colburn, Bruce Tripp, J.B. Livramento, Alan Davis, Hovey Clifford, Bill Martin, John Burke and Dr. Ed Sholkovitz for assistance in sampling and analysis, and Drs. Philip Gschwend, Jay Means, René Schwarzenbach and Ed Sholkovitz for useful discussions. Financial support was provided by a Andrew W. Mellon Foundation grant to the Coastal Research Center, Woods Hole Oceanographic Institution, the Education Office of the Woods Hole Oceanographic Institiution/ Massachusetts Institute of Technology Joint Program in Oceanography, and the Dept. of Commerce, NOAA, National Sea Grant Program under grants NA 83 AA-D-00049 (R/P-13) and NA 84 AA-D-00033 (R/P-17). This is Contribution No. 5870 from the Woods Hole Oceanographic Institution.

REFERENCES

1. Mackay, D., R. Mascarenhas and W.Y Shiu. "Aqueous Solubility of Polychlorinated Biphenyls," *Chemosphere* 9: 257-264 (1980).
2. Rapaport, R.A. and S.J. Eisenreich. "Chromatographic Determination of Octanol-Water Partition Coefficients (K_{ow}'s) for 58 Polychlorinated Biphenyl Congeners," *Environ. Sci. Technol.* 18: 163-170 (1984).
3. National Academy of Sciences. "Polychlorinated Biphenyls," (Prepared by the National Research Council) NAS (1979).
4. Karickhoff, S.W., D.S. Brown and T.A. Scott. "Sorption Kinetics of Hydrophobic Pollutants on Natural Sediments," *Water Res.* 13: 241-248 (1979).
5. Chiou, C.T., L.J. Peters and V.H. Freed "A Physical Concept of Soil-Water Equilibria for Nonionic Compounds," *Science* 206: 831-832 (1979).
6. Means, J.C., S.G. Woods, J.J. Hassett and W.L. Banwart. "Sorption of Polynuclear Aromatic Hydrocarbons by Sediments and Soils," *Environ. Sci. Technol.* 14: 1524-1528 (1980).
7. Schwarzenbach, R.P. and J. Westall. "Transport of Nonpolar Organic Compounds from Surface to Groundwater. Laboratory Sorption Studies," *Environ. Sci. Technol.* 15: 1360-1367 (1981).
8. Chiou, C.T., P.E. Porter and D.W. Schmedding. "Partition Equilibria of Nonionic Organic Compound between Soil Organic Matter and Water," *Environ. Sci. Technol.* 17: 227-231.
9. Gschwend, P.M. and S. Wu. "On the Constancy of Sediment-Water Partitioning of Hydrophobic Organic Pollutants," *Environ. Sci. Technol.* 19: 90-96 (1985).
10. Means, J.C. and R. Wijayaratne. "Role of Natural Colloids in the Transport of Hydrophobic Pollutants," *Science* 215: 968-970 (1982).
11. Carter, C.W. and I.H. Suffet. "Binding of DDT to Dissolved Humic Material," *Environ. Sci. Technol.* 16: 735-740 (1982).
12. Wijayaratne, R.D. and J.C. Means. "Sorption of Polycyclic Aromatic Hydrocarbons by Natural Estuarine Colloids," *Mar. Environ. Res.* 11: 77-89 (1984).
13. Hassett, J.P. and M.A. Anderson. "Association of Hydrophobic Organic Compounds with Dissolved Organic Matter in Aquatic Systems," *Environ. Sci. Technol.* 13: 1526-1529 (1979).
14. Meyers, P.A. and J.A. Quinn. "Factors Affecting the Association of Fatty Acids with Mineral Particles in Sea Water," *Geochim. Cosmochim. Acta* 37: 1745-1759 (1973).
15. Brownawell, B.J. and J.W. Farrington. "Biogeochemistry of PCBs in Interstitial Waters of a Coastal Marine Sediment," Submitted to *Geochim. Cosmochim. Acta*.
16. Weaver, G. "PCB Contamination in and Around New Bedford, Mass.," *Environ. Sci. Technol.* 18: 22A-27A (1984).

17. Farrington, J.W., B.W. Tripp, A.C. Davis and J. Sulanowski "One View of the Role of Scientific Information in the Solution of Environmental-Economic Problems," in *Proceedings of the International Symposium on "Utilization of Coastal Ecosystems: Planning, Pollution, Productivity," 22-27 November 1982, Rio Grande, RS, Brazil, Vol. 1*, L.N. Chao and W.W. Kirby Smith, Eds. (University of Rio Grande do Sul) (in press).
18. Henrichs, S.M. "Biogeochemistry of Dissolved Free Amino Acids in Marine Sediments," PhD Thesis, MIT/WHOI (1980).
19. Sholkovitz, E.R., and D.R. Mann. "The Pore Water Chemistry of $^{239,240}Pu$ and ^{137}Cs in Sediments of Buzzards Bay, Massachusetts," *Geochim. Cosmochim. Acta* 48: 1107-1114 (1984).
20. Duinker, J.C. and M.T.J. Hillebrand. "Characterization of PCB Components in Clophen Formulations by Capillary GC-MS and GC-ECD Techniques," *Environ. Sci. Technol.* 17: 449-456 (1983).
21. Mullin, M.D., C.M. Pochini, S. McCrindle, M. Romkes, S.H. Safe and L.M. Safe. "Synthesis and Chromatographic Properties of All 209 PCB Congeners," *Environ. Sci. Technol.* 18: 468-476 (1984).
22. Eisenreich, S.J., P.D. Capel and B.B. Looney "PCB Dynamics in Lake Superior Water," in *Physical Behaviour of PCBs in the Great Lakes*, D. Mackay, S. Patterson, S.J. Eisenreich, and M.S. Simmons, Eds. (Ann Arbor Science Publishers, Inc., 1983), pp. 181-212.
23. Ballschmiter, K. and M. Zell. "Analysis of Polychlorinated Biphenyls (PCB) by Glass Capillary Gas Chromatography," *Frezenius Z. Anal. Chem.* 302: 20-31 (1980).
24. Albro, P.W. and C.E. Parker. "Comparison of the Compositions of Aroclor 1242 and Aroclor 1016," *J. Chromatography* 169: 161-166 (1979).
25. Albro, P.W., J.T. Corbett and J.L. Schroeder. "Quantitative Characterization of Polychlorinated Biphenyl Mixtures (Aroclors 1248, 1254, and 1260) by Gas Chromatography Using Capillary Columns," *J. Chromatography* 105: 103-111 (1981).
26. Summerhayes, C.P., J.P. Ellis, P. Stoffers, S.R. Briggs and M.G. Fitzgerald. "Fine Grain Sediment and Industrial Waste Distribution and Dispersal in New Bedford Harbor and Western Buzzards Bay, Massachusetts," Woods Hole Oceanographic Instititution Technical Report, WHOI-76-115 (1977).
27. Krom, M.D. and E.R. Sholkovitz. "Nature and Reactions of Dissolved Organic Matter in the Interstitial Waters of Marine Sediments," *Geochim. Cosmochim. Acta* 41: 1565-1573 (1977).
28. Nissenbaum, A., M.J. Baedecker and I.R. Kaplan. "Studies on Dissolved Organic Matter from Interstitial Waters of a Reducing Marine Fjord," in *Advances in Organic Geochemistry*, H. von Gaertner and H. Wehner, Eds. (Oxford, England: Pergamon Press, 1972), pp. 427-440.

29. Elderfied, H. "Metal-Organic Associations in Interstitial Waters of Narragansett Bay Sediments," *Amer. J. Sci.* 281: 1184-1196 (1981).
30. Chiou, C.T., P.E. Porter and T.D. Shoup. "Comment on Partition Equilibria of Nonionic Organic Compounds Between Soil Organic Matter and Water," *Environ. Sci. Technol.* 18: 295-297 (1984).
31. Horzempa, L.M. and D.M. DiToro. "PCB Partitioning in Sediment-Water Systems: The Effect of Sediment Concentration," *J. Environ. Qual.* 12: 373-380 (1983).
32. DiToro, D.M. and L.M. Horzempa. "Reversible and Resistant Components of PCB Adsorption-Desorption: Isotherms," *Environ. Sci. Technol.* 16: 594-602 (1982).
33. O'Connor, D.J. and J.P. Connolly. "The Effect of Concentration of Adsorbing Soleds on the Partition Coefficient," *Water Res.* 14: 1517-1523 (1980).
34. Voice, T.C., C.P.Rice and W.J. Weber Jr. "Effect of Solids Concentration on the Sorptive Partitioning of Hydrophobic Pollutants in Aquatic Systems," *Environ. Sci. Technol.* 17: 513-518 (1983).
35. Karickhoff, S.W. "Sorption Kinetics of Hydrophobic Pollutants in Natural Sediments," in *Contaminants and Sediments, Vol. 2*, R.A. Baker, Ed. (Ann Arbor, MI: Ann Arbor Science Publishers, Inc., 1980), pp. 193-205.
36. Karickhoff, S.W. "Semi-Empirical Estimation of Sorption of Hydrophobic Pollutants on Natural Sediments and Soils," *Chemosphere* 10: 833-846 (1981).
37. McDevit, W.F. and F.A. Long. "The Activity Coefficient of Benzene in Aqueous Salt Solutions," *J. Amer. Chem. Soc.* 74: 1773-1777.
38. Gordon, J.E. and R.L. Thorne. "Salt Effects on Non-Electrolyte Activity Coefficients in Mixed Aqueous Electrolyte Solutions-II. Artificial and Natural Sea Waters," *Geochim. Cosmochim. Acta* 31: 2433-2443 (1967).
39. Bopp, R.F. "The Geochemistry of Polychorinated Biphenyls in Sediments of the Tidal Hudson River," PhD Thesis, Columbia University (1979).
40. Dexter, R.N. and S.P. Pavlou. "Distribution of Stable Organic Molecules in the Marine Environment: Physical-Chemical Aspects. Chlorinated Hydrocarbons," *Mar. Chem.* 7: 67-84 (1978).
41. Tanabe, S. and R. Tatsukawa. "Vertical Transport and Residence Time of Chlorinated Hydrocarbons in the Open Ocean Water Column," *J. Ocean. Soc. Japan* 39: 53-62 (1983).
42. Zsolnay, A. "Coastal Colloidal Carbon: A Study of its Seasonal Variation and the Possibility of River Input," *Estuar. Coast. Mar. Sci.* 9: 559-567 (1975).

CHAPTER 8

SILICONES IN ESTUARINE AND COASTAL MARINE SEDIMENTS

Robert E. Pellenbarg
Combustion and Fuels Branch
Chemistry Division, Code 6182
Naval Research Laboratory
Washington, D.C. 20375-5000, USA

ABSTRACT

Polyorganosiloxanes (silicones) are synthetic, extremely inert surface active organic compounds widely used in consumer products. Silicones were measured in the filter cake, sludge, and aqueous effluent produced at the Blue Plains Wastewater Treatment Facility in Washington, D.C. (a major point source in Potomac estuary, up to 95 ppm, dry wt/wt basis), in Potomac River sediments (0.5-3 ppm), sediments of the heavily impacted New York Bight (0-50 ppm), and in sediments of the Chesapeake Bay (0-30 ppm). These results indicate that silicones are useful tracers of anthropogenic impact on sediments.

INTRODUCTION

Poly(organosiloxanes)(silicones) are synthetic surface active polymers which have been in wide use only since the second World War. Silicones feature a backbone of alternating silicon and oxygen atoms, analogous to a string of beads, with various organic substituents bonded to the silicon atoms in the chain. Most commonly, based on quantities manufactured commercially, methyl or phenyl groups occupy the silicon atoms, yielding dimethyl or diphenyl silicones. Physical properties of the silicone are controlled largely by chain length. The silicone can be tailored to have a water- or oil-like viscosity, to be rubbery, or to be a rigid solid, depending on average molecular weight. For commonly encountered dimethyl silicone oils molecular weight is approximately 10,000 [1].

The dimethyl silicone oils as a class exhibit desirable surface active properties. These silicones find extensive application in consumer products such as water and soil repellants for fabrics, anti-foam additives for cooking oils and beverages, anti-stick formulations for self-adhesive labels, and additives for automotive antifreeze and medical preparations. These products obtain desirable characteristics from only trace amounts of the chemically inert silicone, which is not destroyed during use of the product containing it. Indeed, inertness is one of the more useful properties exhibited by silicones, which pass unaltered through the sewage treatment process [2].

As silicones possess surface active properties, it was hypothesized that these compounds would accumulate at phase

MARINE AND ESTUARINE GEOCHEMISTRY, Sigleo, A. C., and A. Hattori (Editors)

boundaries when released to the environment. The seston-water/sediment-water interface is most significant in this regard. Silicones would ultimately collect in aquatic sediments as coatings or coating constituents on sedimentary particulates. The research summarized briefly here examined, initially, samples of materials produced by the Blue Plains Wastewater Treatment Facility, Washington, D.C. [3], to confirm the sewage plant as a point source for silicones. Then, samples from the adjacent Potomac River [3], as well as sediments from the New York Bight [4] were examined to confirm the environmental dispersal of silicones. Lastly, sediments from the Chesapeake Bay [6] were examined to demonstrate the potential utility of silicones as tracers for anthropogenic additions to sediments.

METHODS

The analytical methodology for silicones in environmental samples is summarized in Figure 1. Briefly, a sample is freeze dried, extracted with diethyl ether, and the extract analyzed for silicon content by atomic absorption spectrophotometry. Atomization is with acetylene-nitrous oxide flame, and standarization with octaphenylcyclotetrasiloxane (Kodak) in methyl isobutyl ketone [3,4]. All data here are reported in terms of organic silicon content of a sample, on a dry weight basis, and dimethyl silicones are 37.9% silicon, wt/wt. The technique outlined here recovers approximately 35% of the silicone in a sample [3].

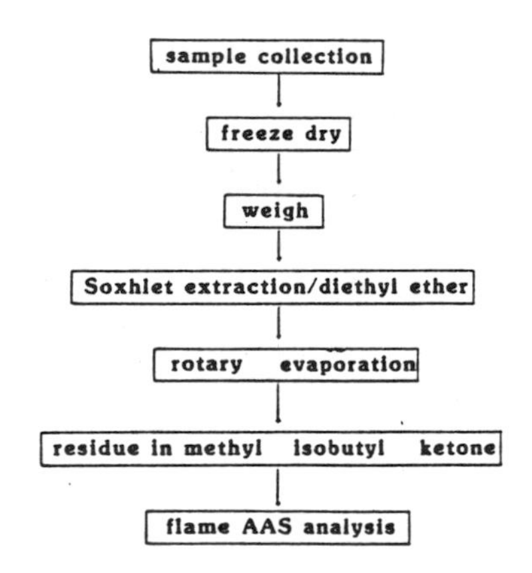

Figure 1. Analytical flow scheme.

RESULTS AND DISCUSSION

The study of the environmental dispersal of silicones began with the confirmation that sewage treatment plants are potential sources for silicones. Samples of materials produced by the Blue Plains Wastewater Treatment Facility in Washington, D.C. were analyzed for extractable silicone. Relevant data are presented in Table I, and show that solids and aqueous effluent from a

TABLE I. Silicones in Materials from Blue Plains Wastewater Treatment Facility

	Sample	Extractable organic Si, ppm
A. Filter Cake	F1	26.3
	F2	42.3
	F3	45.7
	F4	38.6
	F5	32.8
	Average	37.7 7.4
B. Sludge	S1	103.8
	S2	89.1
	S3	96.4
	S4	94.4
	S5	96.7
	Average	96.1 5.3
C. Aqueous Effluent	E1	4.3 ppb
	E2	4.8 ppb
	E3	4.1 ppb
	E4	4.7 ppb
	Average	4.5 0.3

* All samples were freezed dried, then extracted with diethylether. Data are reported as organic silicon content, based on 35% recovery of silicones which are 37.9% silicon (adapted from [3]).

sewage treatment installation contain silicone. The potential impact of such a source of silicones is documented in Figure 2, which reports silicone content of superficial sediments in the Potomac River, adjacent to the Blue Plains Plant. It is seen that riverine sediments can show evidence of direct anthropogenic impact, at least in the immediate vicinity of a well defined source of such a synthetic organic material as silicone.

The observations in the Potomac were extended to sediments of the New York Bight [4]. This region (Figure 3) receives, and had received, vast quantities of urban discards both by direct dumping and by more diffuse input via discharge from the Hudson River. Sedimentary silicone content in the New York Bight shows a concentration gradient away from a major dump site [4], and in a pattern observed independently by following such sedimentary

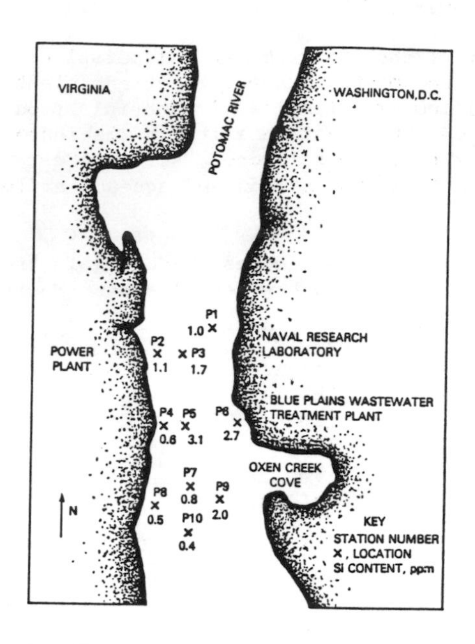

Figure 2. Extractable silicones in the surface sediments of the Potomac River, Washington, D.C.

markers as tomato seeds and steroid derivatives [7,8]. Note, however, that neither of these sedimentary parameters provide the unquestioned anthropogenic origin nor the physical or chemical inertness of silicones.

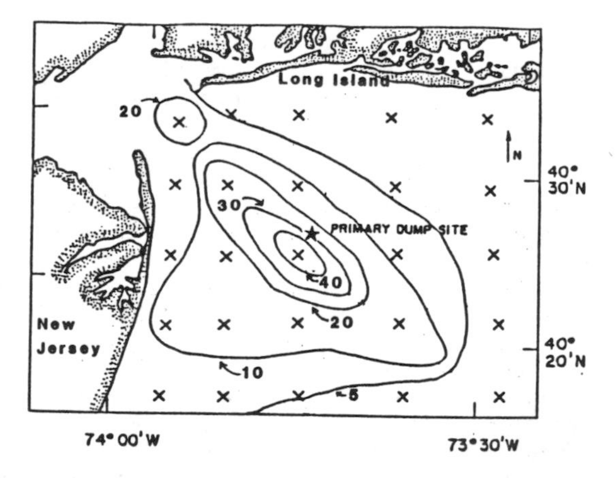

Figure 3. Extractable silicones in the sediments of the New York Bight, USA. Numbered contours refer to organic silicon (ppm, dry wt/wt basis) in the sediments. X marks sampling locations.

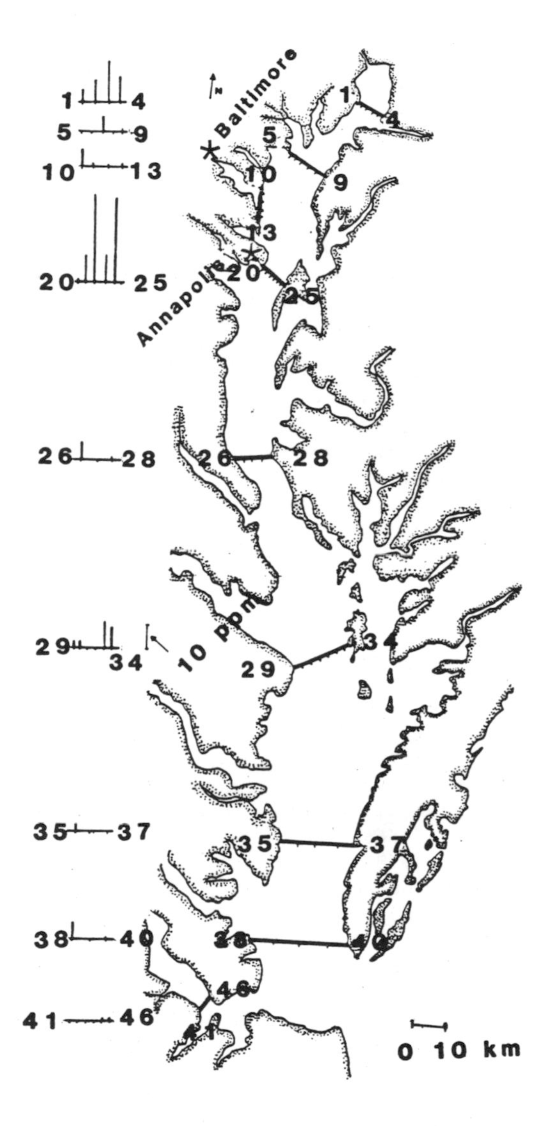

Figure 4. Extractable silicones in the sediments of Chesapeake Bay, USA. Sampling transects are on the map, with station numbers at each end of the transect. Bar graphs on the left of the figure indicate sedimentary silicone content.

As a further examination of the large scale dispersal of silicones in sediments, samples from the Chesapeake Bay [6] were analyzed, with results presented in Figure 4. Two important observations relate to this data. First, in a general sense, sedimentary silicone content was highest near urban centers, such as down stream from Baltimore and Annapolis. The Annapolis sampling transect is especially significant as it is in the vicinity of the turbidity maximum for the Bay, where organic matter flocculates and settles to the bottom. This flocculation process scavenges the surface active silicones, giving the high levels of silicone at stations 20-25. Secondly, closely related to this observation is the fact that down-bay, only near-shore samples exhibit appreciable extractable silicone. These near-shore sediments are relatively high in organic carbon content as determined by combustion [6]. Thus, there is clear evidence from field observations that the hydrophobic, surface active silicones will accumulate in organic rich regimes.

In conclusion, the research briefly summarized in this paper confirms silicones as totally synthetic, highly inert and chemically unique markers for anthropogenic additions to sediments. The movement of impacted sediments has been traced from a clearly defined source, a sewage treatment plant [3], into nearby sediments [3] and, in a larger geographical context, throughout a large estuarine system [6]. Indeed, sedimentary transport may be followed by monitoring silicone content, as reported earlier [4].

LITERATURE CITED

1. Noll, Walter. 1968. Chemistry and Technology of Silicones. New York, Academic Press, 702 p.
2. Hobbs, E.J., Keplinger, M.L., and Calandra, J.C. 1975. Toxicity of polydimethylsiloxanes in certain environmental systems. Environmental Research, 10, p.397-406.
3. Pellenbarg, Robert E. 1979. Environmental poly(organosiloxanes)(Silicones). Environmental Science and Technology, 13, p.565-569.
4. Pellenbarg, Robert. 1979. Silicones as tracers for anthropogenic additions to sediments. Marine Pollution Bulletin, 10, p.267-269.
5. Smith, A. Lee (ed.). 1974. Analysis of Silicones. New York: John Wiley and Sons, 407 p.
6. Pellenbarg, Robert E. 1982. Silicones in Chesapeake Bay sediments. Marine Pollution Bulletin, 13, p. 427-429.
7. Shelton, R.G.J. 1971. Sludge dumping in the Thames estuary. Marine Pollution Bulletin, 2, p.24-27.
8. Hatcher, P.G. and Keister, L.E., and McGillivary. 1977. Steroids as sewage specific indicators in New York Bight sediments. Bulletin Environmental Contamination and Toxicology, 17, p.491-498.

NUTRIENT CYCLES AND TRANSPORT PROCESSES

CHAPTER 9

BUDGETS AND RESIDENCE TIMES OF NUTRIENTS IN TOKYO BAY

Eiji Matsumoto
Marine Geology Department
Geological Survey of Japan
Yatabe, Ibaraki, Japan

ABSTRACT

The total nutrient input to Tokyo Bay may be estimated as the sum of the output, which comprises sedimentation on the bay bottom, and outflow from the bay mouth. Element analyses were combined with Pb-210 dating to enable calculation of sedimentation rates of nutrients. Rates of nutrient outflow were determined from hydrological observation and element analyses. The residence time of nutrients were calculated to be 0.116 year for phosphorus and 0.123 year for nitrogen. The residence times of nutrients were close to the residence time of the bay water (0.129 year), suggesting that nutrients in the bay are rapidly recycled in the water column, and finally may be transported to the open ocean.

INTRODUCTION

Tokyo Bay is heavily contaminated by man's activities, and elevated concentrations of nutrients now show evidence of impacting the ecosystem in the bay. The possible short- and long-range effects of the disposal of nutrients on the ecosystem must be considered, and biochemical research must include not only accurate measurements of nutrients in the water column and the sediment, but also their fate. The residence time of nutrients in the water column is of special importance in connection with nutrient behavior. In order to evaluate the residence time of nutrients in the bay, it is essential to determine concentrations and the rate of input of nutrients to the bay water. Potentially important sources are sewers, rivers, rainfall, aerial fallout, vessel dumping, oceanic water, benthic flux, etc. Unfortunately, accurate information concerning each input is not available. However, the total input to the bay may be estimated as the sum of the output, which comprises sedimentation on the bay bottom and outflow from the bay mouth.

The primary objectives of our investigation are to estimate the total input on the basis of sedimentation and outflow for nutrients, and then, to evaluate the residence times of these nutrients in the Tokyo Bay water.

EXPERIMENTAL

General Features of Tokyo Bay

Tokyo Bay is surrounded by densely populated and heavily

MARINE AND ESTUARINE GEOCHEMISTRY, Sigleo, A. C., and A. Hattori (Editors)

industrialized areas. Therefore the quantities of nutrient and metal inputs are very large, and the effects are easily measured (1).

Tokyo Bay is located on the eastern side of Honshu Island, Japan. Tokyo Bay is 1000 km^2 in area and has an average depth of 17 m, which becomes deeper up to 70 m to the south. The bay is connected with the North Pacific through the Uraga Channel which is up to 800 m in depth. The water exchange between the bay and open ocean is restricted by a narrow bay mouth 7 km in width. The residence time of the bay water is estimated to be 0.129 year (2).

Sampling

Sediment cores were collected at stations located on a 2.5 mile grid during September 1980 and August 1981 aboard the R.V. Kaiko No. 5. The corer was a modified gravity corer which consists of an acryl pipe (11 cm inner diameter), a sediment catcher and a clear vent. We succeeded in obtaining sediment cores with almost integral sedimentary strata by slow lowering of this corer. Immediately after collection, the sediment cores were cut into 5 cm sections, sealed in plastic containers and stored in a refrigerator.

The sediment samples were dried at 110°C. The weight loss was recorded, which gives the water content of the sediments. The residue was pulverized to a fine powder for dating and element analyses. The density of solid phases present in the sediments was determined by using a specific gravity bottle. The sea salt content in dried sediment samples was calculated on the basis of water content and salinity in the interstitial water. The measured concentrations for sediments were normalized to salt-free sediments.

The seawater samples for nitrogen were collected at 9 stations from the R.V. Tansei Maru during January and September 1980. The phosphorus samples were collected at 6 stations from the R.V. Tansei Maru during June 1983.

Analyses

The radioactivity of Pb-210 was measured by counting beta-activity of its daughter nuclide, Bi-210, or gamma-activity of Pb-210 itself. In the case of beta counting, Pb-210 was extracted for the sediments chemically, and counted with a Tracerlab Omni/Guard low-background counter (ICN Tracerlab, Mechelen, Beligium) (3). Also, Pb-210 was measured non-destructively by an ORTEC Gamma-X germanium detector equipped with a DAAS 7053 spectroanalyzer (EG&G ORTEC, Oak Ridge, Tennessee).

For phosphorus analyses, 5 ml each of concentrated reagent grade $HClO_4$ and HF were added to the 0.5 g sediment sample contained in a teflon bottle. The airtight-capped bottle was placed on a 100° hotplate for 6 h, and the mixture allowed to dry in a fume hood. One ml of concentrated HNO_3 and 99 ml of H_2O were added to the bootle. The solution was analyzed for phosphorus by direct aspiration into an argon plasma using a Jobin Yvon Model JY48P inductively coupled plasma emission spectrometer (SA Jobin Yvon, Longjumeau, France).

Nitrogen analysis for sediments was performed by dry combustion in oxygen at 800°C followed by gasmetric analysis of evolved N_2 using a Yanako Model MT-500 nitrogen analyzer (Yanako Ltd, Kyotō, Japan).

Total phosphorus for water was determined from analysis of reactive phosphate by $K_2S_2O_8$ treatment of an unfiltered seawater sample.

Total nitrogen for water was obtained as the sum of dissolved and particulate nitrogens. The dissolved nitrogen was determined by analysis of ammonia by the Kjeldahl treatment of a sample filtered through a glass fibre filter. The particulate nitrogen was determined from analysis of N_2 by dry combustion of the filter residue using a Yanako Model MT-3 nitrogen analyzer.

RESULTS

Accumulation Rates Measured with Pb-210

Accumulation rates were determined from the excess Pb-210 profiles in sediment cores. The excess Pb-210 was calculated by subtracting the supported Pb-210 from the total Pb-210. The supported Pb-210 was determined from Pb-210 measurements using core samples obtained from greater depths and Pb-210 counting in other cores by gamma-spectrometry. Accumulation rate expressed as mass per unit area per time is required for the mass balance consideration and for intercore comparisons because they are independent of porosity differences. The logarithm of excess Pb-210 was plotted against the cumulative weight of sediments, and the accumulation rate was determined from the slope (3).

Surface layers with homogeneous Pb-210 were observed in some cores. The surface homogeneous layer is due to the mixing of sediments by biological and physical activities. If accumulation and mixing are steady, an equal amount of freshly accumulated sediment is displaced below the region of active mixing, and Pb-210 in the displaced sediments decays radiometrically with its characteristic half-life. When a surface mixed layer was observed, an accumulation rate was determined from the logarithmic region below the mixed layer. However, special care for proper application is required, because the thin logarithmic layer is often disturbed with mixing.

The distribution of sediment accumulation rates for Tokyo Bay measured with the Pb-210 technique ranged from 0 at the bay mouth up to 0.60 g/cm^2/yr (Fig. 1). The accumulation rates decrease away from the innermost bay southward and eastward. The bay mouth region is no-depositional and comprises entirely relict sand and Tertiary rock. The annual accumulation rate in the whole Tokyo Bay, except the area shallower less than 5 m in water depth, was calculated to be 1.2 x 10^6 tons.

Element Concentrations in Surficial Sediments and Water

The upper 5 cm of each sediment core collected in 1981 was used for element analysis. As the sediment accumulation rates in

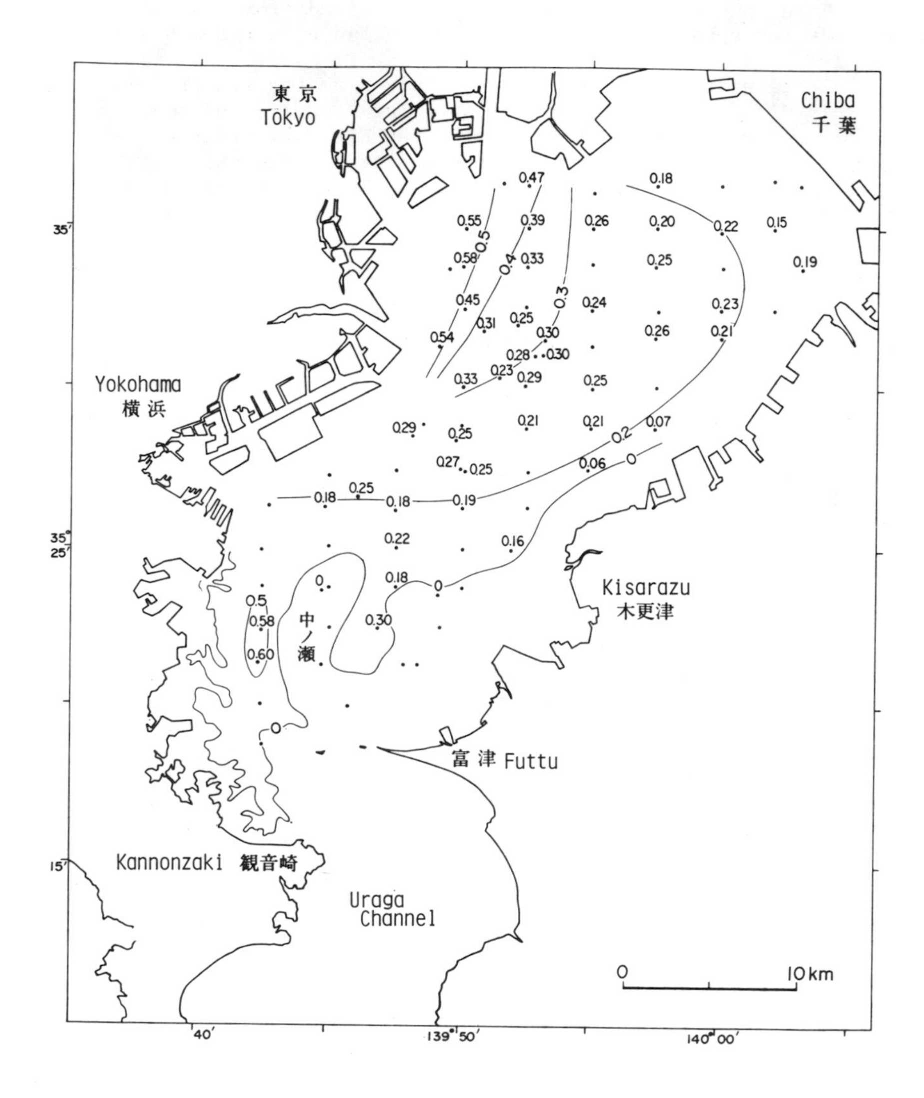

Figure 1. Sedimentation rates in g/cm^2/yr measured with Pb-210.

Tokyo Bay are very fast, the top 5 cm of the cores covers only the past several years. The distributions of surficial concentrations of phosphorus and nitrogen are presented in Figures 2 and 3. The maximum concentrations for phosphorus in these sediments occur near Tokyo (up to 0.55%) and decrease to less than 0.2% near the bay mouth. The range and average concentrations of phosphorus and nitrogen are shown in Table I. The background values were determined from measurements at great depths in the cores. The values reflect the extent of nutrient enrichments in the recently deposited surficial sediments compared to those in deeper sediments accumulated earlier before the advent of the industrial period and dense urban populations.

Table I. Phosphorus and Nitrogen Concentrations in the Surficial Sediments Collected in 1981 from Tokyo Bay.

Element	Number of Samples	Concentration (ppm)		
		Range	Average	Background
P	28	710-1000	840	600
N	29	2000-5500	3600	1500

The range and average nutrient concentrations of the Tokyo Bay water are shown in Table II. A more complete description of the data can be found elsewhere (4, 5). The average nutrient concentrations in the surface water of the North Pacific are also given in Table II. The concentrations in the Tokyo Bay water are higher than those reported for the open ocean, as the results of nutrient input from the surrounding urban areas.

Table II. Phosphorus and Nitrogen Concentrations in Tokyo Bay Water

Element	Number of Samples	Concentration (μg/l)		
		Range	Average	Open Ocean
P	28	50-260	83	4
N	84	280-1200	750	200

DISCUSSION

Model for Residence Time Calculations

The residence time of an element i in the bay water (τ_i) is given by

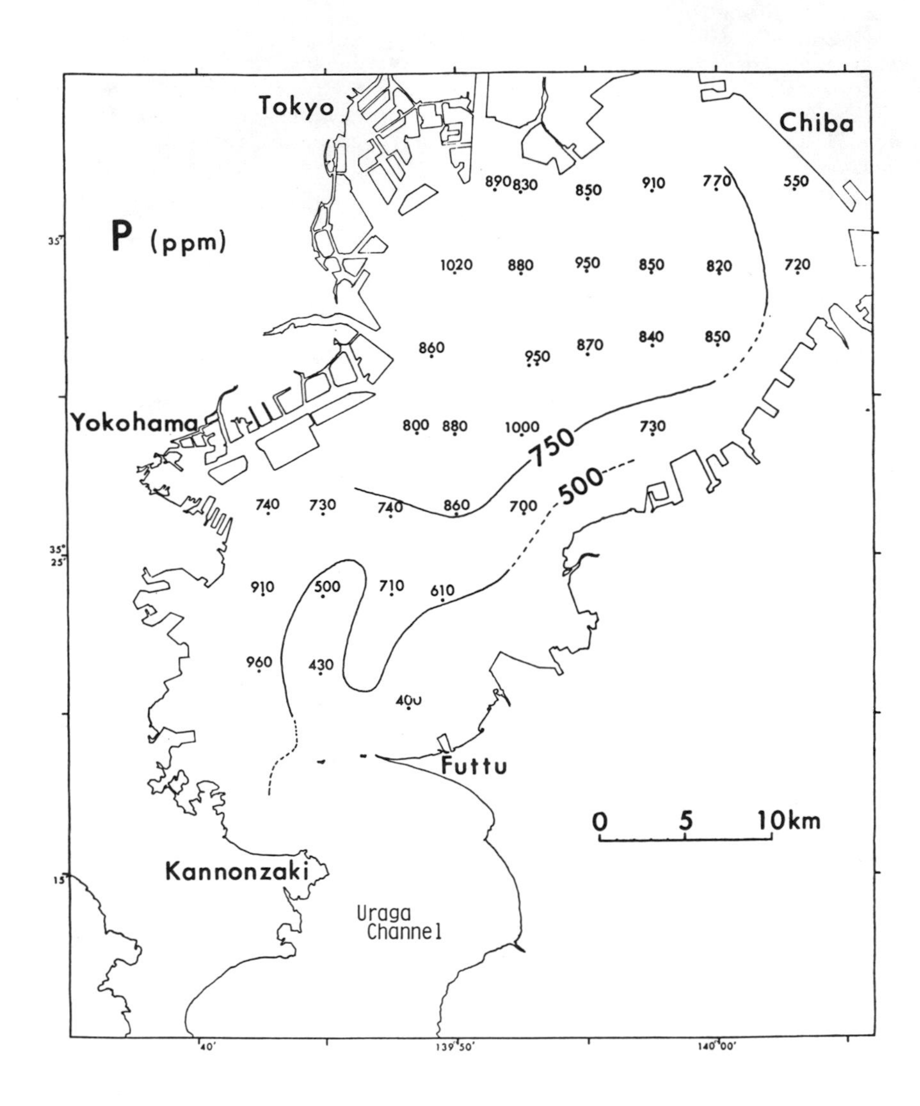

Figure 2. The distribution of phosphorus in the surfical sediments of Tokyo Bay.

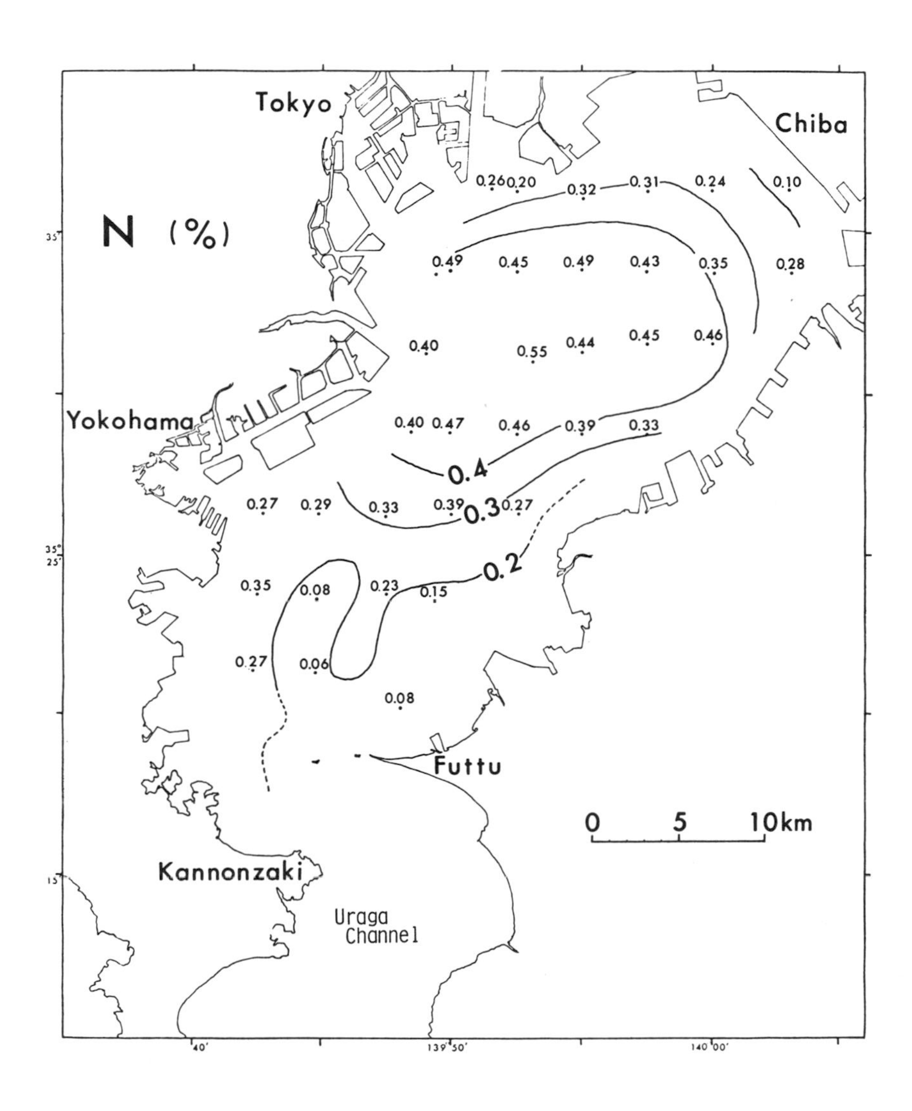

Figure 3. The distribution of nitrogen in the surfical sediments of Tokyo Bay.

$$\tau_i = \frac{C_i V}{I_i}$$

where C_i is the concentration of element i in the bay water, V the volume of the bay water and I_i the input rate of element i into the bay water. Potentially important inputs include sewers, rivers, oceanic water, rainfall, aerial fallout, vessel dumping and benthic flux. Generally, it is not an easy task to evaluate accurately each input. However, the total input to the bay may be estimated as the sum of output which comprises sedimentation and outflow from the bay. Therefore, equation 1 is given by

$$\tau_i = \frac{C_i V}{S_i + C_i v}$$

where S_i is the sedimentation rate of element i into the bay bottom and v is the outward flux of the bay water from the bay mouth. The residence time of the bay water (τ_w) is given by V/v. The value of τ_w in Tokyo Bay was reported to be 0.129 year on the basis of chlorine budget (2). The volume of the bay water is 16 km^3.

Budgets and Residence Times of Nutrients

The annual sedimentation rates of phosphorus and nitrogen to the bay bottom can be calculated by multiplying their average concentrations in surficial sediments by the accumulation rate. The calculation does not take into account the benthic fluxes of nutrients. The outflow of phosphorus and nitrogen from the bay mouth can be calculated by multiplying average concentrations of phosphorus and nitrogen in the bay water by the outflow of bay water (Table III).

Table III. The Budgets and Residence Times of Phosphorus and Nitrogen in Tokyo Bay.

Element	Sedimentation Rate (ton/year)	Flushing Rate (ton/year)	Residence Time (year)
P	1000	10000	0.116
N	4200	90000	0.123

Sedimentation accounts for approximately 9% as much phosphorus as all phosphorus inputs to Tokyo Bay, and 4% of all nitrogen inputs. Callender (1982) reported that the net sedimentation to the bottom accounts for 10% of all phosphorus inputs into the tidal Potomac River (6). This value is comparable to the above value in Tokyo Bay.

The residence times calculated are 0.116 year for phosphorus and 0.123 year for nitrogen. The residence times of phosphorus and nitrogen are close to the residence time of the bay water of 0.13

years, suggesting that the most fraction of nutrients is transported to the open ocean. Hattori et al (1983) measured the rates of uptake and regeneration of inorganic nitrogen in the Tokyo Bay water, utilizing isotope experiment [7]. The results indicate the recycling of nitrogen occurs rapidly in the water column, and only a small fraction of particualte produced in the surface layer may reach the bottom. These findings are consistent with our estimate of nitrogen.

The residence times of heavy metals, however, are quite short relative to the residence time of the bay water [8]. This means heavy metals are generally rapidly deposited after their introduction into the coastal environment.

CONCLUSION

Tokyo Bay is surrounded with a densely populated and industrialized area that introduces a large amount of nutrients into the bay, resulting in eutrophication. The total input of nutrients to the bay was estimated as the sum of the output which comprises the sedimentation on the bay bottom and the flushing from the bay mouth. On the basis of their budgets, the residence time of nutrients was calculated to be 0.116 years for phosphorus and 0.123 years for nitrogen. The values are close to the water residence time of 0.129 years. Nutrients introduced into the bay are utilized by phytoplankton. However, the nutrients once taken up are rapidly regenerated within the water column, and then may be transported to the open ocean.

REFERENCES

1. Matsumoto, E. "Environmental Changes Recorded in Sediments of Coastal Marine Zone to a Big City," *Memoirs Geol. Soc. Japan* 23:91-95 (1983).
2. Unoki, S. and M. Kishino. "Oceanographycal Conditions and Water Exchange in Tokyo Bay," Technical Report of Physical Oceanography Labolatory, Institute of Physical and Chemical Research, Japan (1977).
3. Matsumoto, E. and S. Togashi. "Sedimentation Rates in Funka Bay, Hokkaido," *J. Oceanogr. Soc. Japan* 35:261-267 (1980).
4. Hattori, A. "Distribution and Fate of Pollutants in Coastal Environments: Data Report 1979-1980," Unpublished Report of Special Research Project on Environmental Science, submitted to Ministry of Education, Culture and Science, Japan (1981).
5. Hashimoto, S., K. Fujiwara and K. Fuwa. "Distribution of Alkaline Phosphatase Activity in Tokyo Bay," *Environ. Sci. Tech.* submitted for publication (1985).
6. Callender, E. "Benthic Phosphorus Regeneration in the Potomac River Estuary " *Hydrobiologia* 92:431-446 (1982).
7. Hattori, A., M. Ohtsu and I. Koike. "Distribution, Metabolism and Budgets of Nitrogen in Tokyo Bay," *Chikukagaku* 17:33-41 (1983).
8. Matsumoto, E. "The Behavior of Heavy Metals in Coastal Marine

Environments," in T. Hirano, ed. Marine Environmental Science (Koseishakoseikaku, 1983), pp. 168-177.

CHAPTER 10

SEASONAL AND INTERANNUAL NUTRIENT VARIABLITY IN NORTHERN SAN FRANCISCO BAY

Richard E. Smith
David H. Peterson
Stephen W. Hager
Dana D. Harmon
Laurence E. Schemel
Raynol E. Herndon
U.S. Geological Survey
345 Middlefield Road
Menlo Park, California

ABSTRACT

Nearly two decades (1960-80) of dissolved inorganic nutrient and salinity distribution data for northern San Francisco Bay estuary document seasonal and interannual variations with respect to river flow (a nutrient source) and phytoplankton productivity (a nutrient sink). During winter nutrient sources dominate the nutrient-salinity distribution patterns. During summer, however, the sources and sinks are in close competition. Summers of wet years have characteristics more like winter because sources often dominate the nutrient distributions whereas in summers of dry years sinks dominate.

INTRODUCTION

Information concerning the nature and causes of dissolved inorganic nutrient variability in estuaries on seasonal and interannual time scales is important to a variety of research efforts including studies of the effects of man and climate on estuarine biochemistry [e.g., 1,2], global nutrient cycles [e.g., 3] and estuarine ecology [e.g., 4,5]. The nutrient distributions investigated are dissolved silica, nitrate, phosphate and ammonium in northern San Francisco Bay estuary [cf., 6]. Two characteristic patterns in these distributions are considered. In particular, conservative or near-conservative distributions are associated with periods of high river flow whereas non-conservative distributions are associated with phytoplankton assimilation. The seasonal and interannual variations in river flow are large [Fig. 1]. The seasonal variations in phytoplankton productivity appear to be relatively small [Table I] and their interannual variations have not been determined.

Many factors influence the dissolved inorganic nutrient distributions in estuaries. For this reason and because estuaries are not yet fully understood as complete systems it seems

MARINE AND ESTUARINE GEOCHEMISTRY, Sigleo, A. C., and A. Hattori (Editors)

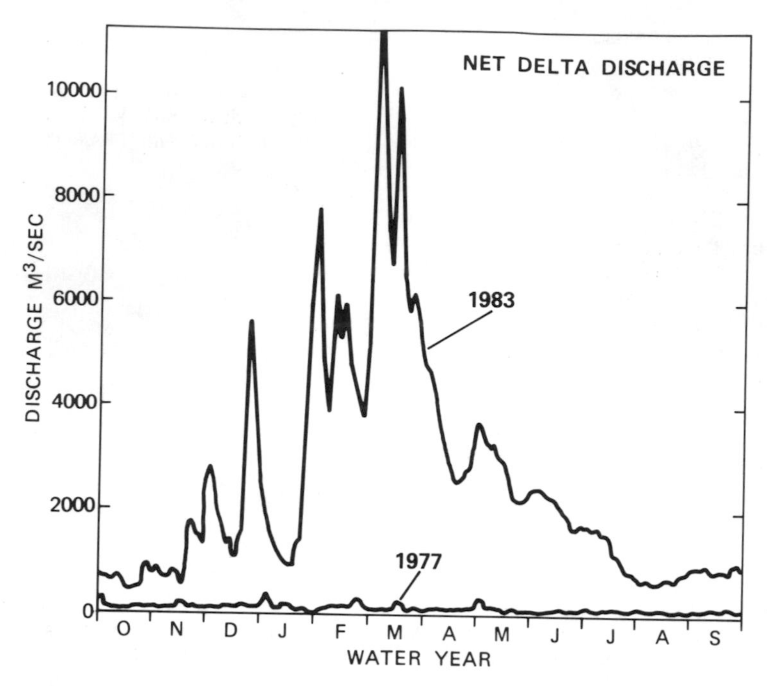

Figure 1. Annual cycle of Sacramento-San Joaquin River flow for the water years 1977 and 1983 (Data from U.S. Bureau of Reclamation).

appropriate to review some examples of estuarine nutrient distributions. What follows is a brief overview given in approximate order of our decreasing knowledge about the behavior of these nutrients in estuarine environments.

Silica

Weathering of rocks and soils is the principal source of silica in rivers [7] and estuaries [8]. If weathering rates have remained relatively constant over the last few hundred years, then riverine dissolved silica concentrations are a relatively constant reference for assessing the changes in concentrations of other dissolved constituents [9]. To document dissolved silica concentrations and other properties, however, requires considerable sampling effort [10] especially in small rivers [11]. In some of the world's largest rivers the variability in dissolved silica is apparently less [12]. In some European rivers increasing phosphate and nitrate loads over the last decades have been accompanied by decreasing dissolved silica concentrations, apparently as a result of increased diatom growth [13]. This is, possibly, a result of the composition of municipal waste which is an important source of nitrogen and phosphate but is not considered a major source of dissolved silica [14].

Table I. Phytoplankton productivity in northern San Francisco Bay

PRODUCTIVITY[1] ($mg\ C\ m^{-2}d^{-1}$)			
Winter		Summer	
Nov-Mar	May-Jun	Jul-Aug	Sep
130	450	350	400

[1]Adapted from 1980 incubation experiments [15]. An average summer productivity is 400 mg C $m^{-2}d^{-1}$, which is equivalent to 6.3, 5.0 and 0.3 mmoles $m^{-2}d^{-1}$ respectively, for silica, dissolved inorganic nitrogen and phosphate based on Redfield's ratio between these elements, and organic carbon [16].

Rivers, then, are a major source of dissolved silica in estuaries and, for many estuaries, the ocean is probably a net sink. A recent literature review concludes most of the riverine silica is ultimately transported to the sea [8]. River flow, however, is not necessarily a major source of silica throughout the year. As river flow diminishes other estuarine sources of dissolved silica emerge and influence silica distributions. In estuaries with large areas and relatively small river inflow, the benthic flux to overlying waters is quantitatively important even exceeding the riverine source [17]. This imbalance between river and estuary implies that in some estuaries riverine dissolved silica can be transformed to biogenic particles and mineralized or recycled within the estuary many times before loss to the sea [17]. Other sources of silica that could help support benthic silica fluxes include riverine and ocean biogenic silica transported to the estuary and ocean dissolved silica transformed to biogenic silica within the estuary. In addition, in instances of extremely large waste sources, waste can be a significant local source relative to other dissolved silica sources, but not relative to phosphate and dissolved inorganic nitrogen [18].

Biotic transformation is the principal sink for dissolved silica in estuaries; whereas the role of abiotic processes, such as adsorption on suspended particles, is considered to be minimal.

Dissolved silica-salinity distributions in estuaries are typically linear or near-linear [8]. In instances of silica depletion to near-zero concentrations [cf., 19] estuaries can be a net sink for both river and ocean silica sources [20]. Estuarine silica concentration maxima have also been observed, presumably due to benthic sources from extensive shoal areas [21].

Nitrate

River sources of nitrate like silica are quantitatively important but, unlike silica, are highly variable. For example, in temperate rivers nitrate concentrations are typically high in winter and low in summer. Ocean sources are also variable. When the continental shelf is narrow, as it is along western North and South America, ocean upwelling is an important nitrate source for estuaries [22]. When the continental shelf is broad, as it is along eastern United States, the cross-shelf transport of ocean water can be sluggish and shelf-water nutrient concentrations are often very low [23]. Municipal waste is a considerable source of nitrate and phosphate in highly populated river basins and estuaries [24]. Also, the benthos has been implicated as both a source and a sink for nitrate but probably less so as a source [cf., 25,26,27].

Biotic transformation is the principal sink for nitrate in estuaries including plants [5] and bacteria [28,29].

Nitrate distributions in estuaries have been recently reviewed [5] and are sometimes similar to dissolved silica (concentrations decrease in a seaward direction [cf., 30, 31]). Nitrate-salinity distributions, however, are probably more variable than silica-salinity distributions. This is expected because source/sink phenomena are more complicated for nitrate than silica.

Phosphate

Riverine phosphate, like nitrate, is influenced by land use (e.g., woodland, agriculture and urban [32]). Phosphate mining rates have exceeded natural weathering rates since about the early 1900s [33] and phosphate is a nonrenewable resources. Waste is a rich source of phosphate [14]. As discussed for nitrate, coastal upwelling is also a potential source of phosphate for estuaries. An evaluation of benthic phosphate sources, however, is complicated by oxic and anoxic phenomena. For instance, iron can serve as an abiotic sink for phosphate when associated with oxidizing sediments and as a source when associated with reducing sediments [25].

Biota are an important sink for dissolved inorganic phosphate in estuaries but probably less so than for nitrate and ammonium [4]. Abiotic adsorption-desorption phenomena are also considered a control on phosphate concentrations in estuaries [34].

Phosphate-salinity distributions in estuaries are commonly linear or near-linear [34]. During the plant growing season, however, phosphate concentrations may be depleted in relatively clear-water estuaries due to phytoplankton uptake [35] or appear in excess due to massive waste sources [18].

Ammonium

Riverine sources of ammonium are generally less than for nitrate except in polluted rivers. Benthic sources of nitrogen, however, are typically in the form of ammonium rather than nitrate. Waste is also a major source of ammonium.

Biota are a major sink for ammonium in estuaries including plants [5] and bacteria [28,29]. Ammonium can be transformed by bacteria to nitrate and vice versa. In any case, phytoplankton generally prefer ammonium over nitrate [30] unless ammonium concentrations are very low (e.g., less than 2μmoles l^{-1}).

In summary, dissolved silica distributions are probably the easiest to interpret for several reasons. Such distributions are generally the result of natural processes and, therefore, are less influenced by human activities. Silica concentrations in rivers are usually high relative to the other nutrients (unless the river basin is highly developed), relative to concentrations in the adjacent coastal ocean, and relative to phytoplankton uptake (Tables I and II). Rivers, in addition to their role as a major source of dissolved silica, are also a forcing function of estuarine transport and circulation. Thus, a knowledge of the dissolved silica distribution can provide a link between estuarine physics, chemistry and biology [cf., 36,37]

Table II. Pristine river and ocean dissolved inorganic nutrient concentrations

	Silica	Nitrate	Phosphate	Ammonium
	(μmole l^{-1})			
River	200[1]	5[2]	0.4[2]	?
Ocean[3]				
5 m	2-30	0-20	0.1-1	0.3
4,000 m	150	40	3	4

[1]Global mean [8].
[2][7].
[3]Pacific Ocean [38].

METHODS

Environmental Setting

Estuaries derive many of their characteristic features from the density (salinity) differences between fresh water and sea water [39]. For example, vertical differences in salinity characterize northern San Francisco Bay as a partially mixed estuary [40]. In this and probably most other western United States estuaries salinity is modulated seasonally by a strong winter (Nov.-Mar.) maximum and summer (May-Sept.) minimum in precipitation [41] and runoff [6].

This study is confined to the northern San Francisco Bay, an estuary with its circulation and salinity responding to variations in Sacramento-San Joaquin River flow. Sacramento-San Joaquin River flow supplies more than 90% of the runoff to the estuary. It averages 600 m^3s^{-1} but ranges from over 2,000 m^3s^{-1} in winter to less than 100 m^3s^{-1} in summer. River flow is the prime seasonal regulator of salinity penetration and water residence time in the northern part of the estuary. At a river flow of 2,000 m^3s^{-1}, salinity penetrates to approximately 40 Km from the seaward end and at a flow of 100 m^3s^{-1} to 90 Km.

Phytoplankton abundance is correlated with variations in water circulation in a general way except during extended periods of extremely low river flow. Phytoplankton biomass decreases in winter and increases in summer, particularly in the landward portions of the estuary. Phytoplankton productivity is a major biological process affecting the chemistry of estuaries in general and nutrient distributions in particular. Furthermore, in general, estuarine nutrient distributions probably follow this annual winter/summer cycle of runoff/productivity. And, as might be anticipated, this seasonality and association with runoff is strongest in northern San Francisco Bay for dissolved silica and perhaps the least for phosphate.

The above is a brief overview, more details about the properties of the estuary, which relate to the discussion of nutrients given here, are given elsewhere: a description of San Francisco Bay environments [42], water circulation [43], the distribution of suspended particulate matter [44] and examples of seasonal phytoplankton [45] and nutrient [46] distributions.

Observations

The data base for this paper spans the period 1960 through 1980 and represents four independent studies [1960 to 1964: 47, 48,49; 1964: 50; 1968: Kramer unpublished data; 1969 to 1980: 51,52,53,54]. Parameters analyzed include salinity, dissolved reactive inorganic phosphate, silica, ammonium and nitrate plus nitrite (hereafter called nitrate). All nutrient analyses from 1971 to 1980 were performed using a Technicon Autoanalyzer. The analyses were usually conducted "on line," with a continuous stream of filtered water (0.45 μ pore diameter) passed to the Autoanalyzer sample tubes. Dissolved silica was determined by an adaptation of the Technician method AII 105-71W [55]; nitrate by AII 100-70W [56]; phosphate by a modification of Atlas <u>et al.</u> [57]; and ammonium is an automated version of the method of Solorzano [58] similar to that of Head [59]. Much of the literature on estuarine nutrient distributions and related processes is difficult to assimilate and interrelate because there are no common procedures of data collection and presentation. Studies in northern San Francisco Bay estuary are no exception [60]. For instance, in some of the earlier studies ammonium observations and nitrate and phosphate were not measured with adequate precision for the purposes here. Data deleted from the figures herein include all ammonium analyses before 1965; all winter (Nov.-Mar.) nitrate analyses before 1965; the Jan. 31, 1961, April 18, 1961 and Jan. 7, 1963 dissolved silica; the 1968 phosphate and all of the nutrients for May 17, 1976. Therefore, the coverage of data is not uniform in time for the four nutrients. Furthermore, spatial coverage is limited to the main channel of northern San Francisco Bay and does not include nutrient sampling in waters overlying the shallow reaches. Despite the above limitations, there are sufficient data in most instances to characterize nutrient variability according to the climatic scenarios of wet, intermediate, dry and very dry years.

SEASONAL VARIABILITY

In general, riverine sources of nutrients are large in winter and small in summer depending on river flow. During winter the cumulative effects of high river flow and associated high suspended sediment concentrations and lower sunlight and temperatures can, more or less, decrease phytoplankton productivity and biomass. For these reasons the effects of phytoplankton are not as strong in winter as summer nutrient concentrations in northern San Francisco Bay are not lowered or depleted as much in winter as summer [cf., 46].

Nutrient concentrations can decrease in summer due to a decrease in rate of supply relative to removal. For example, the summer decreasing concentration of dissolved silica can be explained largely by a decreasing rate of river supply rather than increasing rate of phytoplankton removal (e.g., Fig. 2). However, summer variations in other nutrients are often not as simple. In general, over the course of summer, the decrease in nitrate concentrations is greater relative to phosphate, i.e., the nitrate: phosphate ratio decreases from about 12:1 to 3:1 [unpublished observations]. Unlike silica, variations in river flow are apparently an ineffective control on ammonium concentrations during summer, and summer concentrations are probably controlled more by increased rates of removal (e.g., increased phytoplankton productivity) rather than decreased rates of supply (e.g., river flow). Ammonium sources such as benthic exchange could be greater in summer than winter [27]

INTERANNUAL VARIABILITY

Four annual hydrologic climate conditions are subjectively defined: wet, intermediate, dry, and very dry (Table III, see also Figure 3). Mean river flow during summer is greater than 400 m^3s^{-1} in wet years, and is greater than 120 m^3s^{-1} but less than 200 m^3s^{-1} in very dry years. Winter conditions are categorized according to their following summer flows, that is if the summer was wet, for the purpose of this paper, its preceding winter was also wet. All of the nutrient salinity distributions as per the methods section earlier have been combined according to these four climate conditions and their seasons (Figures 4 to 9).

Table III. Winter-summer variations in Sacramento-San Joaquin River flow during wet, intermediate, dry and very dry hydrologic conditions

Hydrologic Condition	River Discharge (m^3s^{-1}) Winter Nov-Mar	May-Jun	Summer Jul-Aug	Sep
WET[1]	1,600±69%	890±53%	270±36%	470±26%
INT[2]	1,800±78%	360±42%	170±44%	310±24%
DRY[3]	700±46%	180±52%	124±34%	210±33%
V. DRY[4]	230±74%	100±21%	100±27%	92±20%

[1]1963, 1969, 1971, 1974, 1975, 1978
[2]1962, 1970, 1973, 1980
[3]1961, 1964, 1968, 1972, 1979
[4]1976, 1977

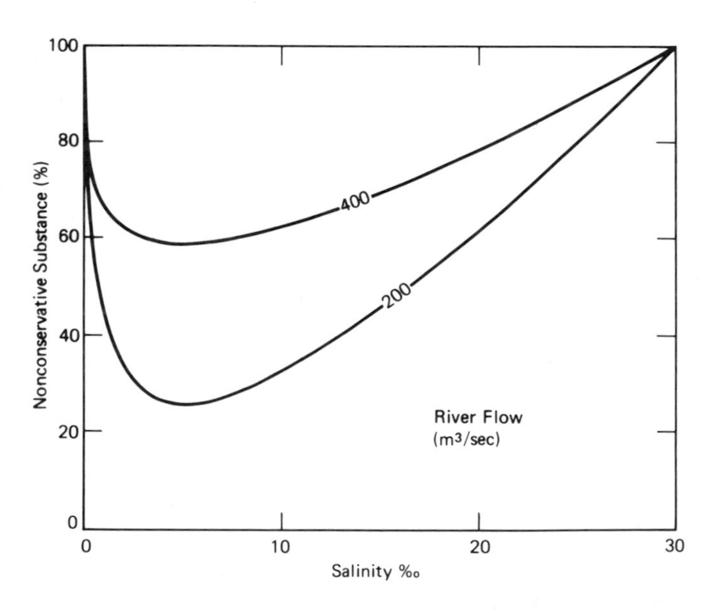

Figure 2. Simulated distribution of a nonconservative substance with salinity as a function of various river flows assuming 100% concentrations of substance at both river and ocean boundaries and a constant uptake of 2% per day. After Peterson et al. (1975).

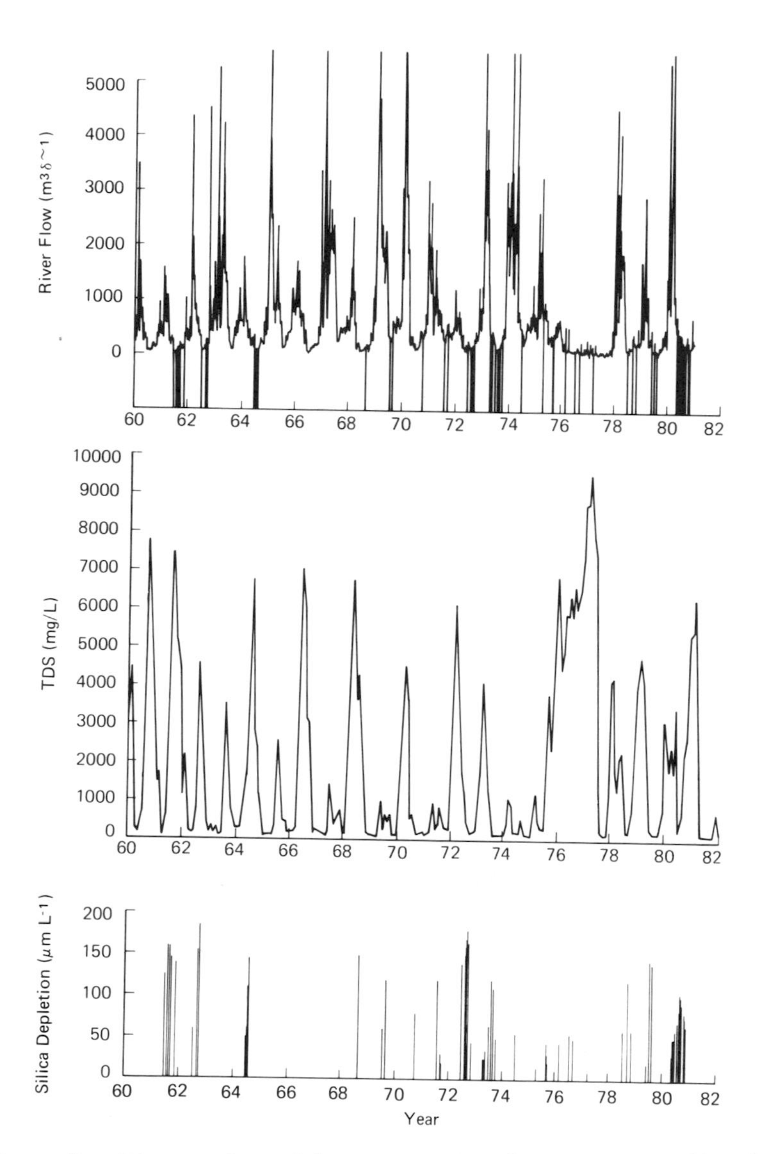

Figure 3. Time series of Sacramento-San Joaquin River flow (upper panel, vertical lines denote cruises when silica depletion was observed), total dissolved solids (middle panel, data from State of California 1983) and silica depletion (lower panel). Silica depletion is the maximum difference in concentration between a conservative mixing value and the observed value at the same salinity.

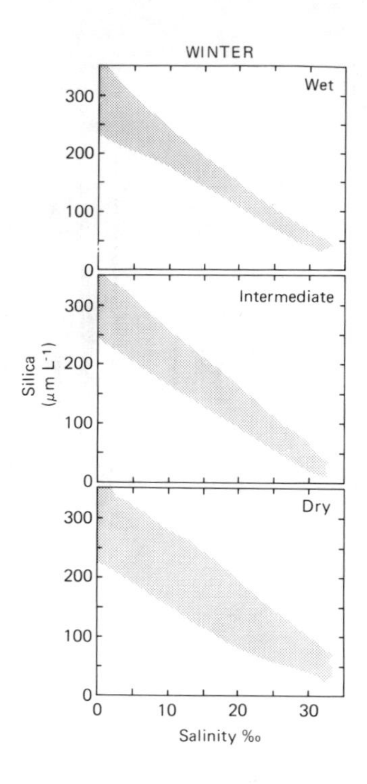

Figure 4. Dissolved silica-salinity field in northern San Francisco Bay estuary during wet, intermediate and dry winters as defined in Table II.

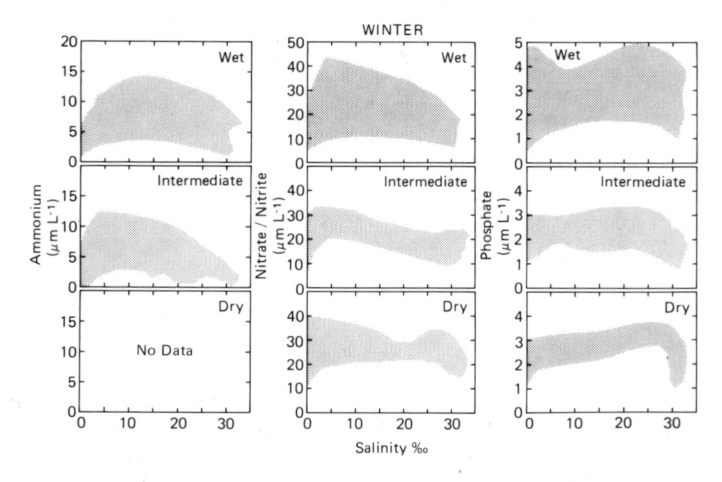

Figure 5. Dissolved ammonium-, nitrate plus nitrite- and phosphate-salinity field in northern San Francisco Bay estuary as in Fig. 4.

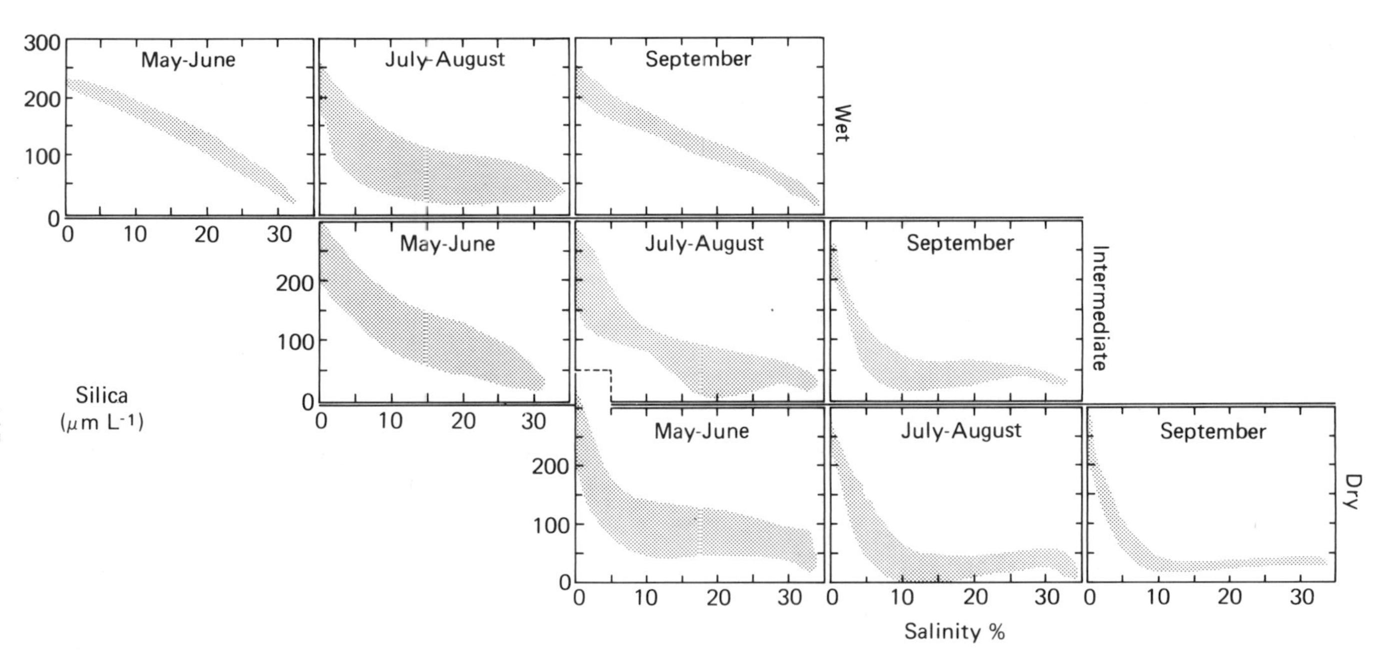

Figure 6. Seasonal dissolved silica-salinity field in northern San Francisco Bay estuary during wet, intermediate and dry summers as defined in Table II. Note panels are offset to illustrate the differences in seasonal behavior.

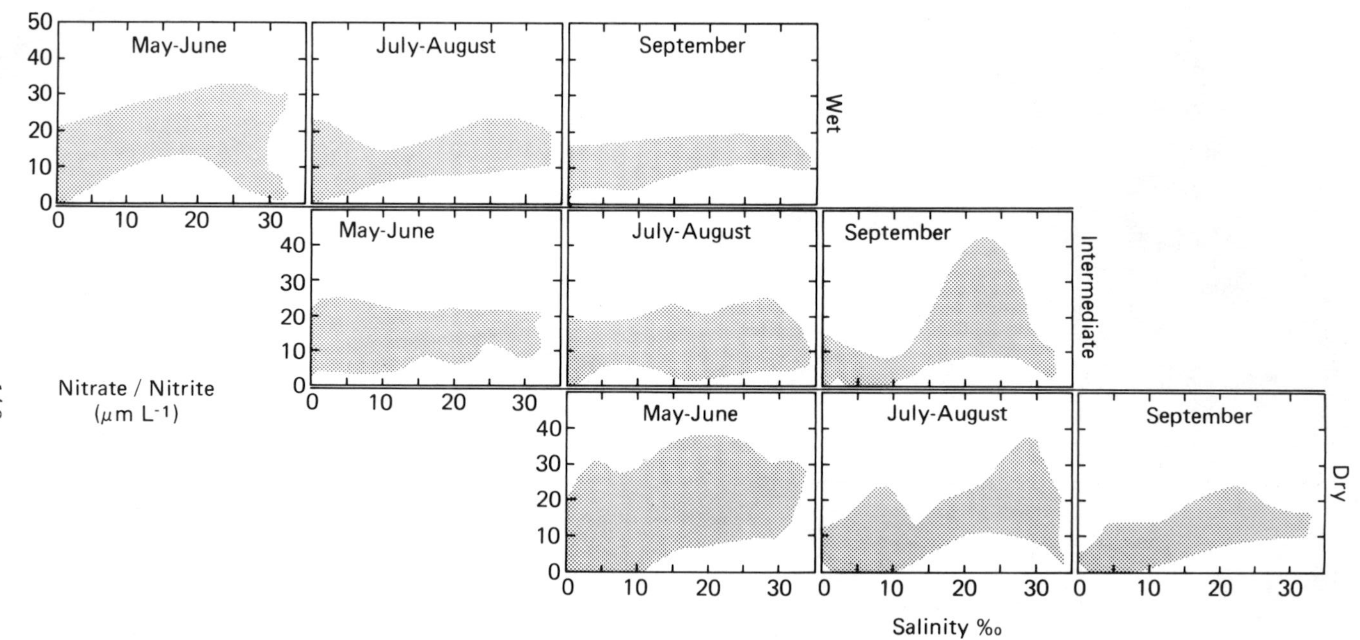

Figure 7. Seasonal dissolved nitrate plus nitrite-salinity field in northern San Francisco Bay estuary as in Fig. 6.

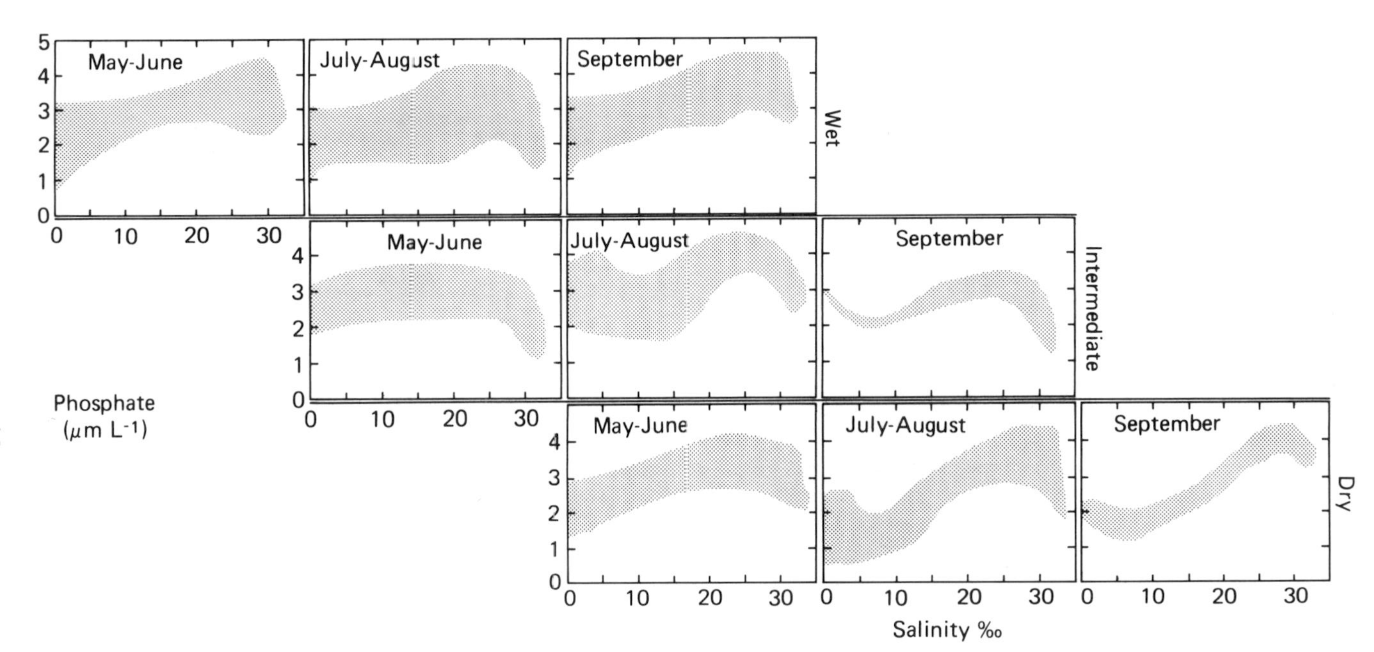

Figure 8. Seasonal dissolved phosphate-salinity field in northern San Francisco Bay estuary as in Fig. 6.

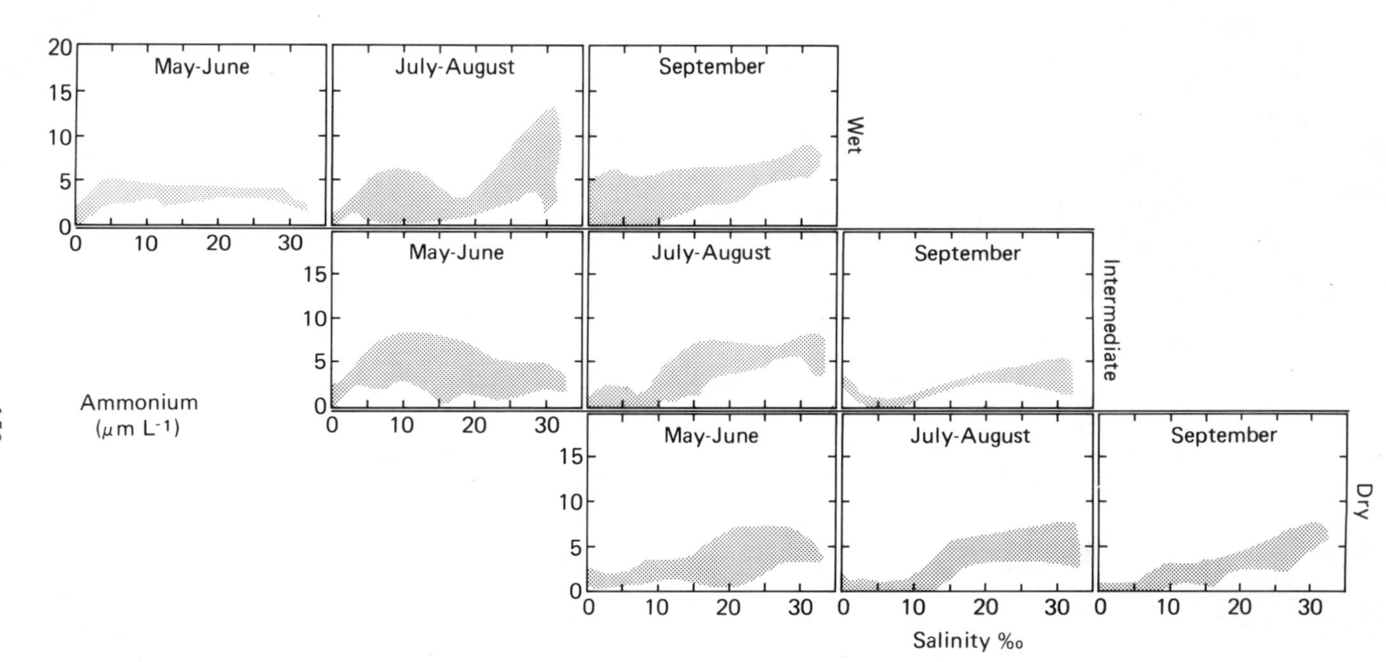

Figure 9. Seasonal dissolved ammonium-salinity field in northern San Francisco Bay estuary as in Fig. 6.

Wet Years

In Winter the sources, external and internal, dominate the dissolved inorganic nutrient distributions during wet winters. Considering all of the dissolved inorganic nutrient distributions and their sources, the phytoplankton productivity sink appears to be relatively weak. The river silica source (the product of the riverine concentration and mean flow) during a wet winter is roughly 50 times greater than the phytoplankton sink when the river source is averaged over $4x10^8$ m^2 (the area of northern San Francisco Bay from San Pablo Bay to Suisun Bay, cf. Fig. 1 p. 116). Therefore, because the phytoplankton sink is relatively weak, the dissolved silica distributions are, as predicted, linear with salinity. Nitrate distributions are similar to those for dissolved silica, but with more variability in riverine concentrations (Fig. 5). Wet winter distributions of phosphate (Fig. 5) show distributions similar to nitrate but are even more variable. Some of this variability is attributed to "internal" sources such as desorption from suspended particulate matter. Ammonium-salinity distributions (Fig. 5) appear to be less linear than the other nutrients suggesting cumulative effects of internal sources including waste and benthic exchange is even more important for ammonium than phosphate [61].

By late spring (May-June), although the river flow and dissolved silica concentrations have probably decreased from winter levels and phytoplankton productivity can increase (Table I), the riverine sources are still about eight times greater than the estimated phytoplankton sink. It is to be expected, then, that dissolved silica concentrations remain linear or near-linear (Fig. 6). However, during July-August, when river flow drops to 270 m^3s^{-1} (or below the wet-summer mean of 400 m^3s^{-1}), dissolved silica sources average only about 2.4 times the phytoplankton sink. Then silica salinity distributions show a slight influence of phytoplankton uptake (a non-linear distribution). By September, however, when river flow is greater than 400 m^3s^{-1} (Table II), the river source is four times greater than the potential phytoplankton sink, and the silica-salinity distributions also return to a near-linear distribution (Fig. 6).

Nitrate is more variable than silica (Fig. 7) because the riverine concentrations, hence, source of silica, is at least 13 times higher than that of nitrate. Although river flow is relatively high, the nitrate source is close to the estimated phytoplankton sink and the distributions can be influenced accordingly. Ocean phenomena can also contribute to variability in nitrate concentrations in the bay. The peak of coastal upwelling occurs around June [62]. The July-August nitrate distributions are similar to dissolved silica distributions in the sense that they show some removal by phytoplankton productivity. By September the river flow increases and effects of phytoplankton productivity on nitrate concentrations are minimized.

The May-June phosphate distributions (Fig. 8) appear to be less influenced by phytoplankton productivity than the nitrate distributions. Riverine concentrations generally vary between 1 and 3 μmole l^{-1}. In the early summer of wet years it appears that the river can be a stronger source of phosphate than nitrate when compared to the strengths of their sinks. In other words, phosphate concentrations should be less sensitive to variations in phytoplankton productivity than nitrate. The July-August and September phosphate distributions could illustrate this difference. The seaward values are often much higher than can be explained solely by upwelling phenomena, therefore, the high phosphate concentrations probably indicate sources other than or in addition to upwelling.

The distribution of ammonium, which was not observed as frequently as nitrate, can show a strong influence of phytoplankton productivity (Fig. 9). It is not surprising that ammonium can be depleted during a wet summer. On the other hand, nitrate is not depleted because the ammonium levels are, in general, much lower in river water than nitrate and because ammonium is preferred over nitrate by phytoplankton as long as ambient ammonium concentrations are above about 2 μ moles l^{-1} [63]. Thus, the river source of ammonium is generally weak even during a wet summer, whereas the phytoplankton sink is especially strong. As noted for phosphate, the frequently high concentrations of ammonium (greater than 5 μ moles l^{-1}) at higher salinities cannot be explained solely by upwelling. It appears that the seaward portion of the estuary as shown or defined is dominated by respiration and mineralization processes during mid- and late-summer, whereas the landward portion of the estuary, Suisun Bay, is dominated by phytosynthesis.

Dry Years

In winter, the dissolved silica distributions (Fig. 4) during dry winters are almost the same as in wet winters. The dry winter nitrate distributions (Fig. 5) are too few to draw any conclusions. The dry winter phosphate distributions (Fig. 5) are similar to their wet winter distributions. There is no ammonium data available for dry winters.

In summer, the dissolved inorganic nutrient distributions during dry summers illustrate the strong role of phytoplankton in controlling nutrient distributions. For instance, the dissolved silica distributions in May-June (Fig. 6) indicate moderate effects of the phytoplankton sink. As summer progresses, this biogenic sink strengthens relative to the river source and, as a result, the silica concentration anomaly (the maximum difference between conservative and observed concentrations) increases in late summer and fall to 80 or 90 percent of the conservative mixing concentration.

As might be expected, nitrate distributions during dry summers (Fig. 7), like silica, also show a strong phytoplankton removal, especially at low salinities. In this way the dry summer nitrate distributions resemble the wet summer ammonium distributions described earlier. This distinction between nitrate and ammonium makes sense because of the differences in strengths of their source/sink terms mentioned earlier.

During dry summers, the phosphate distributions also indicate a phytoplankton sink, especially at low salinities, but perhaps not as strongly as for nitrate. During dry summers ammonium is almost completely removed at low salinities (Fig. 9) showing that phytoplankton, by preferentially using ammonium, are able to keep pace with the ammonium sources.

Intermediate Years

From the above discussion and Figures 4 to 9, it can be seen that for the most part the intermediate patterns lie between the two extremes. The interpretation is left to the reader.

Very Dry Years

Dissolved inorganic nutrient patterns in very dry years (1976 and 1977 Table 3) do not follow the general patterns described above [64]. During very dry years nutrient concentration patterns indicate that nitrate, phosphate and ammonium sources dominate over sinks. Furthermore, dissolved silica remains near-linear throughout summer even with low river inflow. Two explanations have been suggested to account for the unusual nutrient distributions observed during the very dry years. One explanation involves the shift of the salt water regime landward with weak river flow. When this happens, estuarine circulation, the null zone, and the associated turbidity maximum are shifted landward into the deeper water of the river. Hence, the mean water-column light intensity is further decreased and aphotic processes increase relative to photic processes [65,66]. Another explanation is that phytoplankton biomass is suppressed by increased activity of freshwater intolerant benthic filter feeders that migrated landward into Suisun Bay during 1976-77 because of increased salinity [67]. In this case phytoplankton are mineralized and nutrients recycled more rapidly by the increased benthic herbivore activity. it is not yet clear which possible mechanism, water column or benthic, was the most important in controlling dissolved inorganic nutrient distributions during 1976-77. These explanations are not, however, mutually exclusive.

To summarize, estuaries are traditionally studied from an oceanographers perspective--as viewed from the sea. This occurs because sea salt is an integral part of the system. Freshwater inflow is also a major element of an estuary. From the above it seems reasonable to expect that for many estuaries interannual variations in fresh water inflow are a major control interannual estuarine variability in general and dissolved silica distribution in particular. In northern San Francisco Bay interannual riverine influences on phosphate distributions appear to be least relative to the other nutrients. In any case, when all of the nutrient distributions are considered, it is shown that several (many) years of samplings are preferable if not essential to characterize annual nutrient cycles. A study of only one or two annual cycles could be misleading. Our long-range goal is to understand the link between interannual variability and climatic forcing [cf., 2,68].

ACKNOWLEDGEMENTS

We thank W. W. Broenkow and A. C. Sigleo for their manuscript reviews.

REFERENCES

1. Officer, C. B., R. B. Biggs, J. L. Taft, L. E. Cronin, M. A. Tyler and W. R. Boynton. "Chesapeake Bay Anoxia: Origin, Development, and Significance." Science 223:22-27 (1984).

2. Seliger, H. H., J. A. Boggs and W. H. Biggley. "Catastrophic Anoxia in the Chesapeake Bay in 1984." Science 228:70-73 (1985).

3. Bolin, B. and R. B. Cook, Eds. The Major Biochemical Cycles and Their Interactions SCOPE 21 (Chichester, U.K.: John Wiley and Sons, 1983).

4. Boynton, W. R., W. M. Kemp and C. W. Keefe. "A Comparative Analysis of Nutrients and Other Factors Influence Estuarine Phytoplankton Production," in Estuarine Comparisons, V. S. Kennedy, Ed. (NY: Academic Press, 1982), pp. 69-90.

5. Nixon, S. W. and M. E. Pilson. "Nitrogen in Estuarine and Coastal Marine Ecosystems," in Nitrogen in the Marine Environment, E. J. Carpenter and D. G. Capone, Eds. (NY: Academic Press, 1983), pp. 565-648.

6. Conomos, T. J., Ed. San Francisco Bay: The Urbanized Estuary (San Francisco, CA: Pacific Division, Amer. Assoc. Advanc. Sci., 1979).

7. Wollast, R. and F. T. MacKenzie. "The Global Cycle of Silica," in Silicon Geochemistry and Biogeochemistry, S. R. Aston, Ed. (London: Academic Press, 1983), pp. 39-76.

8. DeMaster, D. J. "The Supply and Accumulation of Silica in the Marine Environment. Geochemica et Cosmochimica Acta 45:1715-1732 (1981).

9. Zobrist, J. and W. Stumm. "Chemical Dynamics of the Rhine Catchment Area in Switzerland, Extrapolation to the 'Pristine' Rhine River Input to the Ocean," Inputs to Ocean Systems (NY: UNESCO and U.N. Environment Programme, Unipub, 1982), pp. 52-63.

10. Kennedy, V. C. "Silica Variation in Stream Water with Time and Discharge," Nonequilibrium Systems in Natural Water Chemistry. Advances in Chemistry Series No. 106 (American Chemical Society, 1971), pp. 94-130.

11. Nordin, C. F., Jr. and R. H. Meade. "The Flux of Organic Carbon to the Oceans: Some Hydrological Considerations," Carbon Dioxide Effects Research and Assessment Program, Flux of Organic Carbon by Rivers to the Oceans, NTIS, U.S. Dept. Commerce, 1981), pp. 173-218.

12. van Bennekom, A. J., G. W. Berger, W. Helder and R. T. De Vries. "Nutrient Distribution in the Zaire Estuary and River Plume." Neth. J. of Sea Res. 12:296-323 (1978).

13. van Bennekom, A. J., W. W. C. Gieskes and S. B. Tijssen. "Eutrophication of Dutch Coastal Waters." Proc. R. Soc. Lond. B. 189:359-374 (1975).

14. Officer, C. B. and J. H. Ryther. "The Possible Importance of Silicon in Marine Eutrophication." Mar. Ecol. Progr. Ser. 3:83-91 (1980).

15. Cole, B. E. and J. E. Cloern. "Significance of Biomass and Light Availability to Phytoplankton Productivity in San Francisco Bay." Mar. Ecol. Prog. Series 17:15-25 (1984).

16. Redfield, A. C., B. H. Ketchum and F. A. Richards. "The Influence of Organisms on the Composition of Sea Water," in The Sea, 2, M. N. Hill, Ed. (NY: Wiley and Sons, 1963), pp. 26-77.

17. D'Elia, C. F., D. M. Nelson and W. R. Boynton. "Chesapeake Bay Nutrient and Plankton Dynamics: III. The Annual Cycle of Dissolved Silicon." Geochemica et Cosmochimica Acta 47:1945-1955 (1983).

18. Simpson, J. H., S. C. Williams, C. R. Olsen and D. E. Hammond. "Nutrient and Particulate Matter Budgets in Urban Estuaries," Estuaries, Geophysics and the Environment (Wash. D.C.: National Academy of Sciences, 1977), pp. 94-103.

19. Peterson, D. H., T. J. Conomos, W. W. Broenkow and E. P. Scrivani. "Processes Controlling the Dissolved Silica Distribution in San Francisco Bay," in Estuarine Research. Vol. 1. Chemistry and Biology, L. E. Cronin, Ed. (NY: Academic Press, 1975), pp. 153-187.

20. Officer, C. B. "Discussion of the Behavior of Nonconservative Dissolved Constituents in Estuaries." Estuar. Coast. Mar. Sci. 9:91-94 (1979).

21. Helder, W., R. T. de Vries and M. M. Rutgers van der Loeff. "Behavior of Nitrogen Nutrients and Silica in the Ems-Dollard Estuary." Can. J. Fish. Aquat. Sci. 40 (Suppl. 1):188-200 (1983).

22. Epifanio, C. E., D. Maurer and A. I. Dittel. "Seasonal Changes in Nutrients and Dissolved Oxygen in the Gulf of

Nicoya, a Tropical Estuary on the Pacific Coast of Central America." Hydrobiologia 101:231-238 (1983).

23. Walsh, J. J. "Death in the Sea: Enigmatic Phytoplankton Losses." Prog. Oceanog. 12:1-86 (1983).

24. Wollast, R. "Interactions in Estuaries and Coastal Waters," in The Major Biochemical Cycles and their Interactions, SCOPE 21, B. Bolin and R. B. Cook, Eds. (Chichester, U.K.: John Wiley and Sons, 1983), pp. 385-410.

25. Callender, E. and D. E. Hammond. "Nutrient Exchange Across the Sediment-Water Interface in the Potomac River Estuary." Estuar. Coast. Shelf Sci. 15:395-413 (1982).

26. Fisher, T. R., P. R. Carlson and R. T. Barber. "Sediment Nutrient Regeneration in Three North Carolina Estuaries." Estuar. Coast. Shelf Sci. 14:101-116 (1982).

27. Kemp. W. M., R. L. Wetzel, W. R. Boynton, C. F. D'Elia and J. C. Stevenson. "Nitrogen Cycling and Estuarine Interfaces: Some Current Concepts and Research Directions," in Estuarine Comparisons, V. S. Kennedy, Ed. (NY: Academic Press, 1982), pp 209-230.

28. Kaplan, W. A. "Nitrification," in Nitrogen in the Marine Environment, E. J. Carpenter and D. G. Capone, Eds. (NY: Academic Press, 1983), pp. 139-190.

29. Hattori, A. "Denitrification and Dissimilatory Nitrate Reduction," in Nitrogen in the Marine Environment, E. J. Carpenter and D. G. Capone, Eds. (NY: Academic Press, 1983), pp. 191-232.

30. McCarthy, J. J., W. R. Taylor and J. L. Taft." The Dynamics of Nitrogen and Phosphorus Cycling in the Open Waters of the Chesapeake Bay," in Marine Chemistry in the Coastal Environment, American Chemical Society Symposium Series No. 18 (1975), pp. 664-681.

31. Sharp, J. H., C. H. Culberson and T. M. Church. "The Chemistry of the Delaware Estuary. General Considerations." Limnol. Oceanogr. 27:1015-1028 (1982).

32. Norvell, W. A., C. R. Frink and D. E. Hill. "Phosphorus in Connecticut Lakes Predicted By Land Use." Proc. Natl. Acad. Sci. USA 76:5426-5429 (1979).

33. Bowen, H. J. M. "Natural Cycles of the Elements and Their Perturbation By Man," in Environment and Man Vol. 6, the Chemical Environment, J. Lenihan and W. W. Fletcher, Eds. (NY: Academic Press, 1977), pp. 1-37.

34. Smith, D. J. and A. R. Longmore. "Behavior of Phosphate in Estuarine Water." Nature 287:532-534 (1980).

35. Taft, J. L., W. R. Taylor and J. J. McCarthy. "Uptake and Release of Phosphorus By Phytoplankton in the Chesapeake Bay Estuary, USA." Marine Biology 33:21-32 (1975).

36. Peterson, D. H., J. F. Festa and T. J. Conomos. "Numerical Simulation of Dissolved Silica in the San Francisco Bay." Estuar. Coast. Mar. Sci. 7:99-116 (1978).

37. Rattray, M., Jr. and C. B. Officer. "Distribution of a Non-conservative Constituent in an Estuary with Application to the Numerical Simulation of Dissolved Silica in the San Francisco Bay." Estuar. Coast. Mar. Sci. 8:489-494 (1979).

38. Parsons, T. R. and P. J. Harrison. "Nutrient Cycling in Marine Ecosystems," in Encyclopedia of Plant Physiology New Series Volume 12D Physiological Plant Ecology IV Ecosystem Processes: Mineral Cycling, Productivity and Man's Influence, O. L. Lang, P. S. Nobel, C. B. Osmond and H. Ziegler, Eds. (Berlin: Springer-Verlag, 1983), pp. 47-84.

39. Oey, L. "On Steady Salinity Distribution and Circulation in Partially Mixed and Well Mixed Estuaries." J. Phys. Ocean. 14:629-645 (1984).

40. Pritchard, D. W. "Estuarine Circulation Patterns." Proceedings Am. Soc. of Civil Eng. 81 (separate no. 717), 1-11 (1955).

41. Cayan, D. R. and J. O. Roads. "Local Relationships Between United States West Coast Precipitation and Monthly Mean Circulation Parameters." Monthly Weather Review 12:1276-1282 (1984).

42. Conomos, T. J., R. E. Smith and J. W. Gartner. "Environmental Setting of San Francisco Bay." Hydrobiologia (in press).

43. Walters, R. A., R. T. Chang, and T. J. Conomos. "Time Scales of Circulation and Mixing of San Francisco Bay Waters." Hydrobiologia (in press).

44. Conomos, T. J. and D. H. Peterson. "Suspended-particle Transport and Circulation in San Francisco Bay: An Overview," in Estuarine Processes, Vol. 2, L. E. Cronin, Ed. (NY: Academic Press, 1977), pp. 82-97.

45. Cloern, J. E., B. E. Cole, R. L. Wong and A. E. Alpine. "Temporal Dynamics of Estuarine Phytoplankton: A Case Study of San Francisco Bay." Hydrobiologia (in press).

46. Peterson, D. H. "Sources and Sinks of Biologically Reactive Oxygen, Carbon, Nitrogen, and Silica in Northern San Francisco Bay," in San Francisco Bay: The Urbanized Estuary, T. J. Conomos, Ed. (San Francisco: Pacific Division, Am. Assoc. Adv. Sci., 1979), pp. 175-193.

47. McCarty, J. C., R. A. Wagoner, M. Macomber, H. S. Harris, M. Stephenson and E. A. Pearson. "An Investigation of Water and Sediment Quality and Pollutional Characteristics of Three Areas in San Francisco Bay 1960-61." University of California (Berkeley) Sanitary Eng. Res. Lab. (1962).

48. Storrs, P. N., R. E. Selleck and E. A. Pearson. "A Comprehensive Study of San Francisco Bay, 1961-62: Second Annual Report." University of California (Berkeley) Sanitary Eng. Res. Lab. Rep. No. 63-4 (1963).

49. Storrs, P. N., R. E. Selleck and E. A. Pearson. "A Comprehensive Study of San Francisco Bay, 1962-63: Third Annual Report." University of California (Berkeley) Sanitary Eng. Res. Lab. Rep. No. 64-3 (1964).

50. Bain, R. C. and J. C. McCarty. "Nutrient-productivity Studies in San Francisco Bay." U.S. Public Health Service, Central Pacific Basins Water Pollution Control Admin. (1965).

51. Smith, R. E., R. E. Herndon and D. D. Harmon. "Physical and Chemical Properties of San Francisco Bay Waters, 1969-1976." U.S. Geol. Surv. Open-File Rep. 79-511 (1979).

52. Smith, R. E., and R. E. Herndon. "Physical and Chemical Properties of San Francisco Bay Waters, 1977." U.S. Geol. Surv. Open-File Rep. 80-1191 (1980).

53. Smith, R. E. and R. E. Herndon. "Physical and Chemical Properties of San Francisco Bay Waters, 1978." U.S. Geol. Surv. Open-File Rep. 82-R-0273 (1982).

54. Smith, R. and R. Herndon. "Physical and Chemical Properties of San Francisco Bay Waters, 1980." U.S. Geol. Surv. Open-File Rep. (in press).

55. Technicon Corporation. "Silicates in Water and Wastewater." Technicon Autoanalyzer II, Industrial Method No. 105-71WB (1976).

56. Technicon Corporation. "Nitrate and Nitrite in Water and Wastewater." Technicon Autoanalyzer II, Industrial Method No. 100-70W (1973).

57. Atlas, E. L., Hager, S. W., Gordon, L. I. and Park, P. K. "A Practical Manual for Use of the Technicon Autoanalyzer in Seawater Nutrient Analyses, Revised." Oregon State University, Dept. Ocean., Ref. 71-22.

58. Solorzano, L. "Determination of Ammonia in Natural Waters by the Phenolhypochlorite Method. Limnol. Oceanogr. 14:799-801 (1969).

59. Head, P. C. "An Automated Phenolhypochlorite Method for the Determination of Ammonia in Seawater." Deep Sea Res. 18:531-532.

60. Conomos, T. J., R. E. Smith, D. H. Peterson, S. W. Hager and L. E. Schemel. "Processes Affecting Seasonal Distributions of Water Properties in the San Francisco Bay Estuarine System," in San Francisco Bay: The Urbanized Estuary, T. J. Conomos, Ed. (San Francisco: Pacific Division, Am. Assoc. Adv. Sci., 1979, pp. 115-142.

61. Hammond, D. E., C. Fuller, D. Harmon, B. Hartman, M. Korosec, L. Miller, R. Rea and S. Warren. "Benthic Fluxes in San Francisco Bay." Hydrobiologia (in press).

62. Robinson, S. W. "Natural and Man-made Radiocarbon as a Tracer for Coastal Upwelling Processes," in Coastal Upwelling, F. A. Richards, Ed. (Wash. D.C.: Am. Geophys. Union, 981), pp. 298-302.

63. Peterson, D. H., L. E. Schemel, A. E. Alpine, B. E. Cole, S. W. Hager, D. D. Harmon, A. Hutchinson, R. E. Smith and S. E. Wienke. "Phytoplankton Photosynthesis, Nitrogen Assimilation and Light Intensity in a Partially Mixed Estuary." Estuar. Coast. Shelf Sci. (in press).

64. Peterson, D. H., R. E. Smith, S. W. Hager, D. D. Harmon, R. E. Herndon and L. E. Schemel. "Interannual Variability in Dissolved Inorganic Nutrients in Northern San Francisco Bay Estuary." Hydrobiologia (in press).

65. Peterson, D. H. and J. F. Festa. "Numerical Simulation of Phytoplankton Productivity in Partially Mixed Estuaries." Estuar. Coast. Shelf Sci. 19:563-589 (1984).

66. Cloern, J. E., A. E. Alpine, B. E. Cole, R. L. Wong, J. F. Arthur and M. D. Ball. "River Discharge Controls Phytoplankton Dynamics in Northern San Francisco Bay Estuary." Estuar. Coast. Shelf Sci. 16:415-429 (1983).

67. Nichols, F. H. "Increased Benthic Grazing: An Alternative Explanation for Low Phytoplankton Biomass in Northern San Francisco Bay During the 1976-77 Drought." Estuar. Coast. Shelf Sci. (in press).

68. Strub, P. T., T. Powell and C. R. Goldman. "Climatic Forcing: Effect of El Nino on a Small, Temperate Lake." Science 227:56-57.

CHAPTER 11

SEDIMENT TRAP EXPERIMENTS IN THE ANTARCTIC OCEAN

S. Noriki

K. Harada

S. Tsunogai
Department of Chemistry
Faculty of Fisheries
Hokkaido University
Hakodate 041 JAPAN

ABSTRACT

An array with five sediment traps was deployed at depths of 690, 930, 1330, 2330 and 3130 meters during a period of 24 days from December 1983 to January 1984 in the Antarctic Ocean. The total particulate fluxes were 1.12, 1.10, 0.97, 0.79 and 0.95 $g/m^2/day$ from the shallowest trap. The total fluxes in the Antarctic Ocean were one to two orders of magnitude larger than those in the Pacific and subtropical Atlantic Oceans. The concentrations of major chemical components were fairly constant with depth. The mean composition was 81 % opal, 15 % organic matter, 2 % calcium carbonate. These results reveal that the biological production in the Antarctic Ocean was primarily due to diatoms during this period. By assuming that the flux of Al is equal to the accumulation rate of Al onto the sediment surface, we estimate that the average opal flux was 93 mg-Opal/m^2/day in this period.

INTRODUCTION

Sediment traps have been used for the direct measurement of vertical mass flux and the study of removal processes of chemical elements in the ocean(1-6).

The previous sediment trap studies have demonstrated that the total particulate fluxes and concentrations of major chemical components of settling particles vary widely with a range more than one order of magnitude even in the open ocean, and that the flux is closely related to the biological activity in surface waters.

MARINE AND ESTUARINE GEOCHEMISTRY, Sigleo, A. C., and A. Hattori (Editors)

EXPERIMENTAL

The array with five NH type traps(7) was deployed at depths of 690, 930, 1330, 2330 and 3130 m for 24 days from 20 December 1983 to 13 January 1984 in the Antarctic Ocean(61°33'S, 150°27'E; water depth, 3580 m)(Fig. 1).

Each sediment trap consisted of six cylinders which were 25 cm in diameter and 60 cm high made of polyvinyl chloride. Sodium azide was added to two cylinders of each trap as a bactericide.

The samples from each trap were filtered through preweighed Nuclepore filters(0.6 μm) and used for the determinations of dry weights(8) and chemical components(9). A cylinder sample of each trap was used for the determination of short-lived radioisotopes on board the ship.

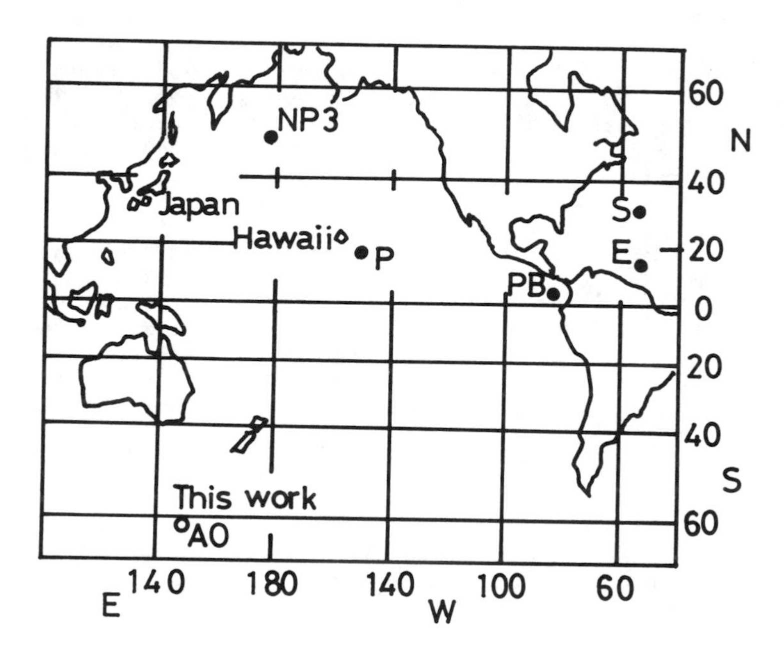

Fig. 1 Map showing locations of the sediment trap stations. Further data are provided in Table 2.

The dried samples were ignited at 450°C for about 24 hours. We have assumed that the ignition loss is equal to the organic matter content. The content of biogenic silica, that is opal, was obtained from the difference in concentrations between total-Si and aluminosilicate-Si in the sample. The concentration of aluminosilicate-Si was calculated by multiplying the Al concentration of the sample by the ratio Si/Al = 28/8 which is the average value of crustal material. The clay content was obtained by assuming that the clay contained 8 % of Al.

RESULTS AND DISCUSSION

Total particulate flux and major chemical composition of settling particles

The total particulate fluxes at five depths are listed in Table 1. The 95 % confidence limits of the total particulate fluxes among the five cylinders were about 10 % of the mean values. The total particulate fluxes observed in the presence of sedium azide in cylinders were not different from those without the bactericide. An average flux of the five depths is calculated to be 0.99 $g/m^2/day$.

The total particulate flux in the Antarctic Ocean is the largest among the previously observed fluxes in deep oceans. The average total particulate flux in the Antarctic Ocean is several times and 20 to 70 times those, respectively, in the northern North Pacific(10) or the Panama Basin(3), and in the Atlantic or the Pacific near Hawaii(3)(Table 2 and Fig. 2).

As mentioned above, the total particulate fluxes vary widely from place to place, where their range is one order of magnitude or more even in the open ocean. The northern North Pacific and the Panama Basin are known to be biologically productive regions. On the other hand, the subtropical Atlantic and the Pacific near Hawaii where small total particulate fluxes are observed are oligotrophic regions. These results suggested that the total particulate flux in the deep sea is related primarily to the biological productivity in the surface water.

The concentrations of major chemical components were fairly constant with depth(Table 3). A striking result is that most of the settling particle consists of opal(Table 3). The large total particulate fluxes are caused by large numbers of opaline shells produced in the surface waters.

Recently, many investigators have suggested that the Antarctic Ocean may not be so productive as previously believed (11, 12). However, the large total particulate flux reported here suggests that the net biological production in the

Table 1. Observed total particulate fluxes
Unit: $g/m^2/day$

	Depth, m				
	690	930	1330	2330	3130
1	1.23*	1.08*	0.84*	0.89*	0.98*
2	1.06*	1.09*	1.05*	0.81*	0.92*
3	1.02	0.96	1.00	0.64	1.06
4	1.26	1.09	0.97	0.83	0.92
5	1.04	1.27	0.99	—	0.88
Mean	1.12	1.10	0.97	0.79	0.95
S. D	0.11	0.11	0.08	0.11	0.07

*The trap cylinders to which sodium azide was added.

Table 2. Mean total particulate fluxes observed in the various deep seas

Location		The shallowest and deepest depths of trap deployed, km (n: No. of traps deployed)	Flux $g/m^2/yr$
Antarctic Ocean(AO)	61.5°S 150.5°E	0.69 - 3.13 (n=5)	360
Northern North Pacific(NP3)	47.9°N 176.3°E	1.04 - 4.38 (n=3)	67*
Panama Basin(PB)	5.4°N 81.9°W	0.67 - 3.79 (n=6)	52**
Tropical Atlantic(E)	13.5°N 54.0°W	0.39 - 5.07 (n=4)	19**
North Atlantic(S)	31.5°N 55.9°W	0.89 - 5.21 (n=4)	6**
Pacific(P)	15.4°N 151.5°W	0.38 - 5.58 (n=5)	5**

* Ref. 10

** Ref. 3

Antarctic Ocean is extremely high at least during the period of this sediment trap experiment.

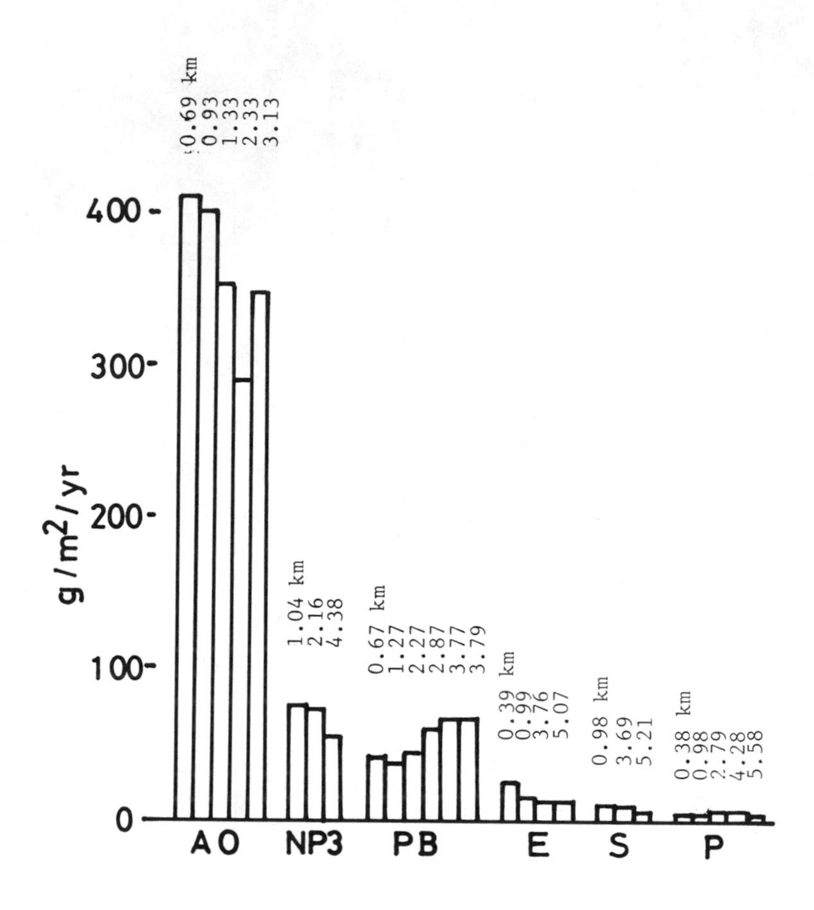

Fig. 2 The total particulate flux at the six stations.

Table 3. Vertical variation of composition of settling particles

Depth km	Organic matter %	Biogenic SiO_2 %	$CaCO_3$ %	Clay(Al, ppm) %
0.69	15.4	77.9	2.3	0.06 (51)
0.93	16.0	80.6	2.0	0.06 (47)
1.33	14.7	80.1	2.0	0.12 (94)
2.33	15.1	82.7	2.1	0.11 (91)
3.13	14.5	83.7	2.2	0.13(107)
Average	15.1	81.0	2.1	0.10 (78)
Surface Sediment	10.5	79.2	2.9	5.9(4700)

Opal flux

The opal flux observed with the sediment trap is normalized to Al by assuming that the true flux of Al(F_{Al}) is equal to the accumulation rate onto the sediment surface(S_{Al}), because the observed particulate flux is apparent and often larger than the accumulation rate at the bottom(5,10). Furthermore, Al having no soluble species in seawater will not be dissolved at the bottom.

The normalized opal flux, F_{opal}, is obtained by

$$F_{opal} = \left(\frac{CP_{opal}}{CP_{Al}} \right) \times F_{Al}$$

where CP_{opal} and CP_{Al} are the concentrations of opal and Al in the settling particles, respectively.

The accumulation rate of opal onto the sediment surface (S_{opal}) is given by

$$S_{opal} = \left(\frac{CS_{opal}}{CS_{Al}} \right) \times S_{Al}$$

where CS_{opal} and CS_{Al} are the concentrations of opal and Al in the surface sediment, respectively.

The accumulation rate of opal is also obtained by

$$S_{opal} = CS_{opal} \times S$$

where S is the sedimentation rate.

According to DeMaster(13), the sedimentation rate of siliceous ooze sediment in the Antarctic Ocean is about 1 cm/1000 yr. By assuming that the bulk density of the surface sediment is 1.4 g/cm^3 and the water content is 70 %, S_{opal} is calculated to be

$$\begin{aligned} S_{opal} &= \frac{79}{100}(\text{g-Opal/g}) \times 0.69(\text{g/cm}^2\ 1000\ \text{yr}) \\ &= 5.5\ \text{g-Opal/m}^2 \cdot \text{yr} \\ &= 15\ \text{mg-Opal/m}^2 \cdot \text{day} \end{aligned}$$

The average normalized opal flux is estimated as follows.

$$\begin{aligned} F_{opal} &= \left(\frac{CP_{opal}}{CP_{Al}} \right) \times \left(\frac{CP_{Al}}{CP_{opal}} \right) \times S_{opal} \\ &= \left(\frac{81}{0.078} \right) \times \left(\frac{0.47}{79} \right) \times 15\ \text{mg-Opal/m}^2.\text{day} \\ &= 93\ \ \text{mg-Opal/m}^2 \cdot \text{day} \end{aligned}$$

The cumulative amounts of opal in the flux of 93 mg-Opal/m^2·day during about two month, 0.093 g-Opal/m^2·day × 59 day = 5.5 g-Opal/m^2, corresponds to the amounts of annual sedimentation of opal.

This result may suggest that the removal of chemical substances in seawater from the surface water to the bottom occurs during a limited period.

As the Opal/Al ratio of the settling particles may vary seasonally due to the change in phytopalnkton species and their activities(10,14), the calculated flux in the Antarctic Ocean may have some uncertainties.

If the observed results are in steady state, on the other hand, the regeneration flux at the sediment surface, RF_{opal}, is obtained as follows:

$$RF_{opal} = F_{opal} - S_{opal}$$

$$= (93 - 15) \text{ mg-Opal/m}^2\text{/day}$$

$$= 78 \text{ mg-Opal/m}^2\text{/day}$$

Tsunogai et al.,(15) have estimated that the opal regeneration in the productive Bering Sea is 12.7 g-SiO_2/m^2/yr at the surface of the sediment. The flux of opal regenerated at the Antarctic sediment surface was about twice that in the Bering Sea.

The percentage of opal regenerated from the settling particle at the sediment surface, R_{opal}, is calculated to be

$$R_{opal}(\%) = \left(\frac{F_{opal} - S_{opal}}{F_{opal}} \right) \times 100$$

$$= \left(\frac{93 - 15}{93} \right) \times 100$$

$$= 84$$

In conclusion, these results indicate that the Antarctic Ocean is extremely productive in biogenic silicate particles and a large number of opaline shells settle onto the sediment surface.

ACKNOWLEDGEMENTS

We wish to thank T. Nemoto and the scientists, officers and crew aboard R. V. Hakuho Maru of University of Tokyo for assistance in the sediment trap experiment. We are also grateful to N. Ishimori, K. Taguchi and T. Kurosaki for their cooperation in the arrangements of sediment trap experiments and the chemical treatment of samples.

REFERENCES

1. S. Honjo, J. Mar. Res. 36, 469(1978).
2. S. Honjo and M. R. Roman, ibid. 36, 45(1978).
3. S. Honjo, S. J. Manganini, J. J. Cole, Deep-Sea Res. 29, 609(1982).
4. R. C. Thunell, W. B. Curry, S. Honjo, Earth Planet. Sci. Lett. 64, 44(1983).
5. S. Tsunogai, M. Uematsu, N. Tanaka, K. Harada, E. Tanoue, N. Handa, Marine Chem. 9, 321(1980).
6. P. H. Wiebe, S. H. Boyd, C. Winget, J. Mr. Res. 34, 341 (1976).
7. S. Tsunogai and S. Noriki, in Kaiyo Kagaku, et., by M. Nishimura(Sanhyo Tosho, Tokyo, 1983), p237.
8. M. Uematsu, M. Minagawa, H. Arita, S. Tsunogai, Bull. Fac. Fish. Hokkaido Univ. 29, 164(1978).
9. S. Noriki, K. Nakanishi, T. Fukawa, M. Uematsu, T. Uchida, S. Tsunogai, ibid. 31, 354(1980).
10. S. Tsunogai, M. Uematsu, S. Noriki, N. Tanaka, M. Yamada, Geochem. J. 16, 129(1982).
11. O. Holm-Hansen, S. Z. El-Sayed, G. A. Franseschini, R. L. Cuhel, in Adaptations within Antarctic Ecosystems, ed., by G. A. Llano(Smithonian Inst., Washington, D. C., 1977), p11.
12. M. Fukuchi, J. Oceanogr. Soc. Japan. 36, 73(1980).
13. D. J. DeMaster, in Antarctic Geoscience, ed., by Campbell Craddock(The Univ. of Wisconsin Press, Wisconsin, 1982), p1039.
14. S. Noriki, N. Ishimori, K. Harada, S. Tsunogai, submitted to Marine Chem.(1985).
15. S. Tsunogai, M. Kusakabe, H. Iizumi, I. Koide, A. Hattori, Deep-Sea Res. 26, 641(1979).

CHAPTER 12

ACCUMULATION OF Cs-137 AND Pu-239,240 IN SEDIMENTS OF THE COASTAL SEA AND THE NORTH PACIFIC.

Kiyoshi Nakamura and Yutaka Nagaya
National Institute of Radiological Sciences
3609, Isozaki, Nakaminato, Ibaraki 311-12 Japan

ABSTRACT

The contents and distributions of Cs-137 and Pu-239,240 were determined in the sediments of the coastal seas of Japan and the North Pacific. The distribution patterns of Cs-137 and Pu-239,240 in the sediment column show rapid vertical penetration of the radionuclides into the sediment. The major mechanism of this movement is thought to be bioturbation. The total inventry of radioactivity in the sediment columns were calculated. Usually, total amounts of Pu-239,240 in the sediments are 7-14 % of those in the overlying water column and the importance of sediment in the cycling of this nuclide in the deep ocean is suggested. A sample from the Seto Inland Sea shows that about 230 % for Pu-239,240 and 40 % for Cs-137 of total global fallout deposits were accumulated in the sediment column. The role of sediments as a sink of radionuclides in the near shore areas is demonstrated.

INTRODUCTION

Substantial amounts of artificial radionuclides have been dispersed as global fallout over the oceans during the past 30 years. In the marine environment, bottom sediment is considered to be one of the most useful indicators of radioactive pollution, because of its slow movement in the sea and large capacity to accumulate radionuclides. In addition, information on the behavior of radionuclides in the deep ocean sediment is also useful for the study of geochemical processes at the sediment-water interface.

The distribution of long-lived fallout radionuclides in the sediments have been widely investigated by V.T.Bowen's group [1-5], especially of those in the Atlantic Ocean. According to their results, detectable amounts of radionuclides have accumulated in deep ocean sediments.

We collected sediment samples from the deep Pacific during the cruise of the R/V Hakuho Maru in 1980 and obtained some other cores thereafter. In this report, the contents and distribution patterns of Cs-137 and Pu-239,240 in the sediments

MARINE AND ESTUARINE GEOCHEMISTRY, Sigleo, A. C., and A. Hattori (Editors)

of the coastal seas of Japan and the North Pacific are discussed.

EXPERIMENTAL

Samples discussed in this report are 5 cores from the North Pacific and 3 cores from the coastal sea of Japan. The locations of sampling stations are shown in Figure 1. The cores of Hakuho-Maru cruise (KH-80-2) and Toyoshio-Maru cruise (Hiuchi Nada) were obtained using gravity corers of J.C.Burke's type. An Aoki-Matsumoto's gravity corer was used for the collection of the sample on the Tansei-Maru cruise (Tokyo Bay and Sagami Bay), and the samples of Syoyo's cruise ("B" area) were collected by a Smith-McIntyre type corer.

The cores were sectioned into 1-5 cm intervals and radiochemical analyse were carried out for each section. When one section was insufficient, some serial sections were combined and analysed as one sample. Twenty to sixty grams of dried sediment were spiked with about 0.8 pCi of Pu-242 tracer (USA-NBS standard solution) and 40 mg of Cs carrier.

After leaching with hot 8N nitric acid, the solution was treated by an anion exchange method for separating Pu and Cs. Plutonium isotopes were purified by repeated anion exchange processes and finally electrodeposited onto a stainless steel disc to measure their alpha-ray activities by a spectrometer.

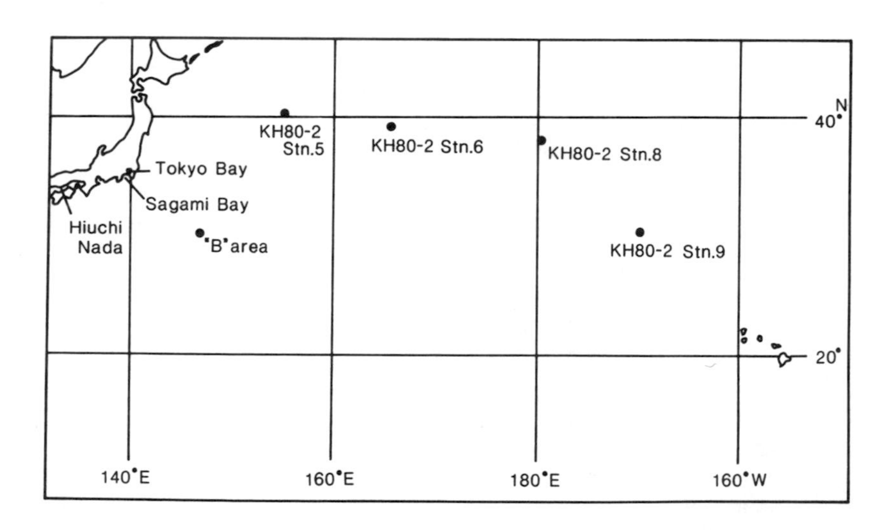

Figure 1. Sampling stations.

Cesium was precipitated by use of ammonium molybdophosphate from the Cs effluent of the anion exchange processes, purified by means of a cation exchange method, and its beta-ray activity was measured in the form of cesium chloroplatinate.

RESULTS AND DISCUSSION

Distribution of Cs-137 and Pu-239,240 in the deep North Pacific sediment.

The distribution profiles of Cs-137 and Pu-239,240 in the open ocean sediments are shown in Figures 2 to 6.

Radioactivities were detectable down to the depth of at least 7 cm for both nuclides in all cores analysed. Considering the fallout history of artificial radionclides and generally accepted sedimentation rates in the deep ocean (less than 10 mm/ 10^3 year), the radionuclides appear to move rather rapidly in the sediment column after deposition.

An area around 30°N and 147°E in the Western Northwest Pacific Basin is a proposed trial dumping site of low level radioactive waste and is called "B" area. Many investigations have been carried out in this area over several years. Recently biological samples were collected and a considerable amount of fish was obtained at the bottom of this area. It suggests the existence of significant biological activity at the bottom of such deep oceans. As Bowen et al.[1] reported, the major mechanism of vertical penetration of these nuclides into deep sediment layers is assumed to be bioturbation.

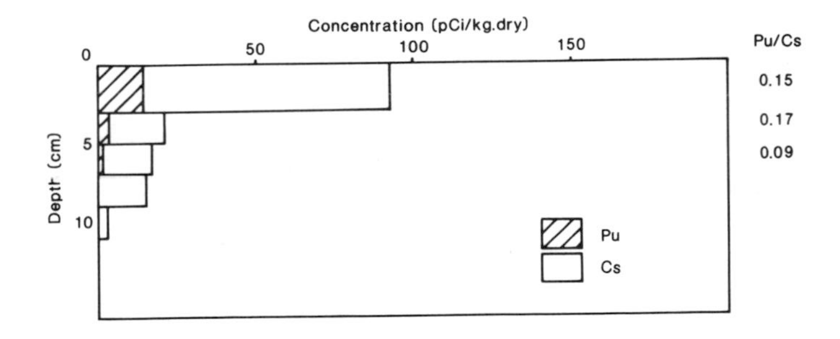

Figure 2. Distribution profile of Pu-239,240 and Cs-137 in the sediment at KH-80-2, Stn.5 (39°58'N, 156°01'E)

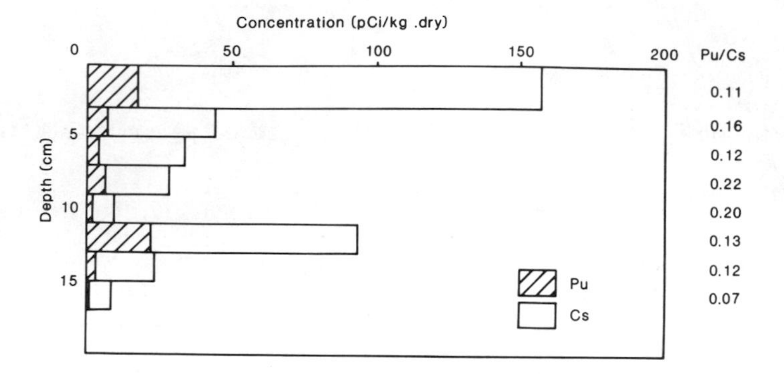

Figure 3. Distribution profile of Pu-239,240 and Cs-137 in the sediment at KH-80-2, Stn.6 (39°00'N, 166°01'E)

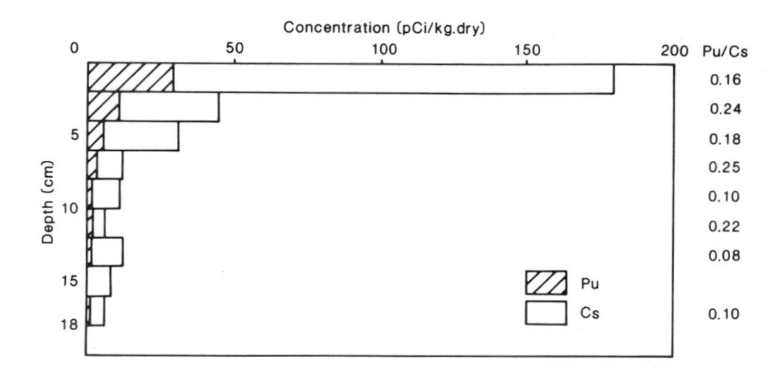

Figure 4. Distribution profile of Pu-239,240 and Cs-137 in the sediment at KH-80-2, Stn.8 (38°03'N, 179°43'E)

The ratios of Pu-239,240 to Cs-137 contents in each sediment column are higher than those in the fallout dust, but lower than the results obtained in the coastal sediments of shallower water depth. After entering the sea, Pu is believed to be removed more rapidly than Cs from the sea surface, but the vertical transport rate of Pu in the water column differs according to the water depth, and a considerable portion of Pu in the water column is retained in the deep layer. Observation of the depth profiles of Pu-239,240 contents in the water column, including those from the same stations where the

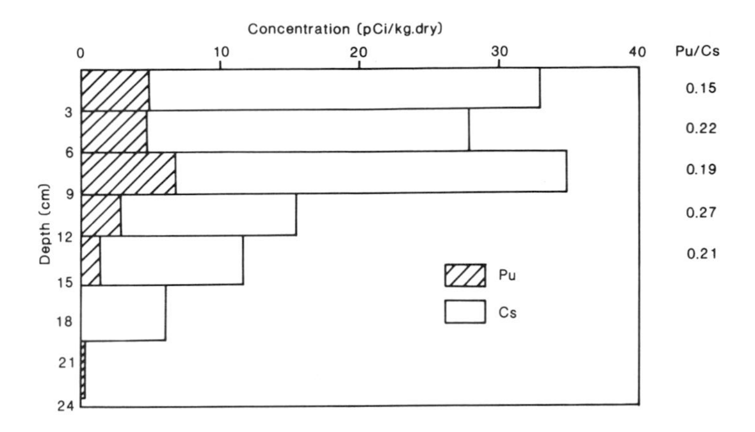

Figure 5. Distribution profile of Pu-239,240 and Cs-137 in the sediment at KH-80-2, Stn.9 (30°00'N, 169°57'E)

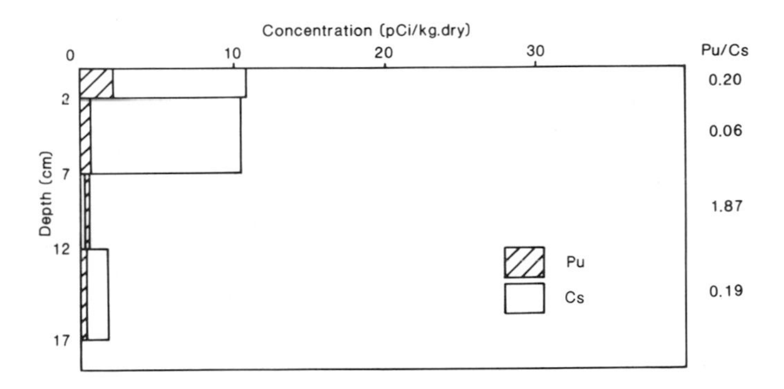

Figure 6. Distribution profile of Pu-239,240 and Cs-137 in the sediment at "B" area. (30°08'N, 146°51'E)

sediments were collected, show this clearly.[6],[7] Near bottom waters usually contained significantly higher Pu-239,240 contents than the overlying water.

Total inventry in the deep sediments.

To evaluate the role of sediments as a sink for nuclides in the ocean, total amounts of Cs-137 and Pu-239,240 in the each sediment column were calculated, and compared with total radioactivities in the overlying water. [6] In the KH-80-2 cruise, a Stn.9 water sample was not obtained so that the data of Stn.11 of the same cruise(30°02'N, 170°40'E) were used for comparison and water data for "B" area were those of sample obtained in 1979. [7],[8] The rest of the water data were from the same stations. Results are shown in Table I. The total deposits in the sediment columns were different between locations for both nuclides. The deposit at the "B" area was significantly lower in comparison with these at the other 4 stations. Noshkin and Bowen [2] found a tendency for the fraction deposited to decrease as the overlying water depth increased in the Atlantic open ocean sediments. In general,

Table I. Comparisons of radionuclides contents in the sediment and in the overlying water column.

Location (Date)	Water Depth	Nuclide	Sediment (mCi/km^2)	Water (mCi/km^2)	Sediment / Water
KH-80-2 Stn.5 (5/1/'80)	5542m	Pu-239,240	0.17	2.3	0.07
		Cs-137	1.24	81	0.015
		(Pu/Cs)	(0.14)	(0.028)	
KH-80-2 Stn.6 (5/2/'80)	5644m	Pu-239,240	0.44	3.2	0.14
		Cs-137	3.46	92	0.038
		(Pu/Cs)	(0.13)	(0.034)	
KH-80-2 Stn.8 (5/13/'80)	5548m	Pu-239,240	0.37	3.6	0.10
		Cs-137	2.24	120	0.019
		(Pu/Cs)	(0.17)	(0.030)	
KH-80-2 Stn.9 (5/16/'80)	5390m	Pu-239,240	0.28	4.0	0.07
		Cs-137	1.52	123	0.012
		(Pu/Cs)	(0.18)	(0.033)	
"B" area (10/29/'80)	6250m	Pu-239,240	0.046	3.0	0.015
		Cs-137	0.296	112	0.003
		(Pu/Cs)	(0.16)	(0.027)	

the depth profiles show that near bottom water contains more radionuclides. It is not clear that the lower amount of radionuclide deposition in the sediment column of "B" area can be explained only by the difference in water depth. There might be a different mechanism of material transport in layers deeper than 5600 m, relative to these in shallower layers. Therefore, more detailed examination of material cycling in near bottom water layers is required. However, due to technical difficulties in sampling and to the fact that considerable time is required for sample collection, there have been no measurements of detailed radionuclide depth profiles of near bottom water layers in deep oceans. And unfortunately, there also remains the possibility that the very thin top layer of sediment probably containing high nuclide contents was lost during sampling.

It is interesting that ratios of Pu-239,240 to Cs-137 contents in the sediment columns are almost equivalent, while total amounts of deposits differ among stations. Plutonium and cesium seem to move in the same manner in the sediments after deposition. The total amount of Pu-239,240 in the sediments was 7-14 % of these in the overlying water column of the same location, except at the "B" area in 1980. In the discussion of material balance and circulation of Pu-239,240 in the deep ocean, sediments must be considered, as well as the distribution of these nuclides in water layers. On the other hand, the amount of Cs-137 in the sediments is not significant in regard to the distribution of total Cs-137 in the ocean ecosystem at present.

Distribution of Cs-137 and Pu-239,240 in the coastal sediment.

The distribution profiles of Cs-137 and Pu-239,240 in the coastal sediments are shown in Figures 7 to 9.

The sampling station for the Tokyo Bay sample is located about 3 miles off the mouth of the Tama River. Effects of materials of terrestrial origin supplied mainly from the river were indicated by high levels of both nuclides. Approximately uniform contents were found down to the depth of 35 cm, probably due to physical and biological mixing. The ratio of Pu-239,240 to Cs-137 contents was more than 10 times higher relative to that of fallout dust. The ratio in the water of the same location was about 0.001(our unpublished data). These data illustrate the higher deposition rate of Pu-239,240 relative to that of Cs-137.

The location of the Sagami Bay sample is also near the mouth of a big river. In spite of a greater water depth, the levels of radionuclides were comparable with those of Tokyo Bay. To explain the irregularity of the Pu-239,240 distribution, the data is still insufficient at present. Considering the complex geographical features around the sampling station, there is a possibility that landslides had occurred. For an adequate discussion, much more information on the horizontal movement of sediment is required.

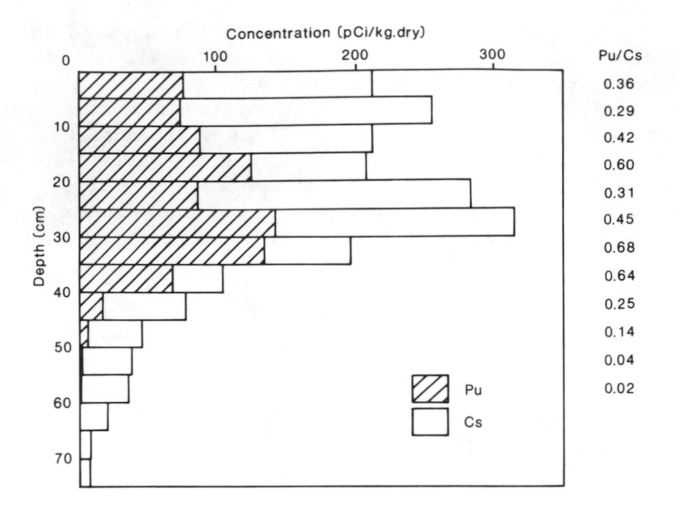

Figure 7. Distribution profile of Pu-239,240 and Cs-137 in the sediment at Tokyo Bay. (35°31'N, 139°52'E) Water depth 22 m Sampling date 6/6/'81

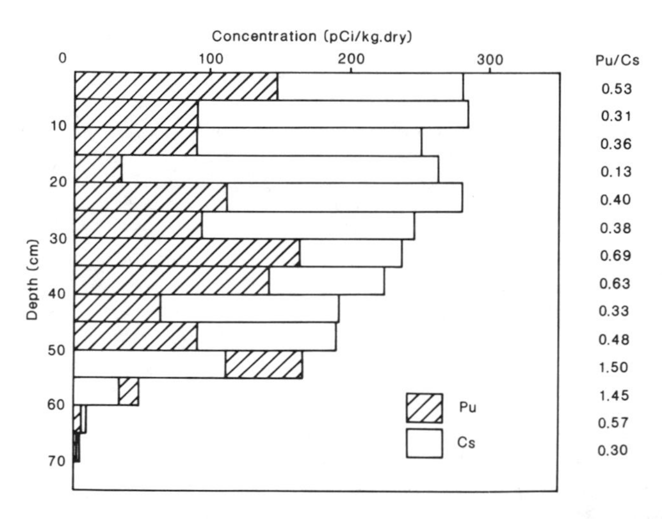

Figure 8. Distribution profile of Pu-239,240 and Cs-137 in the sediment at Sagami Bay. (35°15'N, 139°21'E) Water depth 615 m Sampling date 6/10/'81

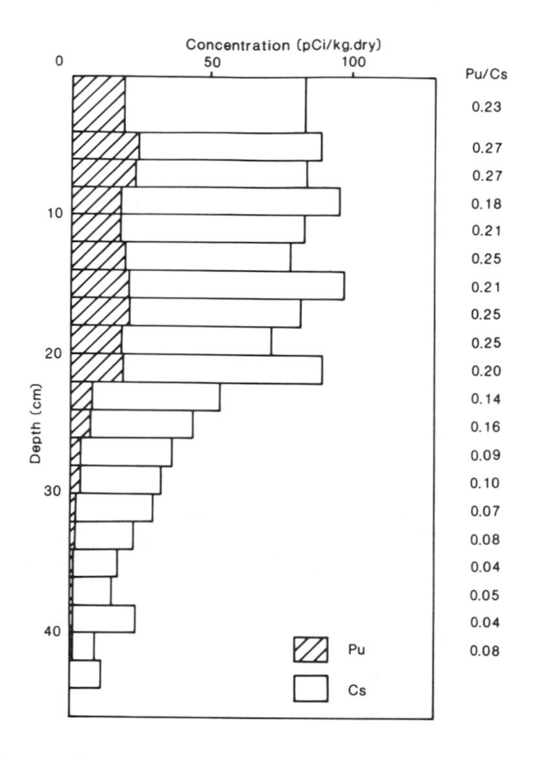

Figure 9. Distribution profile of Pu-239,240 and Cs-137 in the sediment at Hiuchi Nada. (34°03'N, 133°10'E) Water depth 27 m Sampling date 11/18/'83

Hiuchi Nada is located in the central part of the Seto Inland Sea. It has no major freshwater input from large rivers, and the sea water exchange rate is very low [9]. Smooth distribution patterns were observed for both nuclides to the depth of 22 cm, reflecting little physical movement of the overlying sea water. Caluculated amounts of total radioactivities were 2.8 mCi/km^2 and 15.2 mCi/km^2 for Pu-239,240 and Cs-137, respectively. These amounts are about 40 % for Cs-137 and 230 % for Pu-239,240 of the total global fallout deposits. Triulzi [10] also found that a core sample from the Ionian Sea accumulated 3 times greater deposition of Pu-239,240 relative to the estimated Pu deposition. These facts show that the landlocked sea receives great amounts of material of terrestrial origin. The ratio of Pu-239,240 to Cs-137 was 0.18 and was similar to that of the open ocean.

Accumulation of long-lived fallout radionuclides in the investigated sediment of near shore areas indicates the importance of sediment as a sink for pollutants, as previously demonstrated by heavy metal research.

ACKNOWLEDGEMENT

We wish to express our thanks to the staff of the Ocean Research Institute, University of Tokyo and the crew of Hakuho-Maru of the Institute for their cooperation on sample collection. We are also grateful to Dr.H.Tsubota of Hiroshima University, Mr.S.Shiozaki of Hydrographic Department, Maritime Safety Agency and Dr.A.Hattori of the Ocean Research Institute, University of Tokyo, who kindly supplied sediment samples.

REFERENCES

1. Bowen,V.T.,H.D.Livingston and J.C.Burke. "Distribution of transuranium nuclides in sediment and biota of the North Atlantic Ocean," in *Transuranium Nuclides in the Environment* ,(IAEA, Vienna : IAEA, 1976),pp.107-120
2. Noshkin,V.E. and V.T.Bowen. "Concentrations and distributions of long-lived fallout radionuclides in open ocean sediments," in *Radioactive Contamination of the Marine Environment,* (IAEA, Vienna : IAEA, 1973), pp.671-686
3. Labeyrie,L.D., H.D.Livingston and V.T.Bowen. "Comparison of the distributions in marine sediments of the fall-out derived nuclides Fe-55 and Pu-239,240," in *Transuranium Nuclides in the Environment,* (IAEA, Vienna : IAEA, 1976), pp.121-137
4. Livingston,H.D. and V.T.Bowen. "Pu and Cs-137 in coastal sediments," *Earth Planet. Sci. Lett.,* 43 : 29-45 (1979)
5. Bowen,V.T. and H.D.Livingston. "Radionuclide distributions in sediment cores retrieved from marine radioactive waste dumpsites," in *Impacts of Radionuclide Releases into the Marine Environment,* (IAEA, Vienna : IAEA, 1981), pp.33-63
6. Nagaya,Y. and K.Nakamura. "Pu-239,240, Cs-137 and Sr-90 in the Central North Pacific," *J.Oceanogr.Soc.Japan.,* 40 : 416-424 (1984)
7. Nakanishi,T., M.Yajima, M.Senaga, M.Takei, A.Ishikawa, K.Sakamoto and K.Sakanoue. "Determination of Pu-239,240 in sea water," *Nucl.Instr. and Meth.,* 223 : 239-242 (1984)
8. Nagaya,Y. and K.Nakamura. "Artificial Radionuclides in the Western Northwest Pacific (1) Sr-90 and Cs-137 in the Deep Waters," *J.Oceanogr.Soc.Japan.,* 37 : 135-144 (1981)
9. Hoshika,A. and T.Shiozawa. "Sedimentation rates and heavy metal pollution of sediments in the Seto Inland Sea. Part 3. Hiuchi Nada," *J.Oceanogr.Soc.Japan.,* 40 : 334-342 (1984)
10. Triulzi,C., A.Delle Site and V.Marchionni. "Pu-239,240 and Pu-238 in sea water, Marine organisms and sediments of Taranto Gulf (Ionian Sea)," *Estuarine, Coastal and Shelf Science.,* 15 : 109-114 (1982)

CHAPTER 13

EFFECTS OF THE CLAY MINERAL, BENTONITE, ON ACETATE UPTAKE BY MARINE BACTERIA

W. B. Yoon and R. A. Rosson
Marine Science Institute
The University of Texas at Austin
Port Aransas, TX 78373

ABSTRACT

Shallow estuarine waters are frequently turbid due to wind-driven resuspension of surface sediments. Bacterial microheterotrophic uptake (assimilation plus respiration of substrate) in the water column may be affected by resuspended inorganic particles like clay. Effects of clay (bentonite) on bacterial metabolism were studied with cultures of estuarine bacteria in various physiological states, and with samples of natural free-living bacteria, using trace levels of [1-^{14}C] acetate as substrate. All bacteria tested metabolized acetate more efficiently in the presence of clay; either assimilation was increased without increasing respiration, or respiration decreased with little or no change in assimilation. No change in pH was measured during incubation and no adsorption of acetate by clay was detected. The data imply that physiochemical properties of bacterial surfaces may be responsible for the observed effects of clay on bacterial metabolism.

INTRODUCTION

Surface sediments in shallow marine bays and estuaries are frequently resuspended by tidal currents [18], benthic fauna [21,22], or wind-driven currents [4,24]. Resuspended particles normally consist of inorganic silt and clay particles as well as organic detritus [18]. When particulate material is resuspended, the number of attached bacteria in the water column increases [6].

Recent studies have attempted to estimate relative metabolic activities of attached and free-living bacteria [3,6,9,19]. In certain environments, attached bacterial

MARINE AND ESTUARINE GEOCHEMISTRY, Sigleo, A. C., and A. Hattori (Editors)

populations are primarily responsible for the microheterotrophic activity in the water column [6,9].

Laboratory studies concerning the effects of particles on microheterotrophic activity have primarily used clays, silt, or hydroxyapatite as examples of natural inorganic particles for bacterial attachment. Many studies report effects of clays on soil or freshwater bacteria [10,12,15,27]. Only a few studies have addressed effects of inorganic particles such as clays on marine bacteria. For example, Jannasch and Pritchard [8] reported kaolinite increased the efficiency of substrate transformation by isolates of marine bacteria. Little information is available, however, as to the effect of clay on microheterotrophic activity of natural marine bacteria.

The purpose of this study was to determine effects of bentonite on microheterotrophic uptake (assimilation plus respiration) of marine bacteria. The effect of clay on acetate uptake by natural bacteria was compared to uptake by an estuarine bacterial isolate, DG#3, harvested at different growth phases.

MATERIALS AND METHODS

Organisms and Culture Conditions

Strain DG#3, a gram negative isolate from decomposing seagrass was cultured in ZoBell's 2216 broth [17] at 25°C. Log and stationary phase cells were harvested and washed by centrifugation for 15 min at 10,000 x g. Washing solution was 0.2 μm filtered (Gelman GA-8; Ann Arbor, MI) offshore seawater collected about 100 miles from Port Aransas, TX in the Gulf of Mexico. This seawater was filtered to remove bacteria as well as organic and inorganic particulates.

DG#3 was resuspended for uptake assays in various 0.2 μm filtered seawaters: (1) offshore seawater; (2) seawater from the pier-lab at the Aransas Pass Channel; or (3) artificial seawater (salt solution of Kurath and Morita [11]). Uptake assays contained 1.3 x 10^6 cells/ml.

DG#3 was starved: (A) at 4°C for 20 days then shaken (120 rpm) at 25°C for 5 days or (B) at 4°C for 25 days and 5 days with shaking at 25°C. Starved cells were then washed by centrifugation and resuspended in filtered offshore seawater as described above.

Natural bacteria were from estuarine seawater collected at the pier-lab (Aransas Pass Channel). Pier-lab water was filtered through 3.0 μm Nuclepore membranes. The filtrate contained 1.5×10^6 free-living bacteria per ml.

Bacterial concentrations, for both cultured and natural free-living bacteria were determined by epifluorescence direct cell count [7,28]. Suspensions were stained with 4,6-diamino-2-phenylindole DAPI [20] and counted on a Zeiss Universal microscope with a 50W DC HBO mercury burner, G365 excitation, FT395 beam splitter/reflector and LP420 barrier filters.

Bentonite (Clay) Preparation

Bentonite (Sigma Chemical Co., St. Louis, MO) was suspended in distilled water (1 g/100 ml) and sterilized by autoclaving for 20 min.

Acetate Uptake

DG#3 suspensions were assayed in duplicate. Tracer [1-^{14}C]acetate (sodium salt, 1 μCi/ml, 0.69 μCi/μg; New England Nuclear, Boston, MA) was added (50 μl) to a series of Erlenmeyer flasks (125 ml) containing 20 ml of seawater. When added, the final concentration of bentonite was 25 mg/l.

Uptake assays started with addition of isolate DG#3 or, for natural bacteria, with the addition of [1-^{14}C]acetate to 3.0 μm filtered pier-lab seawater. All flasks were sealed tightly with rubber serum caps and incubated with shaking (120 rpm) at 25°C. Incubation was stopped at appropriate times with 1 ml of 4N H_2SO_4.

To trap radioactive carbon dioxide, each flask was sparged for 10 minutes with N_2-gas. The $^{14}CO_2$ was collected in two serially-connected scintillation vials containing 3a70B scintillation cocktail (Research Products International Corp., Mount Prospect, IL) plus Carbosorb (Packard Instrument Co., Downers Grove, IL) plus anhydrous methanol (10:1:1). Greater than 95% of the $^{14}CO_2$ was routinely trapped. The contents of each flask were then filtered through 0.2 μm filters and the filters and 3a70B scintillation cocktail were added to scintillation vials. Radioactivity trapped on filters (assimilated) or collected in $^{14}CO_2$-trapping scintillation cocktail (respired) was determined. Quench was corrected by channels ratio.

Poison Controls

Mercuric chloride (50 μl of a 1:4 diluted saturated solution) or sodium cyanide (0.2 ml of 1M solution) was added, when appropriate, before addition of [1-^{14}C]acetate. Incubations were terminated and processed as described above.

RESULTS

Bacteria and bentonite flocculated when mixed together in seawater. Greater than 90% of the initially free bacterial cells were attached to clay particles. The flocculation in seawater may have resulted from increased attraction between both clay to clay and clay to bacterial cells, primarily due to decrease of the electrical double-layer thickness at the surface of clay particles [14,23].

Control experiments with 0.2 μm-filtered seawater and clay showed there was no measurable adsorption of [1-^{14}C] acetate on bentonite. Poison controls using mercuric chloride or potassium cyanide also showed there was no measurable adsorption of labeled acetate to clay, to bacteria, or to bacteria-clay complexes (Figure 1).

Bentonite stimulated total acetate uptake (assimilation plus respiration) of DG#3 harvested at log phase (Figure 1) and stationary phase (Figure 2). Only assimilation was stimulated by bentonite; there was no measurable effect on respiration. The effect of clay on acetate uptake rates was independent of DG#3 concentration for cell suspensions containing 10^5 to 10^7 cells/ml.

Clay diminished total uptake of starved cells. However, only respiration was diminished by the presence of clay. There was no substantial difference in the assimilation rate with or without bentonite (Figure 3).

The response of natural free-living bacteria to bentonite was similar to that of the starved cells; total uptake of acetate was diminished by clay. Respiration of acetate was diminished in the presence of clay, while assimilation was essentially identical with or without clay (Figure 4).

For all experiments with [1-^{14}C]acetate as the substrate, assimilation rates were linear over the one hour incubation period. In contrast, respiration rates were not linear; these rates increased with time (Figures 1,2,3,4).

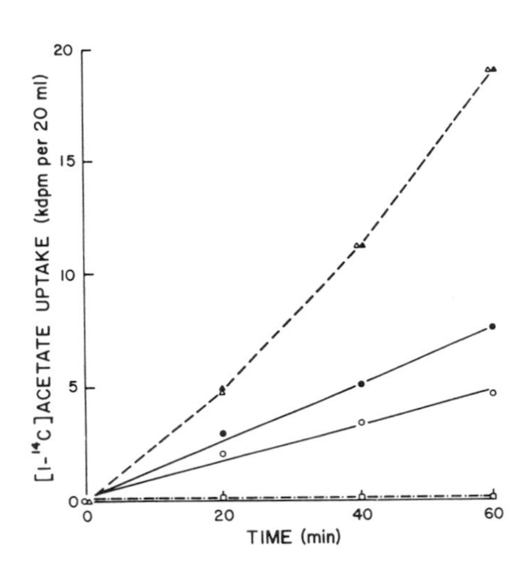

Figure 1. Acetate uptake by log phase DG#3. [1-^{14}C]Acetate (1.1 x 10^5 dpm) was added to 20 ml of 0.2 μm filtered offshore seawater. DG#3 concentration was 1.3 x 10^6 cells/ml. The final bentonite concentration, when added, was 25 μg/ml. Radioactivity assimilated ((●,○); trapped on filters) or respired ((▲,△); $^{14}CO_2$) was determined with (●,▲) or without (○,△) clay after shaking at 25°C. Poison controls (□) were supplemented with mercuric chloride or sodium cyanide.

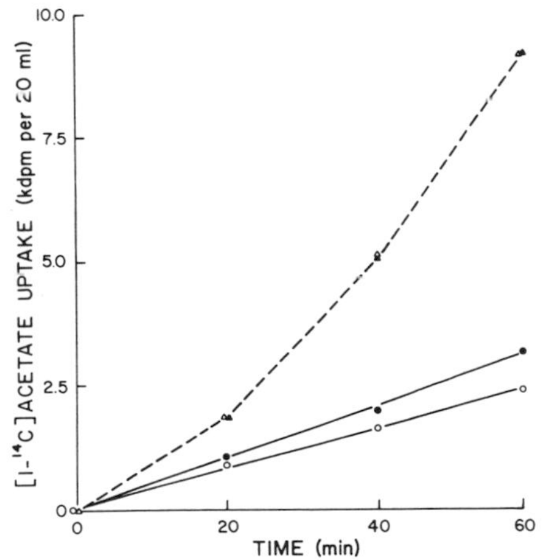

Figure 2. Acetate uptake by early stationary phase DG#3. Symbols and uptake assay conditions as in Fig. 1.

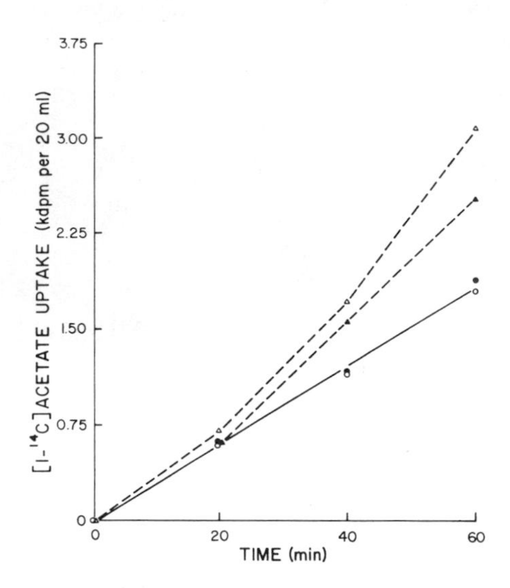

Figure 3. Acetate uptake by starved DG#3. Stationary phase cells were suspended in 0.2 μm filtered offshore seawater and starved for 25 days at 4°C, then 5 days at 25°C (starvation condition "B"). Symbols and uptake assay conditions as in Figure 1.

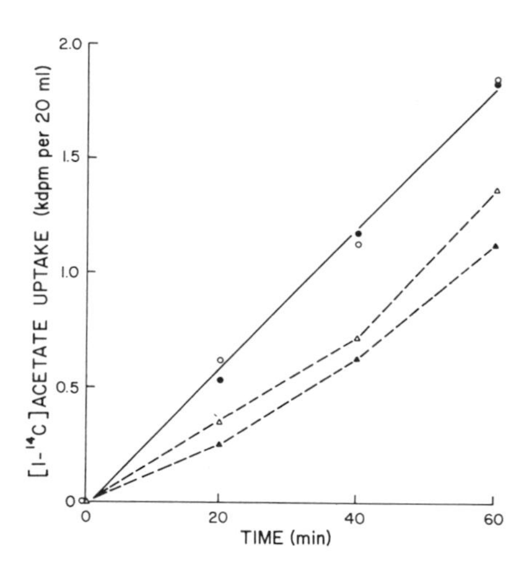

Figure 4. Acetate uptake by natural free-living bacteria. Pier-lab estuarine seawater was filtered through 3.0 μm Nuclepore membranes. The filtrate contained 1.5 x 10^6 free-living (unattached) bacteria per ml. Symbols and uptake assay as in Figure 1.

Laboratory cultures of DG#3 harvested at different growth phases assimilated 20 to 37% of the total uptake (assimilation plus respiration; Table I). For all growth phases studied (log, stationary, or starved cells), bentonite increased the percentage of acetate assimilation of DG#3 by an average of 5.5% (range: 3-8%). Regardless of the presence of clay, log phase cells assimilated the smallest percentage of acetate. The percentage assimilation was greater for stationary cells, and even greater for starved cells. Natural bacteria assimilated a much larger percentage of the acetate than did any DG#3 culture tested, with (62% vs. mean of 35%) or without (57% vs. mean of 30%) bentonite (Table I).

The ratio of assimilation dpm with clay to assimilation dpm without clay is a measure of the effect of clay only on assimilation. Similarly, substitution of respiration for assimilation in the above ratio yields a measure of the effect of clay only on respiration.

The magnitude of stimulation of acetate assimilation in the presence of bentonite decreased as cells aged from log to stationary phase and from stationary phase into starvation (Table I). For log phase cells, bentonite stimulated assimilation of acetate by 57% (ratio = 1.57) compared to a control without clay. Assimilation of acetate by stationary phase DG#3 cells was increased by an average of 33% by bentonite (ratio = 1.30 to 1.36). However, there was no effect of bentonite on respiration of acetate (ratio = 1.0). In contrast, there was little or no increase of assimilation for starved cells by bentonite, while respiration decreased 7 to 19% (ratio = 0.93 to 0.81; Table I).

Natural free-living bacteria responded to clay similar to starved cells. There was almost no effect of bentonite on assimilation (ratio = 1.00 to 1.07), and respiration was diminished by clay (ratio = 0.83 to 0.93). For starved cells, the longer starved, the greater the decrease in respiration when bentonite was added; the ratio decreased from 0.93 to 0.81 (Table I).

Natural seawaters, like offshore seawater, are reported to have a very low concentration of dissolved organic carbon, and an especially low concentration of available organic nitrogen [25]. Estuarine waters, such as our pier-lab seawater, have an elevated level of dissolved organic carbon. We therefore measured the effect of seawater quality on acetate uptake by stationary phase DG#3 suspensions.

Percentage assimilation varied from 25 to 38% without bentonite and from 30 to 45% with bentonite when different seawaters were used. Net increase of percentage assimilation

TABLE I. Effect of bentonite on assimilation and respiration of acetate by marine bacteria harvested at different growth phases[a]

Bacterial Growth Phase	Ratio[b]		Percentage Assimilation[c]		
	Assimilation	Respiration	+clay (A)	-clay (B)	net (A-B)
Log	1.57	1.00	28.6	20.4	8.2
Stationary					
Early	1.30	1.00	25.5	20.9	4.6
Late	1.36	1.04	41.7	35.3	6.4
Starved[d]					
A	1.07	0.93	38.0	35.3	2.7
B	1.00	0.81	41.9	37.0	4.9
Natural					
Free-living	1.00	0.83	61.6	57.0	4.6

[a]Isolate DG#3 was resuspended in 0.2 μm filtered offshore seawater. Radioactivity assimilated and respired by DG#3 was determined.

[b]Ratio = [dpm assimilated (or respired) with clay/dpm assimilated (or respired) without clay].

[c]Percentage assimilation = [assimilation/(assimilation plus respiration)]x100. Net is percentage increase of assimilation with bentonite.

[d]Cells, harvested at the stationary phase, were resuspended in filtered offshore seawater and starved as described in Materials and Methods for 25 days (A) and 30 days (B).

with bentonite was similar for all seawaters tested (mean = 6.1%, range = 5.3 to 6.7%; Table II). The effect of bentonite on assimilation only, or respiration only, also appeared to be independent of seawater quality; mean assimilation ratio was 1.36 and mean respiration ratio was 1.03 (Table II). Therefore, the effect of bentonite on acetate uptake by DG#3 was essentially identical, regardless of seawater quality.

TABLE II. Effect of bentonite on assimilation and respiration of acetate by stationary phase marine bacteria suspended in a variety of seawaters[a]

Seawater	Ratio[b]		Percentage Assimilation[c]		
	Assimilation	Respiration	+clay (A)	-clay (B)	net (A-B)
Artificial	1.34	1.02	30.3	25.0	5.3
Offshore	1.36	1.04	41.7	35.3	6.4
Pier-Lab	1.37	1.04	45.1	38.4	6.7

[a]Isolate DG#3 was harvested in stationary phase and resuspended in various 0.2 μm filtered seawaters.

[b]Ratio = [dpm assimilated (or respired) with clay/dpm assimilated (or respired) without clay].

[c]Percentage assimilation = [assimilation/(assimilation plus respiration)]x100. Net is percentage increase of assimilation with bentonite.

Enrichment of offshore seawater with various levels of acetate (500 to 5,000 μg/l) or with nutrient broth also did not change the effect of bentonite on acetate uptake. Respiration ratios were essentially identical (mean ratio: unenriched = 0.93; + acetate = 0.99; + nutrient broth = 0.96) and assimilation ratios were not substantially different (mean ratio: unenriched = 1.07; + acetate = 1.15; + nutrient broth = 1.16). The net increase of percentage of assimilation ranged from 2 to 4.5% (Table III).

DISCUSSION

These experiments were performed with cultured marine bacteria and with natural free-living marine bacteria from estuarine water. In most experiments, natural seawater was supplemented only with trace levels of acetate. Numbers of cultured bacteria did not exceed bacterial concentrations normally found in seawater. The percentage of assimilation was always increased by clay. The magnitude of this increase was related to the physiological condition of the cells, and

TABLE III. Effect of bentonite on assimilation and respiration of starved marine bacteria suspended in enriched seawater[a]

Seawater	Ratio[b]		Percentage Assimilation[c]		
	Assimilation	Respiration	+clay (A)	-clay (B)	net (A-B)
Control	1.07	0.93	38.3	35.3	3.0
+Acetate					
500 μg/l	1.18	1.00	36.1	33.2	2.9
2,500 μg/l	1.12	1.00	32.2	30.2	2.0
5,000 μg/l	1.15	0.98	31.8	27.7	4.1
+Nutrient Broth[d]					
5.0 ml/l	1.15	0.96	32.7	28.9	3.8
25.0 ml/l	1.16	0.95	34.4	29.9	4.5

[a]Isolate DG#3 was harvested and suspended in 0.2 μm filtered offshore seawater. Starvation was for 20 days at 4°C followed by 5 days at 25°C. Radioactivity assimilated and respired was determined.

[b]Ratio = [dpm assimilated (or respired) with clay/dpm assimilated (or respired) without clay].

[c]Percentage assimilation = [assimilation/(assimilation plus respiration)]x100. Net is percentage increase of assimilation with bentonite.

[d]ZoBell's 2216 broth diluted 1:6.

appeared to be independent of seawater quality. The increase in the percentage of assimilation was due either to stimulation of assimilation with no change in respiration (log and stationary phase cells) or due to a decrease in respiration without considerable change in assimilation (starved or natural free-living bacteria).

It has previously been reported that clay generally stimulates bacterial microheterotrophic activity by either: (1) buffering the system against pH changes [26,27]; by adsorption of toxic metabolites produced by bacterial metabolism [15]; or (3) by concentration of organic nutrients on clay surfaces, thereby increasing the availability of normally dilute nutients [8]. Observations 1 and 2 derive from experiments with high levels of organic nutrients and relatively long incubation times [15,27].

None of these observations can satisfactorily explain our data. No change of pH was observed during incubation in our uptake experiments, either with or without clay. Total organic carbon was very low, only trace levels of [1-^{14}C] acetate were added, incubation times were relatively short, and only 10^6 bacteria/ml were used (typical of natural waters). In the course of a typical experiment, less than 10% of the added [1-^{14}C]acetate was metabolized. It would seem unlikely, therefore, that significant concentrations of toxic metabolites could accumulate in these experiments. Finally, no detectable amount of acetate was adsorbed by bentonite in seawater. Jannasch and Pritchard [8] in their studies with kaolinite also did not detect adsorption of acetate onto clay. However, localized or micro-scale changes in pH, weak adsorption of acetate to bentonite, or removal of low levels of toxic metabolites cannot be completely ruled out on the basis of these experiments, as it is difficult to quantify such micro-scale interaction [8].

Buffering capacity and adsorption of metabolites or nutrients may explain stimulation of bacterial metabolism by clay particles. However it cannot explain the diminished respiration rates of both starved cells and natural free-living bacteria observed in this study. It seems likely, therefore, that clay may also affect microbial activity in an as yet uncharacterized fashion.

All natural and cultured bacteria tested metabolized acetate more efficiently in the presence of clay. Actual rates of acetate assimilation and respiration were affected to different extents for bacteria in different growth phases, and hence in different physiological conditions. However, it should be noted that the percentage of assimilation was always increased by the presence of bentonite. An increase of metabolic efficiency by clay addition is consistent with the results of Martin et al. [15]. Gordon et al. [5], however, observed that metabolic efficiency of attached bacteria was only half of that of free-living bacteria when bacteria were attached to hydroxyapatite.

The mechanism of the increased efficiency in acetate metabolism when bentonite was added remains to be elucidated. It is possible that cell surface characteristics may be involved. Morphological changes are often observed when bacterial growth rates are changed [13] or when bacteria are starved [16]. Baker and Park [1] reported physiological changes in response to substrate depletion resulted in a morphological change from rod to cocccoid shape. DG#3, which was a rod at log phase, also changed its morphology to a coccoid shape when starved. It is possible that such morphological changes indicate physiochemical properties of bacterial surfaces are altered.

Of the cultured bacteria studied, only starved cells exhibited a pattern of the effect of clay similar to natural free-living bacteria. We do not suggest on the basis of these observations that natural free-living bacteria from estuarine waters are in a state of starvation. It is possible, however, that natural free-living bacteria and our starved cultured bacteria may share a common physiochemical characteristic(s) which dictates the observed pattern of response to assimilation and respiration of acetate in the presence of bentonite.

Whatever the mechanism, we postulate from the results of this study that clay may increase metabolic efficiency of bacteria in at least two ways: (1) fast growing bacteria increase assimilation rates without increasing respiration rates, and (2) slow-growing or starved bacteria decrease respiration rates without changing assimilation rates.

SUMMARY

Clay (bentonite), when mixed with a marine bacterial isolate, DG#3 in seawater, formed flocs; greater than 90% of the initially free bacteria were subsequently associated with clay (attached bacteria).

Bentonite stimulated acetate uptake of log or stationary phase DG#3. Only assimilation was stimulated; clay had no effect on respiration. In contrast, clay diminished total uptake in starved bacteria. However, only respiration was diminished, with no substantial effect on assimilation. The effect of clay on natural free-living bacteria showed patterns similar to those of starved DG#3. These results suggested that starved bacteria and natural free-living bacteria may have similar modes of interaction with clays.

Effects of clay on acetate uptake by bacteria were independent of seawater quality and dependent on physiological condition (growth phase). It is likely that physiological state is important for the observed bacterial response to clay. However, the efficiency of acetate metabolism by bacteria was always increased by the presence of clay, regardless of the physiological state of the bacteria and of the seawater quality.

ACKNOWLEDGEMENTS

This research was supported by the University Research Institute Grants (RAR) and Graduate Student Research Grants (WBY) from the University of Texas at Austin.

REFERENCES

1. Baker, D.A. and R.W.A. Park. "Changes in Morphology and Cell Wall Structure that Occur During Growth of *Vibrio* sp. NCTC 4716 in Batch Culture," *J. Gen. Microbiol.* 86: 12-28 (1975).
2. Bell, C.R. and L.J. Albright. "Attached and Free-Floating Bacteria in a Diverse Selection of Water Bodies," *Appl. Environ. Microbiol.* 43:1227-1237 (1982).
3. Bright, J.J. and M. Fletcher. "Amino Acid Assimilation and Electron Transport System Activity in Attached and Free-Living Marine Bacteria," *Appl. Environ. Microbiol.* 45:818-825 (1983).
4. Fanning, K.A., K.L. Carder and P.R. Betzer. "Sediment Resuspension by Coastal Waters: A Potential Mechanism for Nutrient Recycling on the Ocean's Margins," *Deep Sea Res.* 29:953-965 (1982).
5. Gordon, A.S., S.M. Gerchakov and F.J. Millero. "Effects of Inorganic Particles on Metabolism by a Periphytic Marine Bacterium," *Appl. Environ. Microbiol.* 45:411-417 (1983).
6. Goulder, R. "Attached and Free Bacteria in an Estuary with Abundant Suspended Solids," *J. Appl. Bacteriol.* 43:399-405 (1977).
7. Hobbie, J.E., R.J. Daley and S. Jasper. "Use of Nuclepore Filters for Counting Bacteria by Fluorescence Microscopy," *Appl. Environ. Microbiol.* 33:1225-1228 (1977).
8. Jannasch, H.W. and P.H. Pritchard. "The Role of Inert Particulate Matter in the Activity of Aquatic Microorganisms," *Mem. 1st. Ital. Idrobiol.,* 29 Suppl.: 289-308 (1972).
9. Kirchman, D. and R. Mitchell. "Contribution of Particle-Bound Bacteria to Total Microheterotrophic Activity in Five Ponds and Two Marshes," *Appl. Environ. Microbiol.* 43:200-209 (1982).
10. Kunc, F. and G. Stotzky. "Acceleration of Aldehyde Decomposition in Soil by Montmorillonite," *Soil Sci.* 124:167-172 (1977).
11. Kurath, G. and R.Y. Morita. "Starvation-Survival Physiological Studies of a Marine *Pseudomonas* sp.," *Appl. Environ. Microbiol.* 45:1206-1211 (1983).
12. Laanbroek, H.J. and H.J. Geerligs. "Influence of Clay Particles (Illite) on Substrate Utilization by Sulfate-Reducing Bacteria," *Arch. Microbiol.* 134:161-163 (1983).

13. Luscombe, B.M. and T.R.G. Gray. "Effect of Varying Growth Rate on the Morphology of *Arthrobacter*," *J. Gen. Microbiol.* 69:433-434 (1971).
14. Marshall, K.C., R. Stout and R. Mitchell. "Mechanism of the Initital Events in the Sorption of Marine Bacteria to Surfaces," *J. Gen. Microbiol.* 68:337-348 (1971).
15. Martin, J.P., Z. Filip and K. Haider. "Effect of Montmorillonite and Humate on Growth and Metabolic Activity of Some Actinomycetes," *Soil Biol. Biochem.* 8:409-413 (1976).
16. Novitsky, J.A. and R.Y. Morita. "Morphological Characterization of Small Cells Resulting from Nutrient Starvation of a Psychrophilic Marine Vibrio," *Appl. Environ. Microbiol.* 32:617-622 (1976).
17. Oppenheimer, C.H. and C.E. ZoBell. "The Growth and Viability of Sixty Three Species of Marine Bacteria as Influenced by Hydrostatic Pressure," *J. Mar. Res.* 11:10-18 (1952).
18. Oviatt, C.A. and S.W. Nixon. "Sediment Resuspension and Deposition in Narragansett Bay," *Estuarine Coast. Mar. Science* 3:201-217 (1975).
19. Palumbo, A.V., R.L. Ferguson and P.A. Rublee. "Size of Suspended Bacterial Cells and Association of Heterotrophic Activity with Size Fractions of Particles in Estuarine and Coastal Waters," *Appl. Environ. Microbiol.* 48:157-164 (1984).
20. Porter, K.G. and Y.S. Feig. "The Use of DAPI for Identifying and Counting Aquatic Microflora," *Limnol. Oceanogr.* 25:943-948 (1980).
21. Rhoads, D.C. "The Influence of Deposit-Feeding Benthos on Water Turbidity and Nutrient Recycling," *Am. J. Sci.* 273:1-22 (1973).
22. Rhoads, D.C., K. Tenore and M. Browne. "The Role of Resuspended Bottom Mud in Nutrient Cycles of Shallow Embayments," in *Estuarine Research, Vol. 1* L.E. Cronin, Ed. (New York, NY: Academic Press Inc., 1975), pp. 563-579.
23. Santoro, T. and G. Stotzky. "Effect of Electrolyte Composition and pH on the Particle Size Distribution of Microorganisms and Clay Minerals as Determined by the Electrical Sensing Zone Method," *Arch. Biochem. Bisphys.* 122:664-669 (1967).
24. Shideler, G. "Reconnaissance Observations of Some Factors Influencing the Turbidity Structure of a Restricted Estuary: Corpus Christi Bay, Texas," *Texas J. Sci.* 32:59-71 (1980).
25. Sieburth, J. McN. "Sea Microbes," (New York, NY: Oxford University Press, 1979), pp. 24-27.
26. Stotzky, G. "Influence of Clay Minerals on Microorganisms II. Effect of Various Clay Species, Homoionic Clays, and Other Particles on Bacteria," *Can. J. Microbiol.* 12:831-848 (1966).

27. Stotzky, G. and L.T. Rem. "Influence of Clay Minerals on Microorganisms. I. Montmorillonite and Kaolinite on Bacteria," *Can. J. Microbiol.* 12:547-563 (1966).
28. Watson, S.W., T.J. Novitsky, H.L. Quinby and F.W. Valois. "Determination of Bacterial Number and Biomass in the Marine Environment," *Appl. Environ. Microbiol.* 33:940-946 (1977).

CHAPTER 14

FORMS OF SULFUR, CARBON AND IRON IN MARINE SEDIMENTS WITH SPECIAL REFERENCE TO THEIR DEPOSITIONAL ENVIRONMENTS

Takeshi Koma and Yasumoto Suzuki
Geological Survey of Japan

Abstract: Ninety six mudstone and twelve sandstone samples taken from the outcrops in the Kazusa Group were analyzed for sulfur, carbon, chlorine and iron. The following chemical components were determined as multi-indicator components: total sulfur, water-soluble sulfate, HCl-extractable sulfate, disulfide, carbon-bonded sulfur, total carbon, carbonate carbon, water-soluble chlorine and HCl-soluble iron. On the basis of the data obtained, depositional environments of each formation in the Kazusa Group are discussed.

INTRODUCTION

Keith and Degens(1) were the first to estimate the depositional environments of strata, i. e., whether they were deposited in sea water or in fresh water, based on the total sulfur content in mudstones. Many investigations have been conducted in Japan along this line(2-13). To distinguish deposition under oxidized conditions from that under reduced conditions, several other chemical components have also been used(4, 5). We select total sulfur, water-soluble sulfate, HCl-extractable sulfate, disulfide, carbon-bonded sulfur, total carbon, carbonate, organic carbon, water-soluble chloride, and HCl-soluble iron as multi-indicator components for determining depositional redox conditions. We summarize here the analytical data on mudstones and sandstones from the outcrops in the Kazusa Group. Data obtained with some underground samples are also presented. The Kazusa Group is a well known marine sediment stratum in the Boso Peninsula and includes one of the major natural gas reservoirs near Tokyo. It was deposited on the continental shelf and slope from middle Pliocene to Plistocene(14). Geochemical data can provide useful information for the exploration of natural gas.

GENERAL GEOLOGY OF THE KAZUSA GROUP

The Kazusa Group can be divided into the Kurotaki,Katsuura, Namihana, Ohara, Kiwada, Otadai, Umegase, Kokumoto, Kakinokidai, Chonan, Mandano and Kasamori Formations(Fig. 1)(14).

On the basis of the information on the benthonic foraminifera fossils, the Katsuura, Namihana, Ohara and Kiwada Formations have been inferred to be deposited on the lower

MARINE AND ESTUARINE GEOCHEMISTRY, Sigleo, A. C., and A. Hattori (Editors)

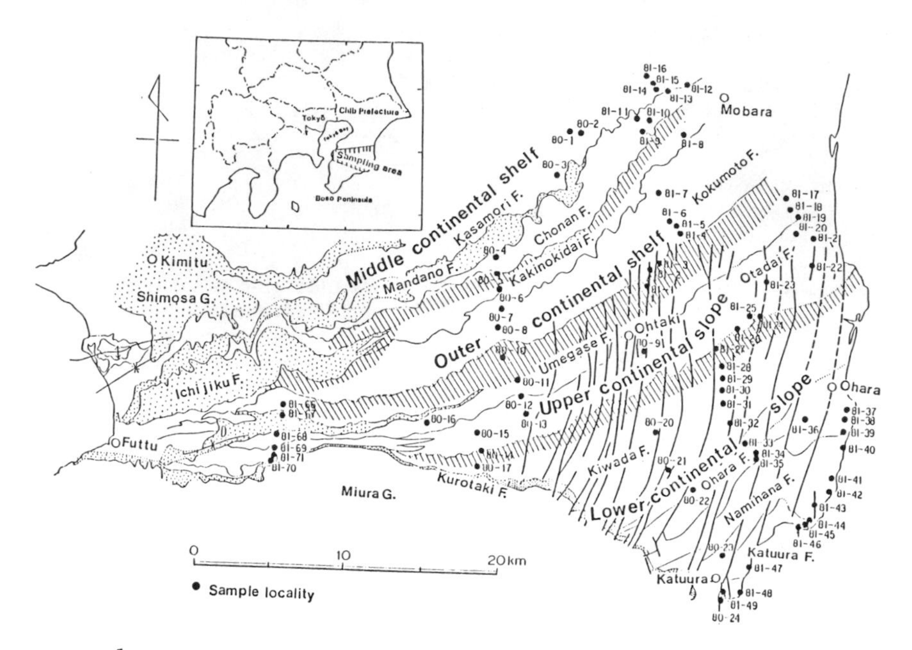

Fig. 1 Outline of geology and sampling locality in the Kazusa Group of the Boso Peninsula.

continental slope, the Otadai and Umegase Formations on the upper continental slope, the Kokumoto and Kakinokidai Formations on the outer continental shelf, and the Chonan, Mandano and Kasamori Formations on the middle continental shelf(14).

The facies of Kokumoto and Kakinokidai Formations laterally change westward into coarse to medium sandstone of the Ichijiku Formation which locally contains pebbly facies. The lateral westward change of the Umegase Formation into sandstone and conglomerate of the Higashi-higasa Formation is also recognized. The Kiwada, Ohara, Namihana and Katsuura Formations show again a facies change to the Kurotaki Formation, which has been interpreted as a marginal formation of the basin. The Higashi-higasa and Kurotaki Formations contain foraminifera of both shallow and deep sea fauna(15).

METHODS

Mudstones and sandstones were collected from the outcrops by the authors during the period from February 1980 to March 1981. The location of the samplings is shown in Fig. 1.

For the determination of total sulfur, the stone sample (0.1g) was placed in a crusible together with 1g iron powder and some tin metal grains, and burned in a high-frequency induction furnace. Sulfur in the sample was converted to sulfur dioxide. The sulfur dioxide was absorbed in sodium sulfate solution containing hydrogen peroxide, and the amount of sulfuric acid produced was determined coulometrically(16, 17).

For water-soluble sulfate, the sediment samples(0.1g) were suspended in 10 ml of water, and ultrasonicated. The insoluble matter was separated by centrifugation. Sulfate in the supernatant was determined by liquid chromatography(19, 20).

Hydrochloric acid extractable sulfate was obtained by heating the sediment samples(1.0g) at 100°C for one hour in 0.12N HCl. The extracted sulfate was determmned by the method of Tabatabai and Bremner(21). The insoluble matter was collected, dried and analyzed for FeS_2 sulfur and carbon-bonded sulfur by the same procedure as described in total sulfur analysis(16, 22, 23). Sulfide sulfur plus elementary sulfur was obtained by subtracting the total amount of sulfate-S and (FeS_2-S) + (C-bonded-S) from the amount of total sulfur(24-26).

Total carbon was determined by combustion to CO_2 and measurement with a thermal conductivity cell. Organic carbon was determined in using the insoluble matter which had been left after the hot HCl treatment(12).

Chloride ion was determined by the liquid chromatographic method of Koma et al.(12). Iron was determined colorimetrically by the o-phenanthrolin method of Sugisaki(27) using the HCl-soluble sulfate fraction.

RESULTS

Total Sulfur(total-S)

The total sulfur in mudstones ranged from 0.202 to 1.103%, and averaged 0.540%(Fig. 2,left). These values fall within the range reported for marine mudstones of Tertiary period in Japan (3-6). The total sulfur in sandstones ranged from 0.027 to 0.327%, and averaged 0.154%(Fig. 2, right). For recent coarse particle sediments at Ariake Sea in Kyushu, an average value of 0.314% has been reported.

Regional variations of the total sulfur content in the Kazusa Group is summarized in Fig. 3, and the vertical variations at two select sites are illustrated in Fig. 4. The total sulfur in the underground samples obtained by a drilling lay between 0.3 and 1.1%(data not shown).

Water-Soluble Sulfate Sulfur(SO_4^{2-}-S:H_2O)

The water-soluble sulfate-sulfur ranged from 0.00 to 0.208%, and averaged 0.020% in mudstones, and from 0.00 to 0.036%, and averaged 0.013% in sandstones(Fig. 2). The percentage of the water-soluble sulfate sulfur relative to the total sulfur, ranged from 1 to 5% for more than 70% of the samples.

Iron Sulfide Sulfur and Elementary Sulfur(FeS-S + element-S)

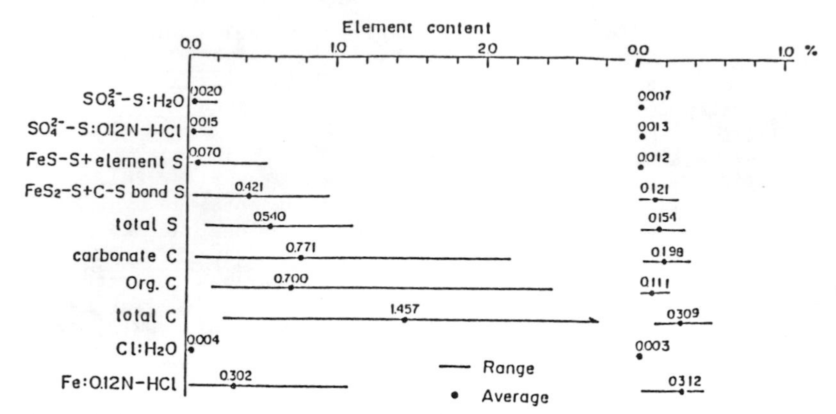

Fig. 2 Ranges and averages of sulfur, carbon, chlorine and iron contents in mudstones(left) and sandstones(right) from the outcrops of the Kazusa Group.

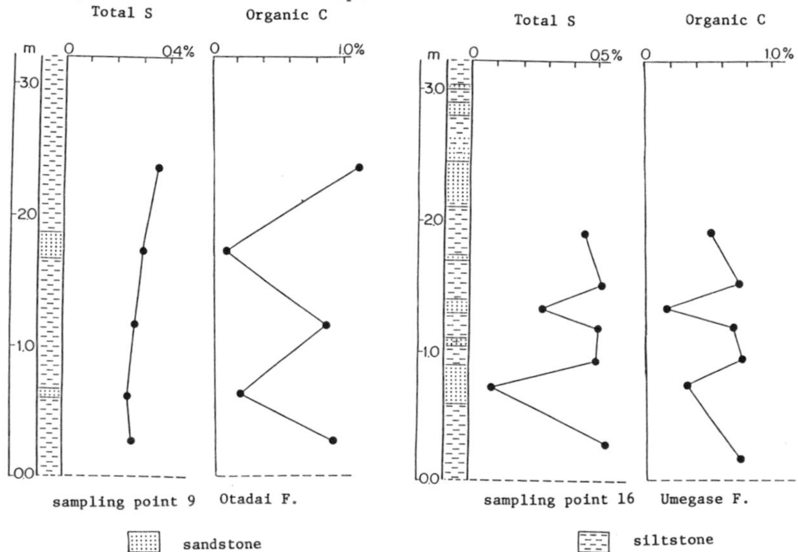

Fig. 3 Distribution of total sulfur and organic carbon in siltstones and sandstones from some outcrops. (left: Otadai F; right: Umegase F)

The sum of FeS-S and element-S in mudstone ranged from 0.00 to 0.054%, and averaged 0.012%. The percentage of this sulfur to that of the total sulfur fell in the range between 0 and 10% for 50% of the samples. About 40% of the samples had the values of this ratio from 11 to 20%. The sum of FeS-S and element-S in the mudstone of the Kazusa Group was lower than that in the mudstones from the Niigata oil and gas field(12).

Iron Disulfide Sulfur and Carbon-bonded Sulfur(FeS_2-S + C-S bond-S)

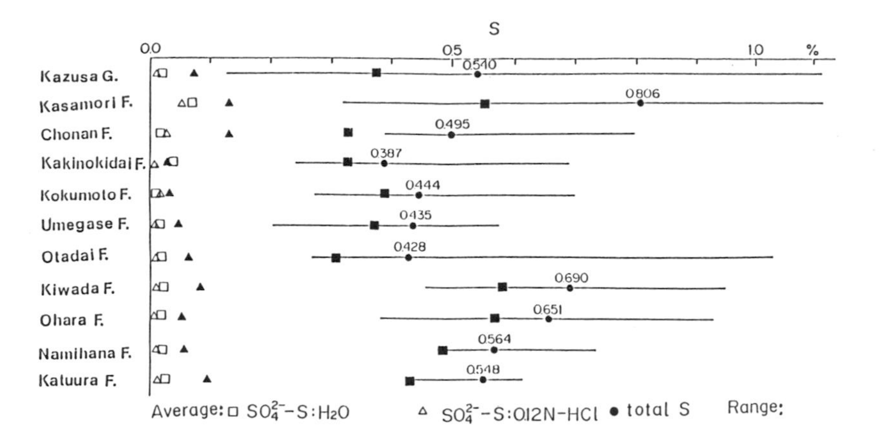

Fig. 4 Averages and ranges of all forms of sulfur in each formation of the Kazusa Group.

The sum of FeS_2-S and C-S bond-S ranged from 0.038 to 0.946%, and averaged 0.421% in mudstones, and from 0.021 to 0.278%, and averaged 0.12% in sandstones(Fig. 2). The percentage of the iron disulfide sulfur relative to the total sulfur exceeded 80% for 76% of the samples. The sulfur of FeS_2 and C-S was correlated with the total-S. According to the morphological observation of these samples under the microscope, there are very fine-grained indefinite shapes and/or spherical shapes which replaced the bodies of dead organisms.

Total Carbon(total-C)

The total carbon in mudstones ranged from 0.245 to 3.46%, and averaged 1.46%. These values are similar to those obtained for mudstones from the Niigata oil and gas field(0.102 to 3.05%), except for mudstones rich in organic matter(3, 28). The total carbon in sandstones ranged from 0.125 to 0.53%, and averaged 0.309%, which is about one-fifth of that in the mudstones.

Carbonate-Carbon(carbonate-C)

The carbonate-carbon ranged from 0.050 to 2.146%, and averaged 0.771% in mudstones, and from 0.214 to 0.375% and averaged 0.198% in sandstones.

Organic Carbon(Org.-C)

The organic carbon ranged from 0.170 to 2.435%, and averaged 0.700% in mudstones and from 0.0355 to 0.226%, and averaged 0.111 % in sandstones. The percentage of the organic carbon relative to the total carbon was between 30 and 60% for 70% of the samples. The carbonate-carbon was generally greater than the organic carbon in both the mudstones and sandstones. The regional and vertical variations of the organic carbon are shown in

Fig. 5 and Fig. 4, respectively.

Water-Soluble Chlorine($Cl:H_2O$) and
Acid-Soluble Iron(Fe:0.12N-HCl)

The water-soluble chlorine was low in both the mudstones and sandstones and their average values were between 0.003 and 0.004%, except for the samples taken at the sea shore. These values were extremely low compared with the values for the drilling core samples(0.2 and 0.4%). Hydrochloric acid soluble iron ranged from 0.017 to 1.075%, and averaged 0.302% in mudstones, and from 0.054 to 0.471%, and averaged 0.312% in sandstones.

DISCUSSION

Water-Soluble Chlorine and Total Sulfur in Marine Sediments

Chlorine in sediments is present as water soluble compounds, and is easily transported with mobile water. On the other hand, sulfur generally is fixed in marine sediments due to the low solubility of sulfide minerals. The water-soluble chlorine content in the underground samples obtained by drilling was actually 100 times greater than that in the outcrops. The low chlorine content in the outcrops can be attributed to wash away of the water-soluble chlorine by rain water. On the other hand, the total sulfur content of the underground samples were similar to that of the outcrop samples. The total sulfur content is not greatly affected by weathering.

Effect of Particle Size on Sulfur and Carbon Contents

In the outcrops, the total sulfur, water-soluble sulfate sulfur, HCl extractable sulfate sulfur, total carbon, organic carbon and carbonate carbon were invariably greater in mudstone

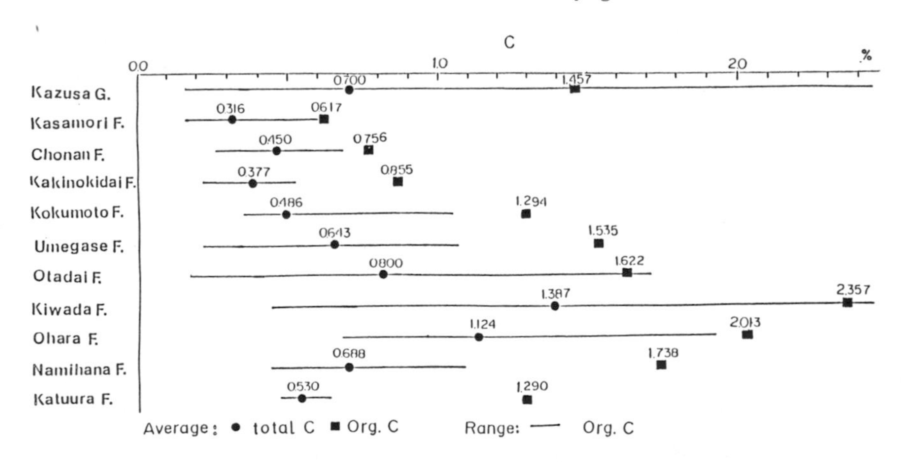

Fig. 5 Distribution of total carbon and organic carbon in each formation of the Kazusa Group.

than in sandstone(Fig. 2). Similar tendencies were found for the total sulfur and total carbon in recent sediments at Ariake Sea in Kyushu and Upper Cretaceous Himenoura Group sediments, Kagoshima Prefecture(8, 17, 32). The total sulfur contents in mudstone and sandstone do not differ substantially when a thin layer of sandstone is interlaminated in thick mudstone strata (Fig. 3, left). However, when the mudstones and sandstone are laminated alternately, the total sulfur contents in both the layers differ greatly from each other(Fig. 3, right). The total sulfur is high in the mudstones and low in the sandstones. On the other hand, the organic carbon is always high in the mudstones and low in the sandstones, irrespective of the thickness of sandstone layers. This tendency is probably controlled by redox environments, since the total sulfur originates primarily from sulfur in iron sulfide. In the strata which are mainly composed of mudstone, even if a thin sandstone layer is interlaminated in them, reduced environments produced by decomposition of organic matter in the mud are extended into the thin layer of sandstone and the reduced environment prevails equally throughout these strata. On the other hand, in the strata in which mudstone and sandstone are laminated alternately in equal thickness, reduced environments are easily produced in the mudstone layer, but in a sandstone layer the decomposition of organic substances may be accelerated by the activity of aerobic bacteria in the more oxidized environment and by benthic organisms.

Total Sulfur and Depositional Environments

On the basis of the total sulfur content and fossil evidence, Berner(33), Berner and Raiswell(34) and Koma et al. (12) suggested that the strata with total-S of more than 0.4% are marine in origin and the strata with 0.6 to 1.0% and with 0.4 to 0.6% were formed under strongly reduced conditions and under weakly reduced conditions, respectively. Judging from the values of the total-S(Fig. 4), the Ohara and Kiwada Formations appear to have been formed under strongly reduced conditions, and the Katsuura and Namihana Formations under stable and weakly reduced conditions. The content of total-S is confined to a relatively narrow range. In the Otadai, Umegase, and Kokumoto Formations, their total-S content is much lower. These sediments are more weakly reducing in state.

The Kakinokidai Formation with the lowest total-S contains much turbidite sandstone accompanied by slump beds and submarine valley deposition beds. It was probably formed under weakly oxidized conditions with mobile water in it. The Chonan Formation containing average total-S of 0.495% seems to be a sediment of weakly reduced state. The Kasamori Formation contains the highest average total-S in the Kazusa Group. Large scattering of the total-S in the Kasamori Formation suggests that this stratum is a sediment of weakly oxidative state in which very strongly reduced sites locally remain.

Carbon and the Depositional Environment

The average organic carbon is correlated with the average total sulfur, except for the Kasamori Formation.

The Katsuura and Namihana Formations are laminated alternately with sandstones and mudstones and appear to have been deposited under the calm environment of the deep sea. In part of these strata, disulfidation of iron sulfide is observed and contains much carbonate carbon and HCl soluble iron (data not shown). These facts indicate that water containing dissolved oxygen moved through this layer.

The average organic and total carbon of the Ohara and Kiwada Formations are higher by a factor of about 2 than the values for the rest of the Kazusa Group(Fig. 5). The total sulfur and disulfide are also high(Fig. 4). Variation of the carbon values, like that of sulfur, seems to be less on the western side and greater on the eastern side. These strata are composed mainly of mudstone and the focus of deposition was in the central east part. Like sulfur, organic carbon appears to have been concentrated in a closed depositional trough. These formations contain the reservoirs of natural gas.

The averages of total carbon of the Otadai, Umegase and Kokumoto Formations are 1.62, 1.54 and 1.29% and those of organic-carbon are 0.80, 0.64 and 0.49%, respectively(Fig. 5). These strata were deposited on a deep trough in the sea(14).

The average values of total and organic carbon of the Kakinokidai, Chonan and Kasamori Formations are the lowest in the Kazusa Group(Fig. 5). These Formations contain much water extractable sulfate sulfur, HCl extractable sulfate sulfur and HCl soluble iron(Fig. 4), suggesting that aerobic bacteria which decompose organic substances were very active in these sediments. The low organic carbon content and high total sulfur content of the Kasamori Formation further suggest that it was deposited in a shallow sea.

CONCLUSION

The geochemical data presented above and other available geological information lead us to the following conclusion:

(1) Among the Kazusa Group, the Katsuura and Namihana Formations were formed under weakly reduced environments and later water movement produced oxidized environments. The Ohara and Kiwada Formations were deposited under strongly reduced environments and later changed into weakly oxidized environments.

(2) The Otadai, Umegase and Kokumoto Formations were deposited under oxidized to weakly reduced conditions.

(3) The Kakinokidai, Chonan and Kasamori Formations were deposited under oxidized to weakly reducing environments. In the Kasamori Formation there are locally strongly reduced environments. After deposition there appears to have been no change in these formations.

(4) Reservoirs of natural gas in localized in the formations (Ohara and Kiwada) which are rich in the total sulfur, disulfide, organic carbon and HCl soluble iron.

Acknowledgements

We are indebted to Prof. N. Nakai, Nagoya University. Prof. K. Taguchi, Tohoku University and Dr. K. Takahashi, the Geological Survey of Japan for their many helpful comments and suggestions to the manuscript.

REFERENCES

1. Keith, M. L. and E. T. Degens. "Geochemical Indicators of Marine and Fresh-water Sediments," in *Researches in Geochemistry*, P. H. Abelson, Ed. (John Wiley and Sons, Inc, New York, 1959), pp. 38-61.
2. Itihara, M. and Y. Itihara. "On Marine Clay and Fresh-water Clay of the Osaka Group," in *Pliocene and Pleistocene series of Chubu district, Japan: Prof. H. Takehara Mem. Vol.*, J. Itoigawa et al. Eds., 1971. pp.173-181.
3. Koma, T. "Chemical Composition of Muddy Rocks from Tertiary Formations in Niigata Oil Field, Central Japan," *Rept. Geol. Surv. Japan*, 250(2):211-227(1974).
4. Koma, T., S. Itoh and S. Yokota. "Chemical Composition of Neogene Muddy Rocks from Chikubetsu Area, Northwestern Hokkaido, Japan," *Jour. Japan. Assoc. Petrol. Technol.*, 39; 95-105(1974).
5. Itoh, S., T. Koma, T. Nemoto, S. Yokota and T. Kimura. "Chemical Composition of Neogene Muddy Rocks from the North Hokkaido Area, Japan," *Bull. Geol. Surv. Japan*, 28:57-67 (1977).
6. Koma T. "Sulfur Content and Its Environmental Significance of Paleogene Muddy Sediments in a Part of the Ishikari Coal Field, Central Hokkaido, Northern Japan," *Jour. Japan. Assoc. Petrol. Technol.*, 43:128-136(1978).
7. Terashima, S., A. Inazumi and S. Ishihara. "Carbon and Sulfur Contents of Pelitic Rocks from Chugoku and Shikoku in Japan," *Bull. Geol. Surv. Japan*, 32:167-181(1981).
8. Tanaka, K., S. Terashima and Y. Teraoka. "Sulfur and Carbon Contents of Mud Rocks from the Upper Cretaceous Himenoura Group, Koshiki-jima, Kyushu," *Bull. Geol. Surv. Japan*, 32:417-431(1981).
9. Terashima, S., S. Nakao and B. Mita. "Sulfur and Carbon Contents of Deep-Sea Sediments from the Central Pacific, GH80-1 Cruise," *Bull. Geol. Surv. Japan*, 33:369-379(1982).
10. Terashima, S., H. Yonetani, E. Matsumoto and Y. Inouchi. "Sulfur and Carbon Contents in Recent Sediments and heir Relation to Sedimentary Environments," *Bull. Geol. Surv. Japan*, 34:361-382(1983).

11. Koma, T., T. Sakamoto and A. Ando. "Total Sulfur Content and Sedimentary Environment of the Late Cenozoic Formations in Ibaraki Prefecture, Japan," *Bull. Geol. Surv. Japan,* 34(6):279-293(1983).
12. Koma , T., Y Suzuki and K. Kodama. "Forms of Sulfur, Carbon, Chlorine and Iron Compounds and Their Depositional Environment, in the Kazusa Group, the Boso Peninsula, Central Japan," *Bull. Geol. Surv. Japan,* 34(4):191-206 (1983).
13. Nakai, N., T. Ohta, H. Fujisawa and M. Yoshida. "Paleoclimatic and Sea-Level Changes deduced from Organic Carbon Isotope Ratios, C/N Ratios and Pyrite Contents of Cored Sediments from Nagoya Harbor, Japan," *Quaternary Research,* 21(3):169-177(1982).
14. Mitsunashi, T., T. Kikuchi and Y. Suzuki. "Geological Map of Tokyo Bay and Adjacent Areas with Expl. Text," *Geol. Surv. Japan. Miscellaneous Maps Series,* 20(1976).
15. Aoki, N. "Benthonic Foraminiferal Zonation of the Kazusa Group, Boso Peninsula." *Trans. Proc. Paleont. Soc. Japan, N. S.,* 70:238-266(1968).
16. Casagrande, D. J., K. Siefert, C. Bershcinski and N. Sutton. "Sulfur in Peat-forming Systems of the Okefenokee Swamp and Florida Everglades Origins of Sulfur in Coal," *Geochim. Cosmochim. Acta,* 41:161-167(1977).
17. Kajima, R. and K. Hoshino. "Determination of Sulphur in Steel by a Modified Combustion Procedure and Coulometric Titration," *Analyst,* 96:835-842(1971).
18. Sugawara, K., T. Koyama and A. Kozawa. "Distribution of Various Forms of Sulphur in Lake-, River- and Sea-muds," *Jour. Earth Sci., Nagoya Univ.,* 1:17-23(1953).
19. Nakae, A., K. Furuya, T. Mikata and M. Yamanaka. "Determination of Sodium Sulfate in Detergents by High Speed Liquid Chromatography," *Jour. Chem. Soc. Japan, Chem. Indust. Chem.,* 11:1655-1659(1977).
20. Nakae, A., K. Furuya and M. Yamanaka. "Determination of Sodium Sulfate in Detergents by Anion-Exchange Chromatography in Nitric Acid-Iron(III) Nitrate Media," *Jour. Chem. Soc. Japan, Chem. Indust. Chem.,* 2:217-221(1978).
21. Tabatabai, M. A. and J. M. Bremner. "Forms of Sulfur, Carbon, Nitrogen and Sulfur Relationships in Iowa Soils," *Soil. Sci.,* 114:380-386(1972).
22. Williams, C. H. and A. Steinberg. "Soil Sulfur Fractions as Chemical Indexes of Available Sulfur in Some Australian Soils," *Aust. J. Agr. Res.,* 10:340-352(1959).
23. Maki, S., S. Nagata, O. Fukuta and S. Furukawa. "Geochemical Study on Organic Matter from Sedimentary Rocks in the Miyazaki Group and the Shimanto Supergroup of Miyazaki Prefecture, Japan," *Bull. Geol. Surv. Japan,* 31(1):1-24(1980).
24. Trask, P. D. and C. C. Wu. "Free Sulphur in Recent Sediments," *Bull. Geol. Soc. Amer.,* 41:89-90(1930).

25. Mapstone, G. E. "Detection of Elemental Sulfur in Gasoline by the Sommer Test," *Ind. Eng. Chem. Anal. Ed.*, 18:498-499(1946).
26. Yagishita, H. "On the Elemental Sulphur Contained in Tertiary Mudstones," *Miscellaneous Reports, Research Institute for Natural Resources*, 41-42:53-61(1956).
27. Sugisaki, R. "A Modified Method of Analysis of Bulk Chemical Composition of Argillaceous Sediments and Data Display with Special Reference to Marine Sediments," *Jour. Geol. Soc. Japan*, 87(2):77-85(1981).
28. Maki, S. "Studies on Organic Matter in the Niigata Gas Field. Part 4. Studies on Organic Matter in Hydrocarbon Deposits," *Bull. Geol. Surv. Japan*, 14:415-430(1963).
29. Hata, Y. "Microbial Production of Sulfides in Marine and Estuarine Sediments," *Contribution from the Shimonoseki University of Fisheries*, 457(14):37-83(1965).
30. Berner, R. A. "Sedimentary Pyrite Formation," *Am. Jour. Sci.*, 268:1-23(1970).
31. Berner, R. A. "Sedimentary Pyrite Formation: An Update," *Geochim. Cosmochim. Acta*, 48:605-615(1984).
32. Katada, M., H. Isomi, E. Omori and T. Yamada. "Chemical Composition of Paleozoic Rocks from Northern Kiso District and of Toyama Clayslates in Kitakami Mountainland: Supplement, Carbon and Carbon Dioxide," *Jour. Jap. Assoc. Miner. Petrol. Econ. Geol.*, 52:217-221(1964).
33. Berner, R. A. "Burial of Organic Carbon and Pyrite Sulfur in the Modern Ocean: Its Geochemical and Environmental Significance," *Amer. Jour. Sci.*, 282:451-473(1982).
34. Berner, R. A. and R. Raiswell. "Burial of Organic Carbon and Pyrite Sulfur in Sediments over Phanerozoic Time: a New Theory," *Geochim. Cosmochim. Acta*, 47:855-862(1983).

CHAPTER 15

SEDIMENTARY PROCESSES OF FINE SEDIMENTS AND THE BEHAVIOUR OF ASSOCIATED METALS IN THE KEUM ESTUARY, KOREA

Chang-Bok Lee
Department of Oceanography
Seoul National University
Seoul 151, Korea

ABSTRACT

The Keum Estuary is characterized by its macro-tidal regime and great seasonal fluctuation in river discharge. The spring-tidal saline water penetrates up to 60 km upstream during the low river discharge period. Concentration of suspended particulate matter(SPM) varies with both velocity and direction of the tidal current, the latter being related to the concentration gradient along the estuary. The development of a turbidity maximum, extending over 40 km along the estuary during the spring-tide and low river discharge period and disappearing during the neap-tide and high river flow, seems primarily related to the tidal range at the mouth. The SPM in the maximum zone is mostly fine-grained and enriched with some heavy metals, especially with Mn, Zn and Cu. Major portions of analyzed metals (Mn, Zn, Cu, Co, Pb and Ni) are associated with the reducible and residual fractions, although some of them show certain relationship with the grain-size and the organic carbon content.

INTRODUCTION

Estuarine sedimentary processes are controlled primarily by the dynamics of the estuary, which in turn depend upon the equilibrium of the river and tidal flow. Numerous previous studies have enabled us to explain the different estuarine circulation characteristics by such basic parameters as river discharge, volume of saline water introduced by tidal current and morphology of the estuary (1-5). Hence, for a given estuary, the major variation in hydrology and related sedimentary processes could be expected to occur following the variations of current velocity over a tidal cycle, of tidal range over a fortnightly spring-neap cycle and of river discharge over a year cycle.

An important phenomenon in the estuarine sedimentary processes of fine sediments is the turbidity maximum, a zone with greater concentrations of suspended particulate matter(SPM) than either the landward or seaward zones. The formation of a turbidity maximum has been attributed both to physico-chemical processes such as flocculation and deflocculation, and to hydrodynamic processes, including net non-tidal circulation and resuspension of bottom sediments (6-8). The SPM in the turbidity maximum is transported almost in a 'closed circle', resulting in both the accumu-

MARINE AND ESTUARINE GEOCHEMISTRY, Sigleo, A. C., and A. Hattori (Editors)

lation of the preferred size range and the increased mean residence time of the SPM in the estuarine zone.

Our recent investigation on the Keum Estuary, a macro-tidal estuary located on the west coast of the Korean Peninsula (Fig. 1), has revealed the formation of turbidity maximum during the spring-tide and low river discharge period. The present paper describes this turbidity maximum and its variation in response to the various fluvial and tidal regimes. In addition the elemental composition, along with that of the nearby bank and bottom sediments is characterized.

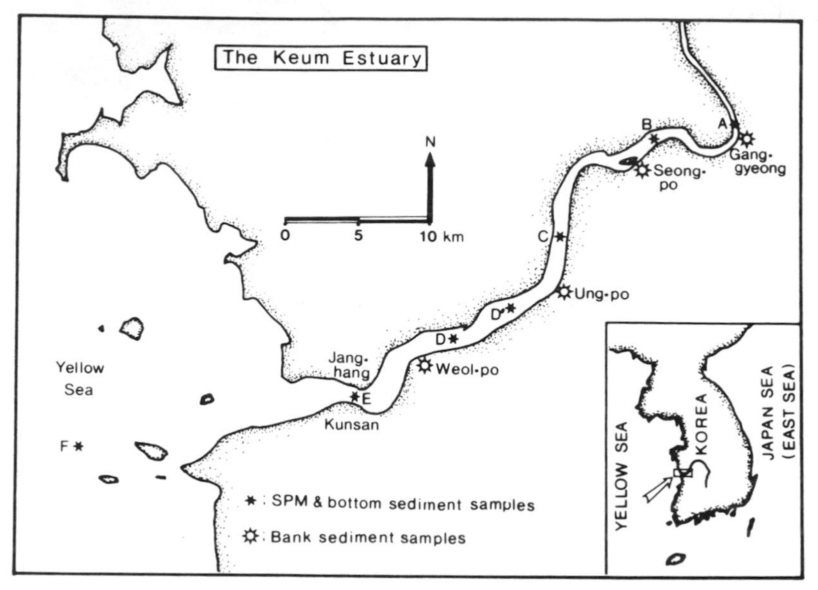

Figure 1. Index map showing the study area and sampling sites.

THE STUDY AREA

The Keum River, the third largest South Korean river next to the Han and Nakdong Rivers, has about 400 km of total length and drains about 10000 km^2 of drainage basin. It is composed mostly of Precambrian metasedimentary complexes and Mesozoic granites with limited extents of sedimentary rocks. The annual fresh water discharge is about 6.4×10^9 m^3, most of it(about two-thirds) being concentrated in the summer rainy period and, consequently, resulting in a great seasonal variation in river flow (Fig. 2).

The coastal zone is characterized by the semi-diurnal, macro-tidal regime, with the mean spring tidal range of about 6 m. The spring-tidal saline water penetrates up to about 60 km upstream from the mouth and forms a well-defined estuarine environment. The dynamic tide, defined as the periodic water level fluctuation related to the tidal cycle, can be observed at about 110 km upstream from the mouth.

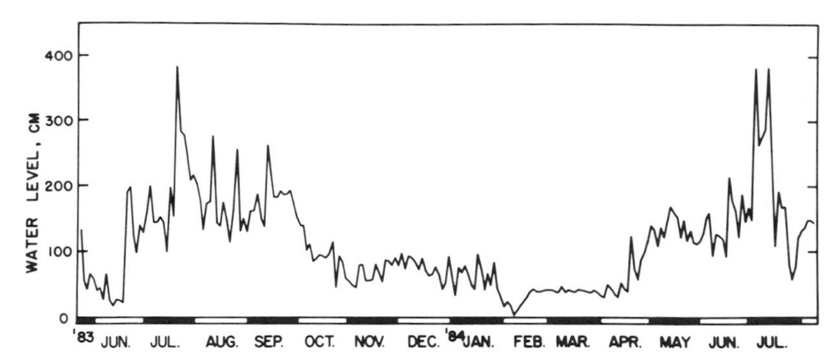

Figure 2. Fluctuation of the Keum River discharge represented by the water level at Ma-am, located at about 110 km upstream from the mouth.

The lower reach of the Keum River shows in general the characteristics of a low gradient, meandering stream. The channel bed near Ganggyeong, about 50 km upstream from the mouth, is lower than the offshore mean sea level and the sinuosity between Kunsan and Buyeo is about 1.7. Although some fluvial islands and channel bars are developed locally, the general feature shows a mono-channel stream character.

SAMPLING AND METHODS

A series of anchor stations along the Keum Estuary were occupied during four periods of field observations, each representing the period of different tidal and fluvial regimes (Fig. 1). During the period of June 10 to 15, 1983, a period of low river discharge and spring-tide, six stations (A-F) were occupied continuously over 1-2 tidal cycles. During August 3-5, 1983, a period of high river discharge and neap-tide, station C was reoccupied over a tidal cycle as was one additional station (D') between stations C and D. During July 15-16, 1984, a period of extremely high river discharge and spring-tide, stations C and D were reoccupied over a tidal cycle. During July 24-25, 1984, a period of high river discharge and neap-tide, stations D and E were reoccupied over a tidal cycle.

At all of these stations, current direction, velocity, and salinity were measured at three depths (surface, mid-depth and 1 m above bottom) at one hour intervals using a CM-2 current meter (Toho Dentan) and a MC-5 temperature-salinity measuring bridge (Hydrobios). Water samples were collected using a Van Dorn sampler and SPM concentration was determined by filtration of these water samples through 0.45 μm Nuclepore filters in the laboratory. Some additional 40 ℓ of surface water samples were collected during the June 1983 field observation within the turbidity maximum zone to obtain an adequate amount of SPM samples for analyses of grain-size and elemental composition. Bottom sediment sampling was done using a grab; and for the bank sediment the uppermost thin layer was collected.

Grain-size analysis was done either using a Sedigraph 5000 (Micromeritics) for SPM samples or by the sieving and pipetting methods for the other sediments. Particulate organic carbon content was determined by back titration after treatment of the sediment with a sulfo-chromic mixture (9,10). Heavy metal contents were analyzed using an IL 251 Model atomic absorption spectrometer. Prior to analysis the sediments were digested with a mixed solution of HF, HNO_3 and $HClO_4$ and leached with dilute HCl (11). The speciation of metals into four chemical fractions (adsorbed, reducible, oxidizable and residual) was determined on the SPM samples by sequential extraction using a modified procedure of Burrows and Hulbert (12) and Kitano and Fujiyoshi (11).

HYDRODYNAMIC CHARACTERISTICS OF THE KEUM ESTUARY

Figure 3 shows the variations of residual velocity, salinity and SPM concentration along the Keum Estuary during the spring-tide and low river discharge period. On the basis of the salinity distribution pattern, the Keum Estuary could be classified as a well-mixed estuary, where neither vertical nor lateral salinity gradients were observed (1,2,5). The mean upstream limit of the salinity intrusion (0.5‰) occurred at about 50 km from the mouth during this period. Being a well-mixed estuary, the residual velocities were all oriented downstream through the entire water column in most of the estuary, except near the inlet where two-layered residual flow was observed. The null point appeared to be located near about 10 km upstream from the mouth.

As the river discharge increased, the limit of the salinity intrusion shifted to the downstream direction and was located near station E, about 8 km upstream from the inlet during the high river discharge period. The variation in tidal range also affected the extent of the salinity intrusion but on a smaller scale than that of river discharge. These influences of fluvial and tidal regimes on the hydrodynamic characteristics of the Keum Estuary can be seen in the residual velocity variation, as shown in Figure 4.

The velocity distribution pattern over a tidal cycle in the Keum Estuary seems primarily to depend on the fluvial regime. We do not have enough data covering the whole spectrum of tidal and fluvial regimes, but our results indicate that the maximum velocity over a tidal cycle occurred during the ebb current when the river discharge was high and during the flood current when the river discharge was low. This change in the flood- and ebb-dominant current pattern may have a great implication on the SPM dynamics of the Keum Estuary while the turbidity maximum, which will be discussed in detail in the following section, is fed and maintained mostly by the resuspension of the bottom sediments.

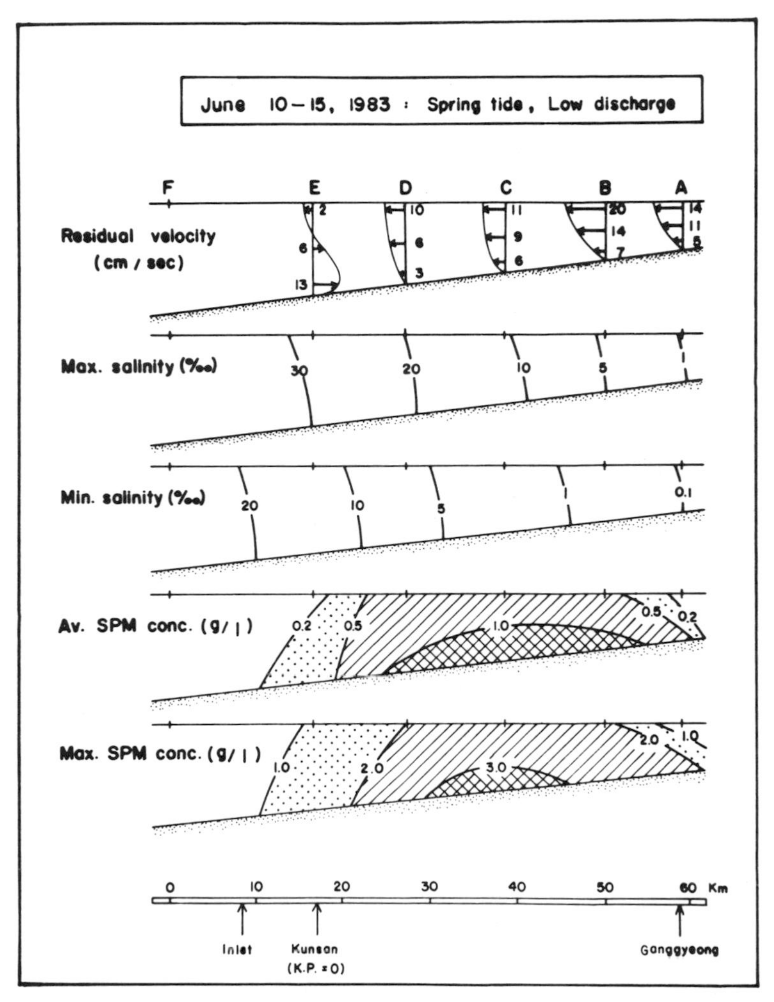

Figure 3. Distribution of the residual velocity, salinity and SPM concentration along the Keum Estuary during the spring-tide and low river flow.

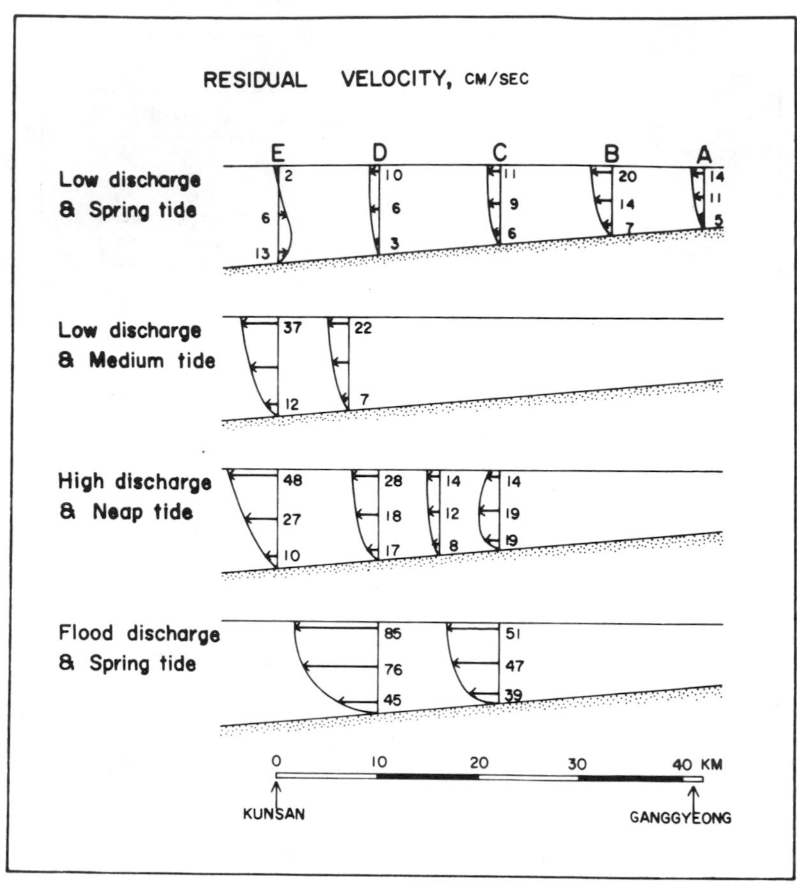

Figure 4. Distribution of the residual velocity under the various fluvial and tidal regimes.

TURBIDITY MAXIMUM AND SUSPENDED SEDIMENT DYNAMICS

The development of the turbidity maximum in the Keum Estuary was most obvious during the low discharge and spring-tide period. This turbidity maximum, extending over about 40 km in length between Kunsan and Ganggyeong, showed a mean SPM concentration (tidal-averaged) of more than 0.5 g/ℓ and a maximum SPM up to 3.5 g/ℓ, while, both to the upstream and downstream directions from this extremely turbid zone, the SPM concentrations decreased rapidly and reached values less than 100 mg/ℓ. The landward limit occurred near the limit of salinity intrusion and the seaward limit coincided roughly with the residual velocity null point (Fig. 3).

The turbidity maximum exhibits a periodic variation in its extent with the tidal cycle, related to the variation in current velocity (Fig. 5). Both at high slack water and low slack water

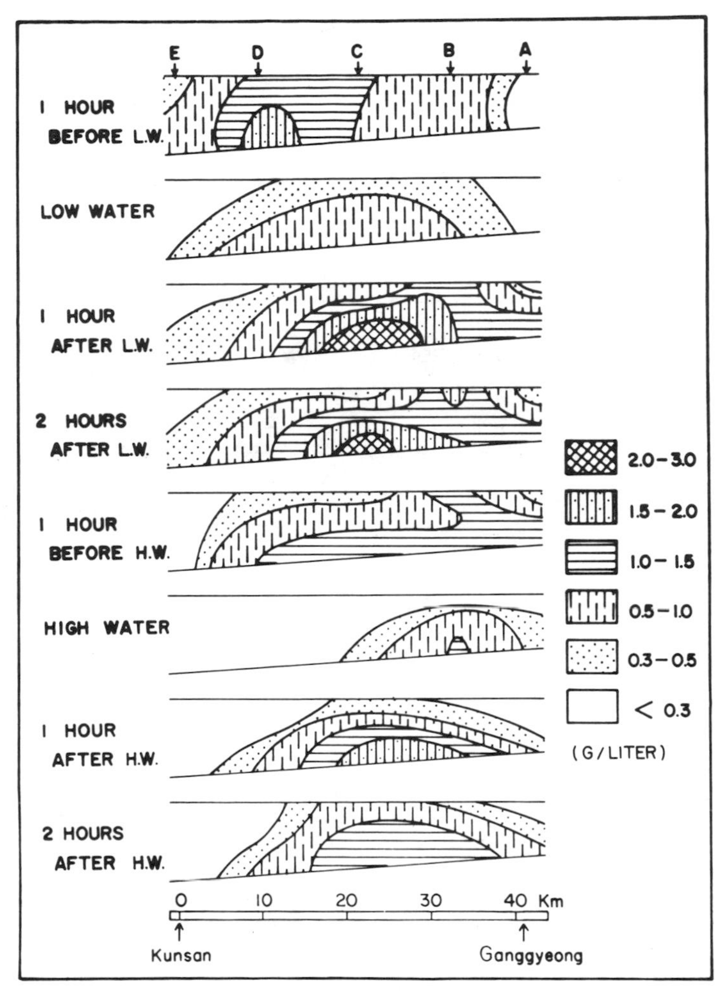

Figure 5. Variation of the SPM concentration over a tidal cycle within the Keum Estuary turbidity maximum.

it shows a minimum extent, and with the increase in both the ebb and flood current velocities it enlarges in extent, fed by the resuspension of bottom sediments. The flood-dominant current pattern during this period is represented by SPM levels whose maximum occurs during the flood current. This, along with the shift of the turbid core between the high water and low water slacks, appears to inhibit the seaward escape of the SPM and therefore permits the turbidity maximum to be maintained during the low discharge period.

A tentative estimation of the SPM transport through unit width of each measurement station was made by integrating the current velocities and SPM levels over a tidal cycle (Fig. 6). Though it may not be considered a quantitative result of the SPM transport in this environment, it shows clearly that the dynamics of SPM is controlled primarily by the tidal regime and supports further the idea that the turbidity maximum is fed and maintained by the resuspension of the bottom sediment by the strong tidal current. The ebb-dominant current pattern during the high river discharge period, being less effective in trapping the SPM within the estuarine zone, may facilitate the seaward transport of resuspended sediments and consequently, the turbidity maximum could not be formed during this period.

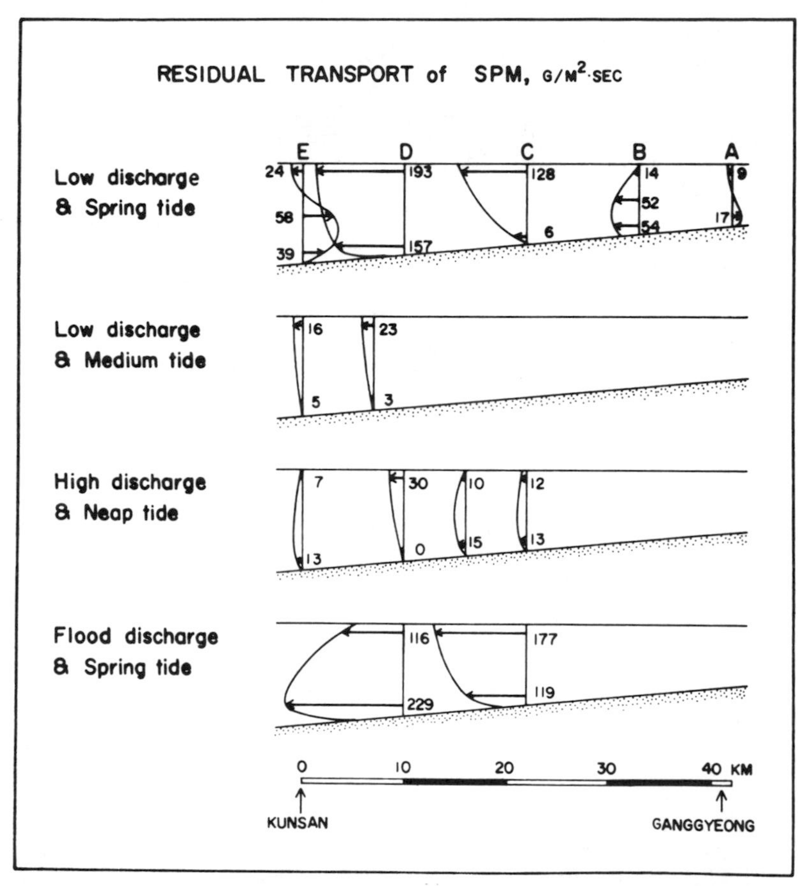

Figure 6. Distribution of the estimated SPM transport through unit width of channel.

GRAIN-SIZE CHARACTERISTICS OF SEDIMENTS

The grain-size of SPM is extremely fine with a mean size of 8.9ϕ. Sand grains form about 2% of the total SPM while the less than 2 μm fraction is on the average 45% of SPM (Table I). Compared with the SPM, the bottom and bank sediments are relatively coarse with a mean size of 3.5ϕ and 5.7ϕ respectively.

Table I. Textural Properties of SPM and Deposited Sediments

	Samples	K.P.*	>62 μm (%)	62-16 μm (%)	16-2 μm (%)	<2 μm (%)	Mean Size (ϕ)
SPM	Station A	42	2.3	24.7	34.9	38.1	8.5
	B	33	2.7	14.3	38.1	44.9	8.9
	C	22	1.5	8.2	39.6	50.7	9.4
	D	10	3.8	14.9	44.7	36.6	8.6
	E	0	0	5.5	40.7	53.8	9.5
Bottom	Station A	42	64.1	23.1	10.3	2.5	4.0
	B	33	90.5	4.5	3.7	1.3	1.2
	C	22	61.8	22.9	12.5	2.8	3.6
	D	10	18.8	56.2	19.4	5.6	5.2
Bank	Ganggyeong	41	69.0	20.3	6.6	4.1	3.2
	Seongpo	31	13.4	48.5	22.8	15.3	6.1
	Ungpo	19	15.4	35.1	30.2	19.3	6.5
	Weolpo	8	10.5	40.1	26.8	22.6	6.8

* K.P. = Kilometric Point which represents the distance along channel (in Kilometer) in the upstream direction from Kunsan Harbour which is located at about 8 km upstream from the estuary inlet.

METAL CONTENTS IN THE SEDIMENTS

The manganese content varied between 975 and 1262 ppm in the SPM, between 368 and 1333 ppm in the bank sediments, and between 491 and 848 ppm in the bottom sediments. The zinc content had a range of 360-1530 ppm in the SPM, 161-619 ppm in the bank and 157-627 ppm in the bottom sediments. The copper level was between 37 and 379 ppm in the SPM, 16 and 30 ppm in the bank sediments, and 8 and 28 ppm in the bottom sediments. The cobalt content was between 40 and 52 ppm in the SPM, 20 and 32 ppm in the bank sediments, and 19 and 27 ppm in the bottom sediments. The lead content had a range of 52-64 ppm in the SPM, 38-59 ppm in the bank sediments, and 31-54 ppm in the bottom sediments. The nickel content had a range of 31-44 ppm in the SPM, 21-39 ppm in the bank sediments, and 13-30 ppm in the bottom sediments.

As we can see from the above results, all the analyzed metals showed their maximum contents in the SPM and minimum contents in the bottom sediments. Although the level of particulate organic carbon content in this environment was generally low compared with other coastal areas, it was highest in the SPM (average 1.03%) and lowest in the bottom sediments (average 0.39%). This difference in organic matter contents together with that of grain-size may partly have affected the different level of metallic contents in the three types of sediments considered. We can demonstrate these effects by pair diagrams relating the metal contents with the granulometry (grain size) and the organic carbon contents of sediments (Fig. 7, 8). Except for copper, the relationships of the metallic contents with the percentages of fine fractions (clay and fine silt) are obvious. With the organic carbon contents also, these relationships can be found again, especially for cobalt, lead and nickel.

The average metal contents in the SPM, bank and bottom sediments of the Keum Estuary have been compared with other coastal areas around the Korean Peninsula and around the world (Table II). The SPM is enriched in all the analyzed metals compared with the bottom and nearby bank sediments, most significantly in copper and zinc. When compared with other coastal areas, the highly enriched nature of zinc, copper and cobalt in the Keum Estuary becomes evident. The zinc and copper levels, in particular, in the SPM are comparable to those of the Rhine River, one of the most polluted estuaries (16). A smelter located near Janghang, in the vicinity of the inlet, may have been partly responsible for the contamination of the Keum Estuary sediments with these metals.

Table II. Average Metal Contents in the Keum Estuary Sediments as Compared with Sediments from other Areas (in ppm)

Areas		Mn	Zn	Cu	Co	Pb	Ni
Keum Estuary:	SPM	1082	897	283	44	58	40
	Bank	770	410	25	27	49	30
	Bottom	618	313	17	24	43	22
Gyeong-gi Bay, Korea (13)		455	167	12	7	32	26
Jin-hae Bay, Korea (14)			76	34		28	
Southeastern Coastal Area, Korea (15)		499	111	18	13	28	29
Rhine River, Germany (16)			903	192	26	251	164
New York Bight, USA (17)			254	141		170	24
Average Shale (18)		850	95	45	19	20	68

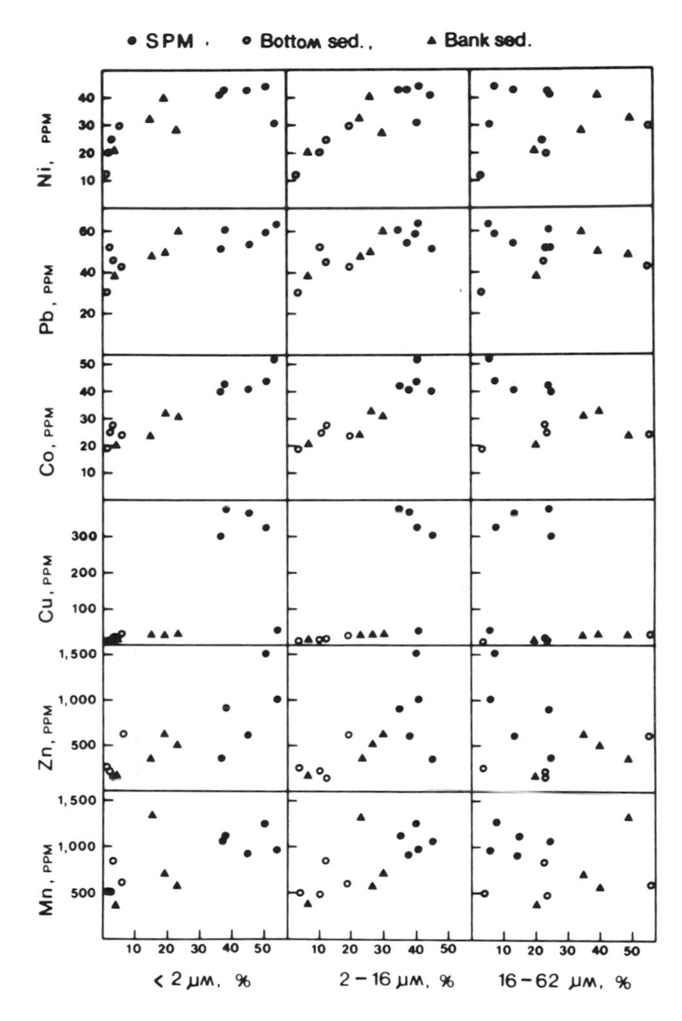

Figure 7. Relationships between the metal contents and the granulometry of sediments.

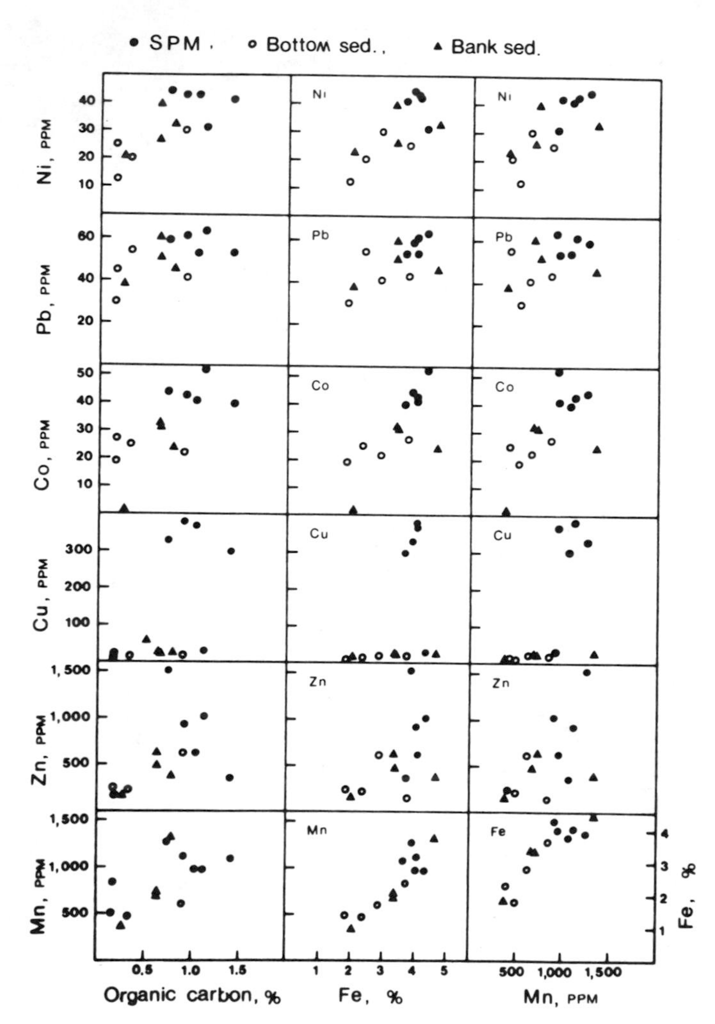

Figure 8. Relationships between the metal contents and the organic carbon, iron and manganese contents of sediments.

SPECIATION OF THE PARTICULATE METALS IN THE SPM

Table III shows the results of the sequential extraction of the six analyzed metals from the SPM in the maximum turbidity zone. The exchangeable fraction, extracted by 1 M ammonium acetate solution (11), consists of the labile adsorbed metals. The reducible fraction was dissolved with 0.25 M hydroxylamine hydrochloride in 25% acetic acid (12) and was regarded as carbonate and ferro-manganese oxide phases. Hydrogen peroxide was used to extract the metal species associated with organic complexes. The residual fraction was digested at about 175°C with a mixed solution of nitric acid, hydrofluoric acid and perchloric acid, and extracted with dilute nitric acid (11).

Most manganese in the Keum Estuary SPM was associated with reducible fraction whereas for zinc, cobalt and nickel the residual fraction comprised more than 70% of the total content. For copper and lead, both the reducible and residual fractions were equally important although the two fractions showed different degrees of contribution to the metal contents. For all the analyzed metals, the exchangeable fraction was of minor importance and was almost nondetectable except manganese, cobalt and copper. The minor contribution of the exchangeable phase in the estuarine particulate metals was also reported from the Jinhae Bay, Korea (19), the Weser Estuary, Germany (20) and the Yellow River, China (21). The contribution of the organic fraction, greatest for copper and non-detectable for lead and zinc, was much lower than that of reducible and residual fractions in the Keum Estuary SPM.

The preponderance of the reducible fraction in the particulate managanese content, as well as the residual fraction in the cobalt case, was also shown in the Amazon and Yukon Rivers (22). The total contents of iron and manganese show a fairly good correlation (Fig. 8) which suggests that the iron and manganese oxides or hydroxides are important metal scavengers in this environment.

The different level of cobalt, lead and nickel contents among the SPM, bank and bottom sediment can be attributed mostly to the difference in grain-size and organic content. For manganese and zinc also, these effects must have played a significant role, as shown by the relationships in Figures 7 and 8. But, for copper, there must have been other distributing mechanisms which have led this metal to be enriched in the SPM. Although the major part of the particulate copper is associated with the reducible and oxidizable fractions, the unusually great enrichment of copper in the SPM has the effect of masking the relationships that the copper content has with the grain-size, organic carbon and manganese contents as shown in Figures 7 and 8.

Hong (14) reported from the Jinhae Bay, Korea, that the oxidizable fraction of the particulate copper constitutes on the average about 59% of the total in the SPM and 48% in the deposited sediment. Kitano and Sakada (23) showed from the Nagoya Port sediment study that a significant amount of copper was trapped in the estuarine sediment in the form of iron sulfide phase. On the other hand, Jouanneau et al. (24) discussed the general impover-

Table III. Partitioning of Particulate Metals in the Keum Estuary SPM

Metal	Station	Exchangeable Fraction (%)	Reducible Fraction (%)	Oxidizable Fraction (%)	Residual Fraction (%)	Total Content (ppm)
Mn	St. A	2.1	68.3	4.7	24.4	1126
	B	5.1	62.9	4.4	27.6	976
	C	3.9	79.7	5.0	11.4	1262
	D	5.6	75.6	5.8	13.0	1073
	E	2.5	75.2	6.3	16.0	975
	(Average)	(3.8)	(72.4)	(5.3)	(18.5)	(1082)
Zn	St. A	0.2	40.8	2.7	56.3	935
	B	--	27.2	0.9	71.9	634
	C	--	11.5	0.7	87.8	1530
	D	--	36.9	2.1	61.0	360
	E	--	6.1	0.7	93.2	1024
	(Average)	(--)	(24.5)	(1.5)	(74.0)	(897)
Cu	St. A	1.8	56.0	16.5	25.7	379
	B	1.9	58.0	15.3	24.8	371
	C	1.2	58.3	15.0	25.5	326
	D	1.1	53.4	12.5	33.0	302
	E	--	12.8	--	87.2	37
	(Average)	(1.2)	(47.7)	(11.9)	(39.2)	(283)
Co	St. A	0.5	14.0	6.5	79.0	43
	B	4.1	21.1	5.7	69.1	41
	C	4.0	20.8	7.3	67.9	44
	D	4.3	20.5	7.8	67.3	40
	E	5.7	16.4	5.7	72.2	52
	(Average)	(3.7)	(18.6)	(6.6)	(71.1)	(44)
Pb	St. A	--	29.4	--	70.6	61
	B	--	31.6	--	68.4	54
	C	--	31.0	--	69.0	59
	D	--	32.9	--	67.1	52
	E	--	32.3	--	67.7	64
	(Average)	(--)	(31.4)	(--)	(68.6)	(58)
Ni	St. A	--	10.2	--	89.8	43
	B	3.5	20.8	--	75.7	43
	C	--	18.4	--	81.6	44
	D	--	22.4	--	77.6	41
	E	0.8	22.5	--	76.7	31
	(Average)	(0.9)	(18.9)	(--)	(80.2)	(40)

ishment of metals associated with SPM in the Gironde Estuary, France, relating its cause to the consumption of organic matter and the subsequent solubilization of associated metals. In the Keum Estuary, it is very difficult at this stage to infer any mechanisms involved in the large enrichment of copper in the SPM. However, it is probable that the organic matter has taken part in these mechanisms.

CONCLUSIONS

The results of the present study demonstrate that the formation of a turbidity maximum is an important phenomenon in the sedimentary processes of fine sediments in the Keum Estuary. Its spatial and temporal evolution depends almost entirely on the estuarine hydrodynamic conditions, which in turn depends upon the balance between the tidal and fluvial regimes. The turbidity maximum is caused by the resuspension of bottom sediments and the flood-dominant current pattern, and is most extensively developed during the spring-tide and low river discharge period. The ebb-dominant current pattern during the high river flow period appears not to be favorable for its development.

The suspended particulate matter of the Keum Estuary turbidity maximum is enriched with heavy metals, as compared with the bottom and bank sediments. Hence, the formation of a turbidity maximum may play a significant role in the metal budget in this environment. Some of this enrichment is closely related to the difference in grain-size and organic load, as in the cases of cobalt, lead, nickel, and possibly manganese and zinc. Other undefined mechanisms may also be involved, especially for copper, resulting in the unusual enrichment in the suspended particulate matter.

Among the four phases of metal associations occurring in the particulate form, the reducible and residual fractions are the dominant phases for all the six analyzed metals in the Keum Estuary turbidity maximum. The exchangeable fraction, which is generally considered as the most bioavailable among the particulate phases, is of minor importance. The metallic association with the oxidizable fraction is also small, except copper, where it forms about 12% of the total content. The small proportion of the oxidizable metal fraction is probably related to the well-mixed character of the Keum Estuary and also to the increased residence time of the estuarine suspended matter.

ACKNOWLEDGEMENTS

This study was supported by the Korea Science and Engineering Foundation. Prof. Dr. S.J. Kim of the Department of Geology, S.N.U., kindly arranged the use of AAS in his department laboratory. I would also like to thank Mr. T.I. Kim, M.S. Choi and Miss H.J. Cha for their help in the field and laboratory work.

REFERENCES

1. Pritchard, D.W. "Estuarine Circulation Patterns", *Proc. Am. Soc. Civ. Engin.,* New York, 81: 1-11 (1955).
2. Pritchard, D.W. "Observations of Circulation in Coastal Plain Estuaries", in *Estuaries,* G.H. Lauff, Ed. (Am. Assoc. Adv. Sci., Publ. No.83), pp. 37-44 (1967).
3. Simmons, H.B. "Field Experience in Estuaries", in *Estuary and Coastline Hydrodynamics,* A.T. Ippen, Ed. (New York: McGraw-Hill, 1966), pp. 673-690.
4. Schubel, J.R. "Estuarine Circulation and Sedimentation", in *The Estuarine Environment, Estuaries and Estuarine Sedimentation* (Washington, D.C.: Am. Geol. Inst., 1971).
5. Allen, G.P. "Etude des processus sédimentaire dans l'estuaire de la Gironde", Thèse Doc. ès-Sciences, Univ. de Bordeaux I (1972).
6. Postma, H. "Sediment Transport and Sedimentation in the Estuarine Environment" in *Estuaries,*G.H. Lauff, Ed. (Am. Assoc. Adv. Sci., Publ. No.83), pp. 158-179 (1967).
7. Schubel, J.R. "Turbidity Maximum of the Northern Chesapeake Bay", *Science* 161: 1013-1015 (1968).
8. Allen, G.P., G. Sauzay, P. Castaing and J.M. Jouanneau. "Transport and Deposition of Suspended Sediment in the Gironde Estuary, France", in *Estuarine Processes,* M. Wiley, Ed. (New York: Academic Press, 1977), vol.2, pp. 63-81.
9. Johnson, M.J. "A Rapid Micromethod for Estimation of Non-Volatile Organic Matter", *J. Biol. Chem.,* Baltimore, 181: 707-711 (1949).
10. Etcheber, H. "Etude de la répartition et du comportement de quelques oligo-éléments métalliques (Zn, Pb, Cu et Ni) dans le complexe fluvio-estuarien", Thèse Doc. 3e Cycle, Univ. de Bordeaux I (1978).
11. Kitano, Y. and R. Fujiyoshi. "Selective Chemical Leaching of Cadmium, Copper, Manganese and Iron in Marine Sediments", *Geochemical J.* 14: 113-122 (1980).
12. Burrows, K.C. and M.H. Hulbert. "Release of Heavy Metals from Sediments: Preliminary Comparison of Laboratory and Field Studies", in *Marine Chemistry in the Coastal Environment,* T.M. Church, Ed. (Washington, D.C.: Am. Chem. Soc., ACS Symp. Ser. 18, 1975), pp. 382-393.
13. Korea Ocean Research and Development Institute, "Coastal Marine Environment: Marine Geology of the Bay of Gyunggi",(1981).
14. Hong, G.H. "A Study on the Geochemistry of Suspended and Bottom Sediments in the Jinhae Bay, Korea", Master's Thesis, Seoul National Univ. (1981).
15. Lee, D.S. and S.J. Han. "The Contents of Heavy Metals in Sediments from the Southeastern Coastal Area of Korea", *J.Oceanol. Soc. Korea* 13: 11-16 (1978).
16. Förstner, U. and G. Müller. "Heavy Metal Accumulation in River Sediments: a Response to Environmental Pollution", *Geoforum* 14: 53-61 (1973).
17. Carmody, D.J., J.B. Pearce and W.E. Yasso. "Trace Metals in

Sediments of New York Bight", *Mar. Poll. Bull.* 4: 132-135 (1973).

18. Turekian, K.K. and K.H. Wedepohl. "Distribution of Elements in Some Major Units of Earth's Crust", *Bull. Geol. Soc. Am.* 72: 175-192 (1961).
19. Hong, G.H., Y.A. Park and K.W. Lee. "Partitioning of Heavy Metals in Sediments from Jinhae Bay, Korea", *J. Oceanol. Soc. Korea* 18: 180-184 (1983).
20. Förstner, U., W. Calmano, K. Conradt, H. Jaksch, C. Schimkus and J. Schoer. "Chemical Speciation of Heavy Metals in Solid Waste Materials (Sewage Sludge, Mining Wastes, Dredged Materials, Polluted Sediments) by Sequential Extraction", in *Proc. Intern. Conf. on Heavy Metals in the Environments, Amsterdam,* pp. 698-704 (1981).
21. Hong, Y.T. and U. Förstner. "Speciation of Heavy Metals in Yellow River Sediment", in *Proc. Intern. Conf. on Heavy Metals in the Environment, Heidelberg,* pp. 872-875 (1983).
22. Gibbs, R.J. "Mechanisms of Trace Metal Transport in Rivers", *Science* 180: 71-73 (1973).
23. Kitano, Y. and M. Sakada. "Bahaviour of Heavy Metals in River Input to the Ocean System", in *Proc. Aquatic Environm. Pac. Region, Taipei* (SCOPE/Academia Sinica, 1978), pp. 66-76.
24. Jouanneau, J.M., H. Etcheber and C. Latouche. "Impoverishment and Decrease of Metallic Elements Associated with Suspended Matter in the Gironde Estuary", in *Trace Metals in Sea Water,* C.S. Wong et al., Eds. (New York: Plenum Press, 1983), pp. 245-263.

ORGANOMETALLICS AND TRACE METALS

CHAPTER 16

BIOLOGICAL MEDIATION OF MARINE METAL CYCLES: THE CASE OF METHYL IODIDE

F. E. Brinckman and G. J. Olson
Chemical and Bioprocesses Group
National Bureau of Standards
Washington, D. C. 20234

J. S. Thayer
Department of Chemistry
University of Cincinnati
Cincinnati, OH 45221

ABSTRACT

Exocellular biogenic metabolites solubilize and methylate heavy metals and may be important in global metal cycling. Methyl iodide, ubiquitous in marine environments, solubilizes bulk metals and refractory binary and ternary metal sulfides, possibly represented by oceanic suspended particulates, producing methylated sulfur coproducts. With tin, an element which forms water stable methyl derivatives, we report that stannous sulfide and chloride react with MeI to produce methyltin(IV) species and tin(IV) as cassiterite, a major tin ore (SnO_2), is solubilized but not methylated by MeI. Dimethyl-β-propiothetin, a common algal metabolite, reacts with cell-permeable I^- to produce MeI and with OH^- to form Me_2S. Based on these results, we construct a model for environmental heterogeneous methylation reactions mediated by intracellular or extracellular MeI and methylsulfonium compounds which may bear on the frequently reported methylmetal(loid) species reported in marine environments. Further **in vitro** and **in vivo** experiments are underway to test this model.

INTRODUCTION

The global distributions and persistence of methylmetals and methylmetalloids in the marine environment (Table I) suggests that in addition to anthropogenic activities, natural cycling mechanisms for the formation and degradation of these compounds occur and should be investigated. The organotins provide an example where anthropogenic introduction of varied tin(IV) species, R_nSnL_{4-n}, bearing long-lived, covalently bonded organic functions R differing in numbers and kinds, along with familiar inorganic, labile geganions, L, is now estimated to be comparable to total natural biogeochemical tin fluxes [1,2],

MARINE AND ESTUARINE GEOCHEMISTRY, Sigleo, A. C., and A. Hattori (Editors)

Table I. Examples of Methylmetals and Methylmetalloids Reported in Marine Environments

Methyl Species	Typical Matrix or Source	Concentration Range, nmol/kg	Refs.
$MeHg^+$	fish tissue	4-110	26
	estuarine sediment	0.3-1	27
$MeAsO(OH)_2$	algae	0.2-0.3	28
Me_2AsOOH	algae	1.3-2.4	28
$MeSbO(OH)_2$	marine and estuarine	0.004-0.1	7
$Me_2SbO(OH)$	waters	0.005-0.03	7
Me_4Pb	tidal flats	0.002-1.1[a]	29
Me_4Sn	polluted harbor waters	0.08-2.5	20
$Me_nSn^{(4-n)+b}$	coral and sea shells	0.3-3.8	30
$Me_nGe^{(4-n)+b}$	oceanic water column	0.1-0.3	9

[a] nmoles m^{-3}; [b] n = 1-2.

On the other hand, the biological mediation of certain molecular forms of heavy elements may control marine concentrations of these species. Based on thermodynamic information for arsenic in aerobic oceanic environments, we would predict that As(V) concentrations should far exceed those of As(III) by many orders of magnitude. However, recent direct measurements indicate that <As(V)TOTAL>/<As(III)TOTAL> is small [3]. Both the relatively long half-lives of As(III) species in aerobic waters [4] and widespread microbial reduction of As(V) species [3,5] appear to account for the environmental situation observed. Moreover, the facile uptake and production of methylarsenical species in both (III) and (V) oxidation states by primary trophic levels creates a pool of the metalloid in oceanic surface waters. Similar considerations certainly also apply to some other heavy elements in the marine environment, of which methyltin [6], methylantimony [7,8], and methylgermanium [8,9] are notable examples. Methylelement formation in the sea might result from one or a combination of three basic chemical processes [10]:

$$E^{++} + CH_3CO_2^- \longrightarrow CO_2 + CH_3E^+ \qquad (1)$$

$$E^{++} + CH_3^- \longrightarrow CH_3E^+ \qquad (2)$$

$$E^{++} + CH_3^+ \longrightarrow CH_3E^{3+} \qquad (3)$$

Though the free-radical process (eq. 1) may involve a cellular metabolite, such as acetate, the methylation step requires photoactivation [10] and will not be considered here. Processes 2 or 3, however, can also involve cellular metabolites, respectively producing methylelements by either metathetical or oxidative methylation mechanisms which are influenced by metal availability and local marine conditions (e.g., pH, pCl, Eh, etc.). Depending upon intrinsic stability of metal- or metalloid-carbon bonds in saline aqueous solutions, prospects for discovering new methyl species in marine cycles may be assessed in terms of relatively simple chemical paths, albeit biologically controlled. Among these, both the carbanionic or carbonium ion reactions, respectively 2 and 3, suggest that field searches for methylthallium, methylphosphorus, methylsilicon, methylplatinum, and the like, will prove profitable as analytical methods become increasingly available [11].

Previously, it was shown [12] that metals and metalloids occurring in higher oxidation states in common mineral forms, including those found in marine environments, could be solubilized by an exocellular metabolite, methylcobalamin, $Me\text{-}B_{12}$. Presumably, the dissolution of otherwise insoluble mineral forms proceeded via carbanionic methylation (Equation 2, above), even though the methylated products might not survive. Wood et al. first described production of $MeHg^+$ via interaction of inorganic mercury with microbially-produced $Me\text{-}B_{12}$ [13]. Thayer has investigated many metal and metalloid oxides with similar success, and shown that refractory, insoluble oxides of Sn(IV), Pb(IV), Au(III), In(III), Sb(V), Bi(V), Mn(IV), Pt(IV), among others [12,14] were solubilized by a complex mechanism sensitive to metal particle surface areas. It is not yet clear whether methylmetal intermediates are alone responsible for dissolution into aqueous media.

Methylcobalamin evidently is not as widely dispersed in the environment as methyl iodide which is a labile, volatile metabolite capable of methylating metals or metalloids [15,16], or at least solubilizing their refractory mineral forms [17]. The high degree of reactivity between aquated MeI and heavy elements observed by ourselves and others is summarized in Figure 1. Moreover, there is a lack of understanding of the origin and wide distribution of MeI in the oceans [18]. White [19] has recently proposed that reactions between dimethylsulfonium compounds prevalent in marine algae, specifically dimethyl-β-propiothetin (DMPT) and iodide (or Cl^- and Br^-) accumulated by the algae may account for global occurrence of MeI (or MeCl and MeBr) in the sea and atmosphere.

On these bases, and in consideration of new evidence gathered in our laboratories, we have undertaken studies on the potential involvement of both exocellular and endocellular biological processes in marine metal cycling, especially with respect to methylmetals and methylmetalloids. In this paper we report on the occurrence of DMPT in estuarine plankton, its reaction with inorganic iodine to produce MeI, and, based on previous work in our laboratory and others, propose a scheme whereby MeI may participate in environmental methylation processes.

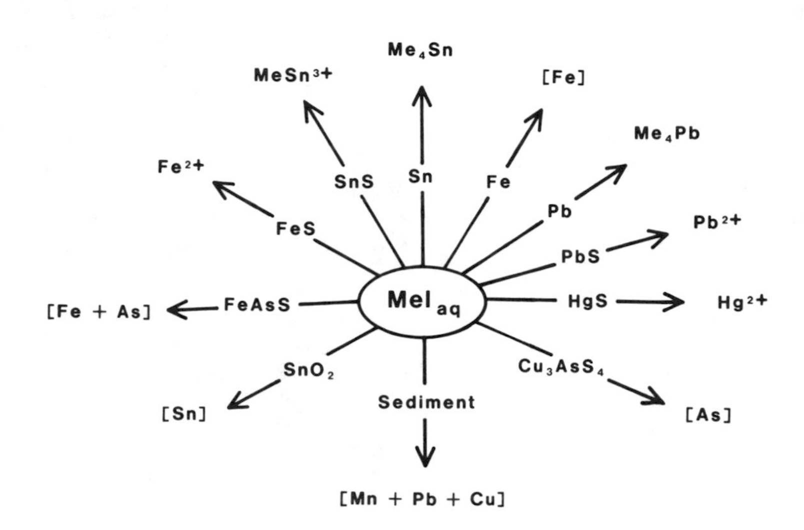

FIGURE 1 Observed heterogeneous reactions between methyl iodide in water and bulk metals (Sn, Fe, Pb), binary metal sulfides, ternary minerals enargite (Cu_3AsS_4) and arsenopyrite (FeAsS), an oxide ore cassiterite (SnO_2), and estuarine sediment obtained from the Chesapeake Bay are compared. Speciated forms of methylated products and the oxidation states of some metal ions are shown; the products in brackets were quantitated by GFAA but not speciated. For the metal sulfides, dimethyl sulfide was the common coproduct, sometimes associated with lesser amounts of dimethyl disulfide.

EXPERIMENTAL SUMMARY

Sample Collection and Assay for DMPT in Plankton

Chesapeake Bay water samples were collected aboard the research vessel Ridgeley Warfield. Samples were collected at approximately 1 m depth while the ship was moving, with the exception of the station at lat. 38° 51' 59" long. 76° 24' 14" which was collected on station at a depth of 15 m, with the temperature 22 °C and salinity 11 ^{0}/oo. Temperature and salinity data were not obtained for samples collected while moving. However, such data was collected at nearby station sites at anchor. Temperatures for these sites ranged from 22-24 °C in June and 22-25 °C in July and salinities ranged from 4-8 ^{0}/oo in June and 4-6 ^{0}/oo in July. Relative fluorescence (chlorophyll) levels were determined using a flow-through fluorometer which was zeroed with deionized water. Plankton were examined for the presence of dimethylsulfonium compounds using the method described by White (19) involving decomposition of DMPT by alkali to produce dimethylsulfide. Plankton were collected by filtration of 300 mL of water samples through 0.8 µm pore size, 47 mm diameter membrane filters (Millipore Corp.). The filters were placed in 9 mL glass vials containing 1.0 mL of either deionized water, KI (1.0 M) or NaOH (1.0 M). Control vials containing clean filters and reagents were also run. Vials were sealed with teflon septum caps and after 24-48 hours incubation at 20-25° C, headspace gases were analyzed for volatile S- and I-containing compounds using a gas chromatograph (Hewlett Packard 5730A) equipped with a flame photometric detector (GC-FPD, refs. 20,21). The glass column was 6' x 1/4" (2 mm i.d.) co-packed with 3% SP2401 and 10% SP2100 on 80/100 mesh Supelcoport (Supelco, Inc.) and was operated isothermally at 25 °C with N_2 as a carrier gas (20 mL min^{-1}) The detector was maintained at 200 °C and the injection port at 150 °C. The flame photometric detector was equipped with a 460 nm "cut-off" filter (Ditric Optics, Hudson, MA) to select for sulfur and iodine emissions.

Reactions of DMPT and MeI

DMPT (Research Plus, Inc.) and KI (1.0 mL of 10 mM aqueous each) were mixed and incubated aerobically in teflon septum capped 20 mL vials. After 18 hours at 30° C headspace gas samples (0.05 - 0.25 mL) were withdrawn using a gas tight syringe and analyzed by GC-FPD (19,20). MeI (20 µL) was mixed with 0.5 g of 50 mesh cassiterite ore (SnO_2) in 2.0 mL deionized water. After 15 days at 30° C approximately 1.0 mL of solution was removed, centrifuged at 12,000 x.g. for 2 min, and soluble tin was determined by graphite furnace atomic absorption spectrophotometry.

RESULTS AND DISCUSSION

Marine Algal Sulfur Metabolite In Methylation of Metals

As summarized in Figure 2, we found that DMPT reacts with iodide in water to produce MeI, as predicted by White. Methyl iodide was confirmed at 1.35 ug/mL in headspace gas, but no dimethylsulfide (DMS) was observed (detection limit, 0.1 ng [22]). Addition of 10 µL 2N NaOH to the vial resulted in DMS evolution. A control vial containing DMPT but no iodide or hydroxide showed no MeI (detection limit, 2.5 ng) or DMS by GC-FPD, neither was MeI nor DMS detected above KI or NaOH solutions alone. In these preliminary experiments we also examined SnS (particulate) and aqueous $SnCl_2$ for reactions with DMPT. No volatile methyltin products were detected by GC-FPD (detection limits < 1 ng for Me_4Sn [22]), nor were solvated methyltins released by borohydride treatment observed using purge-and-trap GC-FPD [20,21]. After 5 days at 30° C, no methyltin species formed, but dimethylsulfide (DMS) evolved from all samples containing DMPT treated with hydride, consistent with well-characterized [19] cleavage of DMS from DMPT by nucleophilic hydroxyl ion. Moreover, methyltin products were not detected in a mixture of DMPT, KI, SnS (1.5 mg), though MeI was produced (1.42 µg/mL) in headspace gas. This is surprising in view of direct reaction observed between MeI and SnS to produce aquated $MeSn^{3+}$ [16]. Presumably, particulate SnS unfavorably competes with dissolved MeI in the presence of unreacted DMPT or its demethylated acidic product [19]. The comparable result with cellular DMPT in vivo remains to be tested for tin and other metal compounds.

Methyl iodide also undergoes reaction with Sn(IV) in its common oxide ore, cassiterite SnO_2, to solubilize tin although no methyltins form. In those experiments cassiterite ore in rock was aerobically incubated with MeI in water. After 15 days at 30° C, the vial contained 0.021 µg/mL soluble tin as determined by graphite furnace AA, whereas a control vial containing cassiterite in water with no MeI showed no detectable (<0.010 µg/mL) soluble tin.

The occurrence of DMPT in Chesapeake Bay plankton and its correlation with chlorophyll levels was also investigated. The Chesapeake Bay plankton trapped on filters evolved DMS after treatment with NaOH (Table II). There was not a strong correlation between _in vivo_ chlorophyll levels and DMS production. However, the sample containing the highest chlorophyll level on each of the two sampling trips evolved the greatest amount of DMS. Evolution of small amounts of DMS from water-treated control filters from two sites may reflect enzymatic cleavage of DMPT [25] or possibly bacterial protein decomposition during incubation. Challenger and coworkers

[23,24] and, later, White [19] showed that DMS is evolved from DMPT upon treatment of the latter compound with alkali. These investigations showed that several marine and some fresh water algae contained DMPT by this method [19,24].

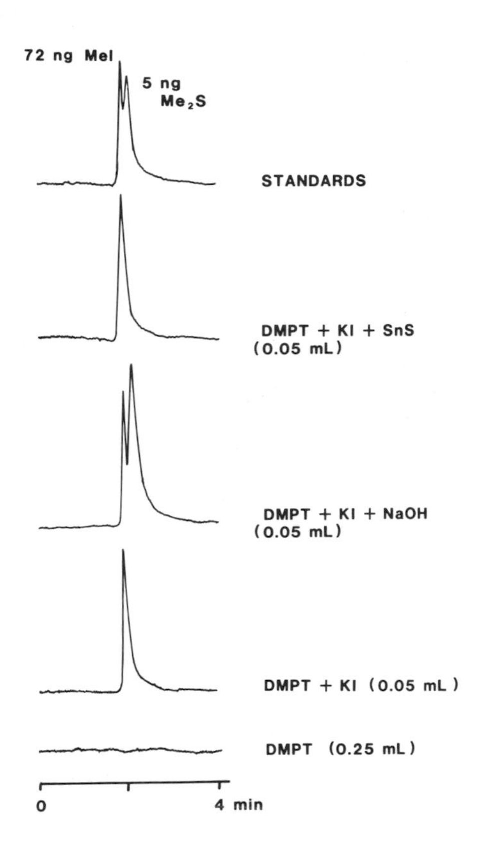

FIGURE 2 Gas chromatograms showing the production of methyl iodide and dimethyl sulfide from aqueous DMPT solutions (5 mM) spiked with potassium iodide (5 mM final concentrations), NaOH (10 µL of 2 N), and stannous sulfide (1.5 mg). Numbers in parentheses indicate volume of headspace gas injected into gas chromatograph equipped with element-selective flame photometric detector.

Table II. Release of DMS from Chesapeake Bay Plankton[a]

Date	Location	DMS Evolved (ng/vial)		Relative fluorescence[c]
		NaOH treated	H_2O control	
June 1984	39° 08' 44" 76° 25' 20"	1.36	ND[b]	80
	39° 04' 40" 76° 23' 35"	0.91	ND	45
	38° 54' 30" 76° 24' 58"	22.4	2.1	90
	38° 51' 59" 76° 24' 14"	10.0	0.84	26
July 1984	39° 09' 47" 76° 25' 26"	0.22	ND	10
	38° 57' 50" 76° 27' 17"	0.84	ND	17
	39° 03' 36" 76° 23' 34"	23.6	ND	26
	39° 10' 43" 76° 26' 23"	0.35	ND	4
Blank filter		ND	ND	-

[a] samples collected from just north of Annapolis Bay Bridge to mouth of Choptank River at 1 m depth.

[b] ND denotes not detectable. Detection limits were 0.1 ng

[c] *in vivo* chlorophyll fluorescence (365 nm excitation) in water samples, arbitrary units

A suggested methylation sequence of iodine and tin by metabolites of marine algae is depicted in Figure 3. Our preliminary findings show that DMPT does in fact react with membrane-permeable iodide to produce MeI which can in turn abiotically or exocellularly react with Sn^{2+}(aq) or SnS(solid) to produce $MeSn^{3+}$(aq) and DMS (16). We have no evidence yet for direct production of methyltins by isolated DMPT, and so can make no deductions about endo- or exocellular metal methylations directly involving this important metabolite. These studies are underway, as are related investigations for methylation of cell-bound tin by algal metabolic methyl iodide and its methylsulfonium precursors.

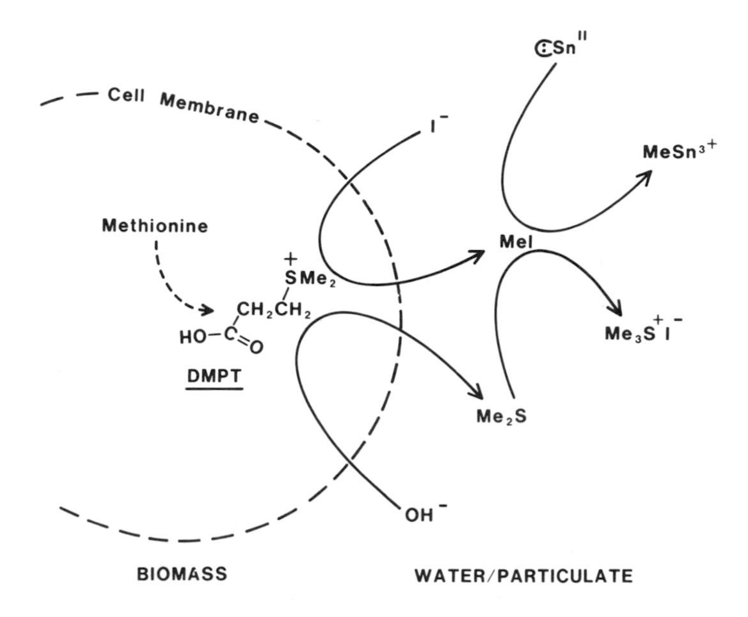

FIGURE 3 Model proposed for either **endo-** or **exo**-cellular methylation of metals and metalloids involves initial production of methyl iodide via reaction between DMPT and membrane-permeable iodide followed either by **intra**- or **extra**-cellular transfer of methyl carbonium ion to nucleophilic center. Shown here is **exo**-cellular methylation of stannous ion following excretion of precursor MeI from a cell. Direct methylation by the DMPT metabolite of a nucleophile is possible but not yet demonstrated. Removal of "active" methyl by environmental sulfur compounds (Me_2S) and OH- probably strongly competes with metal(loid) methylations.

CONCLUSIONS

In view of the propensity for aquated MeI to either methylate or solubilize bulk metals, refractory minerals, and dissolved metal ions at suprisingly rapid rates [16,17], summarized in Figure 1, additional considerations of the relation between biogenic molecules and the suspended mineral particulate load in oceanic water columns must be made. If, as is modelled in Figure 3, sufficient production rates of exocellular MeI occur in favorable marine or estuarine locales, then we may in fact better understand the remarkable ubiquity and uniform concentrations of the methylelements cited in Table I. DMPT apparently occurs in estuarine plankton (this study) as well as in fresh water and marine algae (Table II). Additionally, the solubilization by MeI of many metals or metalloids noted in Figure 1, though not producing stable methyl derivatives, could be significant to their marine cycling and bioavailability.

ACKNOWLEDGEMENTS

We thank Cheryl Matthias, University of Maryland, and Randy Loftus, Chesapeake Bay Institute, for technical assistance. We also thank the University of Maryland for making shiptime available to us aboard the R/V Warfield. This work was supported in part by the Office of Naval Research.

REFERENCES

1. Byrd, J. T. and M. O. Andreae. "Tin and methyltin species in seawater: concentrations and fluxes," Science 218: 565-569 (1983).
2. Brinckman, F. E. "Environmental organotin chemistry today: experiences in the field and laboratory," J. Organometal. Chem. Library 12: 343-376 (1981).
3. Andreae, M. O. "Arsenic speciation in seawater and interstitial waters: the influence of biological-chemical interactions on the chemistry of the trace element," Limnol. Oceanogr. 24: 440-452 (1979).
4. Tallman, D. E. and A. U. Shaikh. "Redox stability of inorganic arsenic(III) and arsenic(V) in aqueous solutions," Anal. Chem. 52: 196-199 (1980).
5. Johnson, D. L. "Bacterial reduction of arsenate in sea water," Nature 240: 44-45 (1972).
6. Eisler, R. Trace Metal Concentrations in Marine Organisms, Pergamon Press, New York (1981).
7. Andreae, M. O., J. F. Asmonde, P. Foster, and L. Van't dack. "Determination of antimony(III), antimony(V), and methylantimony species in natural waters by atomic absorption spectrometry with hydride generation," Anal. Chem. 53: 1766-1771 (1981).
8. Andreae, M. O. and P. N. Froelich. "Arsenic, Antimony, ad Germanium biogeochemistry in the Baltic Sea," Tellus 36B: 101-117 (1984).
9. Hambrick, G. A., P. N. Froelich, M. O. Andreae and B. L. Lewis. "Determination of methylgermanium species in natural waters by graphite furnace atomic absorption spectrometry with hydride generation," Anal. Chem. 56: 421-424 (1984).
10. Brinckman, F. E., G. J. Olson and W. P. Iverson. "The production and fate of volatie molecular species in the environment: metals and metalloids," in Atmospheric Chemistry E. D. Goldberg, Ed. (Berlin: Springer-Verlag, 1982), pp. 231-249.

11. Thayer, J. S. and F. E. Brinckman. "The biological methylation of metals and metalloids," in Advances in Organometallic Chemistry, F.G.A. Stone and R. West, Eds., Vol. 20, (New York: Academic Press, 1982), pp. 313-356.
12. Thayer, J. S., "Demethylation of methylcobalamin: some comparative rate studies," in Organometals and Organometalloids: Occurrence and Fate in the Environment, F. E. Brinckman and J. M. Bellama, eds., (Washington, D.C.: American Chemical Society, 1978), pp. 188-204.
13. Wood, J. M., F. S. Kennedy, and C. G. Rosen. "Synthesis of methyl-mercury compounds by extracts of a methanogenic bacterium," Nature 220: 173-174 (1968).
14. Thayer, J. S. "The reaction between lead dioxide and methylcobalamin," J. Environ. Sci. Health A18: 471-476 (1983).
15. Craig, P. J. and S. Rapsomanikis. "A new route to tris(dimethyltin sulfide) with tetramethyltin as co-product; the wider implications of this and some other reactions leading to tetramethyl-tin and -lead from iodomethane," J. Chem. Soc. Chem. Comm. 1982: 114 (1982).
16. Manders, W. F., G. J. Olson, F. E. Brinckman and J. M. Bellama. "A novel synthesis of methyltin triiodide with environmental implications," J. Chem. Soc. Chem. Comm. 1984: 538-540 (1984).
17. Thayer, J. S., G. J. Olson, F. E. Brinckman. "Iodomethane as a potential metal mobilizing agent in nature," Environ. Sci. Technol. 18: 726-729 (1984).
18. Rasmussen, R. A., M. A. K. Khalil, R. Gunawardena and S. D. Hoyt. "Atmospheric Methyl Iodide (CH_3I)," J. Geophys. Res. 87: 3086-3090 (1982).
19. White, R. H. "Analysis of dimethyl sulfonium compounds in marine algae," J. Mar. Res. 40: 529-536 (1982).
20. Jackson, J. A., W. R. Blair, F. E. Brinckman and W. P. Iverson. "Gas chromatographic speciation of methylstannanes in the Chesapeake Bay using purge and trap sampling with a tin-selective detector," Environ. Sci. Technol. 16: 110-119 (1982).
21. Olson, G. J., F. E. Brinckman, and J. A. Jackson. "Purge and trap flame photometric gas chromatography technique for the speciation of trace organotin and organosulfur compounds in a human urine standard reference material (SRM)," Int. J. Environ. Anal. Chem. 15: 249-261 (1983).
22. Parris, G. E, W. R. Blair, and F. E. Brinckman. "Chemical and physical considerations in the use of atomic absorption detectors coupled with a gas chromatograph for determination of trace organometallic gases," Anal. Chem. 49: 378-386 (1977).
23. Challenger, F. and M. I. Simpson. "Studies on biological methylation. Part XII. A precursor of the dimethyl sulfide evolved by _Polysiphonia fastigiata_. Dimethyl-2-carboxyethylsulphonium hydroxide and its salts," J. Chem. Soc. 1948: 1591-1597 (1948).

24. Challenger, F., R. Bywood, P. Thomas, and B. J. Hayward. "Studies on biological methylation. XVII. The natural occurrence and chemical reactions of some thetins," Arch. Biochem. Biophys. 69: 514-523 (1957).
25. Cantoni, G. L., and D. G. Anderson. "Enzymatic cleavage of dimethylpropiothetin by _Polysiphonia lanosa_," J. Biol. Chem. 222: 171-177 (1956).
26. Westoo, G. "Determination of methylmerury compounds in foodstuffs," Acta Chem. Scand. 20: 2131-2137 (1966).
27. Andren, A. W. and R. C. Hariss. "Methylmercury in estuarine sediments," Nature 245: 256-257 (1973).
28. Andreae, M. O. and D. Klumpp. "Biosynthesis and release of organoarsenic compounds by marine algae," Environ. Sci. Technol. 13: 738-741 (1979).
29. Harrison, R. M. and D. P. H.Laxen. "Natural source of tetraalkyllead in air," Nature 275: 738-739 (1978).
30. Braman, R. S. and M. A. Tompkins. "Separation and determination of nanogram amounts of inorganic tin and methyltin compounds in the environment," Anal. Chem. 51: 12-19 (1979).

CHAPTER 17

TIN METHYLATION IN SULFIDE BEARING SEDIMENTS

C.C. Gilmour
J.H. Tuttle
J.C. Means
University of Maryland
Center for Environmental & Estuarine Studies
Chesapeake Biological Laboratory
Solomons, Maryland 20688-0038 USA

ABSTRACT

Metals in anaerobic sulfide rich sediments are often considered unavailable for biological transformation due to the extreme insolubility of metal sulfides. We found that sediment tin methylation rates are highest under anaerobic conditions. The potential for methylation of Sn was examined in anoxic, sulfidic Chesapeake Bay sediment slurries spiked with 50 mg/l $SnCl_4$. Over 61 days, live slurries produced ug/l levels of mono- (MMT) and dimethyltin (DMT), determined as their corresponding hydride derivatives by GCMS. Whole sediments incubated with inorganic tin also produced ug/l quantities of MMT and DMT in 3 weeks, even at sulfide concentrations in excess of inorganic tin concentration. Production of MMT in sediments was significantly correlated with numbers of both sulfate reducing and sulfide oxidizing bacteria. Desulfovibrio spp. isolated from the sediments were able to methylate Sn in culture medium at rates similar to whole sediment slurries. Sulfate reducing organisms seem to have a dominant role in sediment tin methylation, possibly allowing the mobilization of tin sulfides. Although the relative importance of tin transformation in sulfidic and nonsulfidic sediments is unknown, there is potential for production of highly toxic organotin species in sulfide bearing aquifers and in estuarine and marine sediments.

INTRODUCTION

The methylation of metals in the environment yields organometals of significantly greater toxicity than their inorganic counterparts [1]. These methylation reactions may be biotic, abiotic, or a combination of the two. While methylation of certain heavy metals, e.g. mercury, has been studied in detail, tin methylation has only recently been examined. Methyltins are extremely toxic to mammals [2-5] and aquatic organisms [6], while inorganic tin hardly affects either [7, 8].

MARINE AND ESTUARINE GEOCHEMISTRY, Sigleo, A. C., and A. Hattori (Editors)

Tin methylation occurs in slurries prepared from both freshwater and estuarine sediments [9, 10]. However, the significance of biomethylation relative to abiotic methylation is unknown. Likewise, the inorganic tin species undergoing methylation have not been determined; and the optimal physiochemical conditions, e.g. redox potential and salinity, are unclear.

Methyltin concentrations in estuarine water increase with the surface to volume ratio of the estuary, suggesting a benthic origin for methyltins [11]. The distribution of methyltins in sediments varies with sediment type. Monomethyltin (MMT) predominates in anoxic, polluted sediments, while trimethyltin dominates in aerated, non-impacted sediments [12]. This distribution pattern suggests that inorganic tin methylation is dominant in anoxic sediments while degradation of higher organotins, such as tributyltin, is the dominant aerobic process. MMT is the major methyltin measured in Chesapeake Bay sediments, with levels up to 0.8 ng/gm dry weight of sediment [12]. MMT is the first product of stepwise inorganic tin methylation. Although each subsequent methylation step is thermodynamically less favorable, a biological tin cycle based on the stepwise microbial methylation of inorganic tin to mono-, di-, tri-, and tetramethyltin has been proposed [13].

Unlike tin, mercury methylation has been studied extensively, more so than any other environmental metal methylation process. Mercury methylation may occur abiotically [14, 15, 16], but is intensified by bacterial activity [17]. Methylation is favored under anoxic conditions [18, 19] even though sulfide, commonly found at high concentrations in anoxic estuarine and marine sediments, decreases methylation rates [17, 18, 20], probably due to formation of unavailable HgS. _In situ_ studies with estuarine sediments have shown that the Hg methylation process occurs more rapidly in deeper sediments, which are anoxic and sulfidic, relative to more oxidized surface sediments [21].

Sulfide and sulfide-containing organics are known to cause redistribution reactions among organometallics [22-26], and to catalyze the methylation of inorganic compounds by methyl iodide [27]. However, organic thiols inhibit the reaction of Hg with methylcobalamin [28]. Sulfides may cause the redistribution of trimethyltin to tetra- and dimethyltin (DMT) through an organometallic sulfide intermediate [29], and catalyze similar reactions with methylmercury [25, 26] and trimethyllead [30, 31]. The trimethyltin redistribution reaction occurs in anaerobic estuarine sediments both abiotically and biotically, with an enhanced rate in biologically active sediments [32]. Sulfide and a microbial carbon source both increase the biological rate, although sulfide does not catalyze the production of tetramethyltin in sterile sediments.

In one proposed mechanism for Sn methylation, reduced Sn(II) is the species which undergoes methylation by the biological methylating agent, methylcobalamin [33]. Available data on both Hg and Sn strongly suggest that methylation occurs most rapidly in anoxic sediments as the result of bacterial activity. In this report we show that this is indeed the case for tin methylation in the estuarine environment, and demonstrate the interrelation of tin methylation and the chemistry and bacteriology of the sulfur cycle.

METHODS

Analysis of Methyltins

Methyltins were analyzed as their volatile hydrides with a Hewlett-Packard Model 5985B GCMS system equipped with an HP Model 7675A purge and trap (P/T) sampler [34]. Methyltins were derivitized using $NaBH_4$ at pH 8.0, trapped on Tenax-GC, and desorbed onto the GC column (3% SP2401 copacked with 10% SP2100 on 80/100 mesh Supelcoport) by ballistic heating to 200^oC. The GC separation was performed isothermally at 35^oC. Identification and quantitation of the methylstannanes were performed by monitoring selected ions (nominal m/e 116, 118, 120, 133, 135, 148, 163, 165, and 178). These nine ions were selected from isotope multiplets of the four methylstannanes [35]. Ions 116, 118, and 120 represent inorganic tin ions, while ions 133 and 135 represent tin plus one CH_3 group; 148 represents $Sn(CH_3)_2$ ions, etc. Detection limits were 0.4 ng MMT and 0.07 ng DMT, or 0.08 ppb and 0.02 ppb, respectively, in a 5 ml sample.

In some experiments, the methyltin trapping system was modified and diethyltin (DET) added to samples as an internal standard. Rather than trapping the methylstannanes on a Tenax-GC trap, the organotin hydrides were purged directly onto the head of a chromatographic column packed with 0.2% carbowax 1500 on 80/100 Carbopak C which was held at -40^oC. After a 5 min. purge, the GC oven temperature increased at 15^oC/min for 7 min, then 30^oC/min until the final oven temperature, 150^oC, was reached. Quantitation and identification by mass spectrometry was as described above, except that the selected ions monitored were changed to detect the DET internal standard (nominal m/e 116, 118, 120, 133, 149, 150, 165). Detection limits with this system were near 5 pg for each methyltin. Representative standard and sample chromatograms are shown in Figure 1.

Aerobic vs. Anaerobic Methylation by Sediment Slurries

Slurries of sediments from two sites in Chesapeake Bay were incubated with inorganic anhydrous $Sn(IV)Cl_4$ under aerobic,

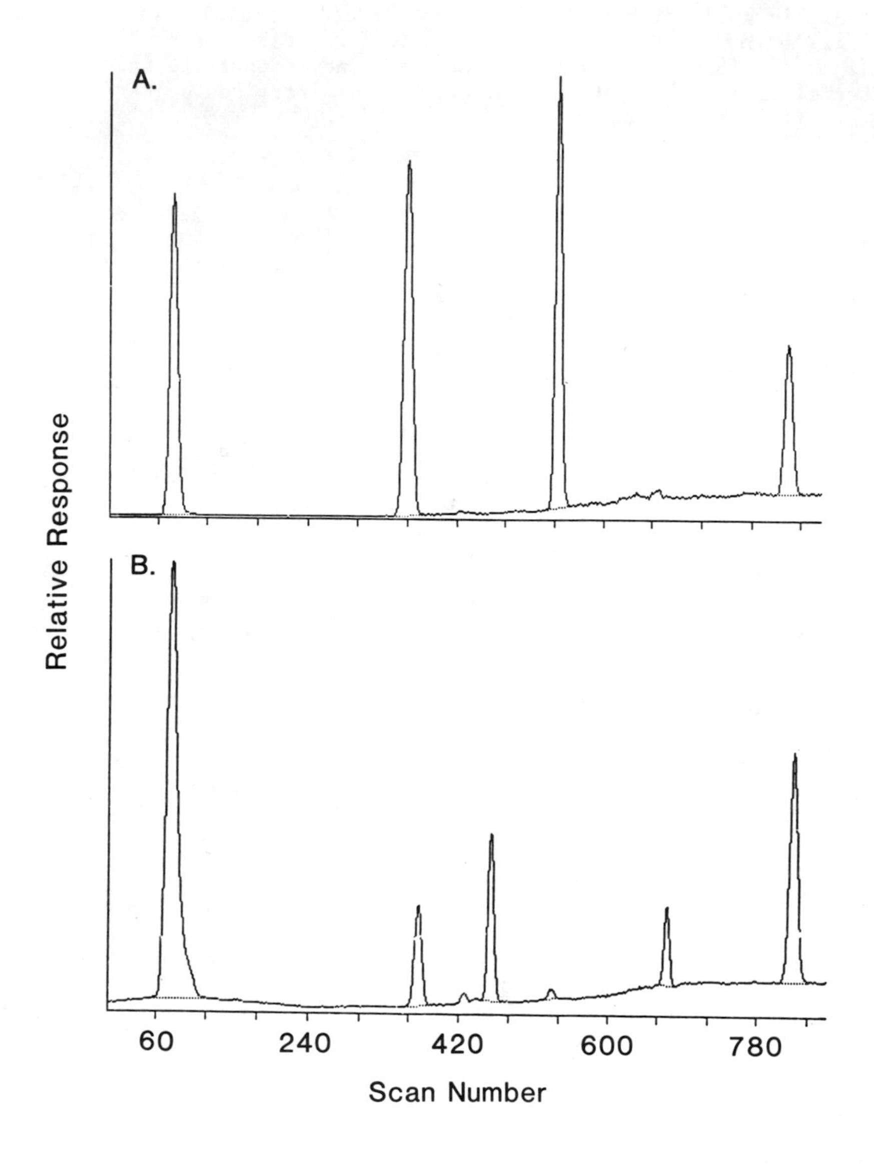

Figure 1. Total ion chromatograms, P/T analysis of methyltins. a. Standard, 10 ng each MMT, DMT, TMT, plus 7 ng DET. b. Sediment, 3 g, after 3-week incubation with inorganic tin.

anaerobic, or microaerophilic conditions in order to assess the potential for tin methylation under different redox conditions. Sediments were collected from highly contaminated Baltimore Harbor, site 1, with a Van Veen grab and held covered in 10 gal. containers until used. Site 2, at the mouth of the Patuxent River, was chosen as a relatively pristine location. Sediments were collected from this site with a 5 cm I.D. Lexan core tube and used immediately. Baltimore Harbor sediments were primarily silty clays, while Patuxent sediments were sandy silts. Both sediments were black, sulfidic, and highly reduced (-150 mV) below 1 cm depth. Slurries consisted of a 1:10 ratio of sediment to culture medium (composed of 2.5 g casamino acids, 0.5 g yeast extract, 1.0 g glucose, 10.0 g NaCl, 2.8 g $MgSO_4 \cdot 7H_2O$, 0.5 g NH_4NO_3, 0.7 g K_2HPO_4, and 0.3 g KH_2PO_4 per liter of distilled water and adjusted to pH 7.0). For each site, three flasks at each redox condition contained sediment slurries. Aerobic slurries were maintained in flasks with loose stoppers and microaerophilic slurries in flasks sealed under nitrogen. To prepare strictly anaerobic slurries, the above medium plus 1 g thioglycolate, 0.03 g cysteine·HCl, and 1mg resazurin per liter was prereduced. The anaerobic mixed slurries were then sealed under oxygen free nitrogen. All cultures were incubated in the dark with shaking (150 rpm) at 30°C. Aliquots of mixed slurry were periodically removed from the cultures and analyzed for methyltins.

Sediment slurries in prereduced culture medium were incubated anaerobically with inorganic tin and assayed for methyltins periodically during a two month incubation in order to examine anaerobic methyltin production over time. Medium GAM consisted of, per liter of medium: 2.5 g casamino acids, 0.5 g yeast extract, 1.0 g dextrose, 0.5 g NH_4NO_3, 2.1 g 31-N-morpholino propanesulfonic acid (MOPS), 1.0 g Na thioglycolate, 10 g NaCl, 2.8 g $MgSO_4 \cdot 7H_2O$, 0.3 g KCl, 0.03 g cysteine, 0.015 g KH_2PO_4, and 0.001 g resazurin. The pH was set at 7.0. The medium was prereduced by boiling under O_2-free N_2. Tin chloride was added (50 ppm for site 1, 10 ppm for site 2) after 70 ml of medium was dispensed into gassed-out, 100 ml serum bottles. The bottles were then sealed and autoclaved. After cooling, triplicate cultures were prepared by adding, under O_2-free N_2, about 7 g of wet sediment to each serum bottle. One of the triplicates, which served as a control, was sterilized by autoclaving the sediment/medium mixture. The cultures were incubated stationary at 27°C. Resazurin in the slurries remained colorless throughout the incubation in all bottles, indicating anaerobic conditions with an Eh below -80mv.

Bacterial isolates from the above sediments were then tested for their ability to methylate $SnCl_4$ in culture medium without added sediment. Sulfate reducing bacteria (SRB) were isolated according to Postgate [36] using the agar shake tube method. Other anaerobes were isolated using shake tubes, Hungate roll tube technique [37], and agar plates incubated in an anaerobic atmosphere. Isolates were tentatively identified using methods and criteria outlined in Bergey's Manual [38]. Fifteen cultures were tested for their ability to produce methyltins in GAM and SRM with 10 mg Sn/l added as $SnCl_4$. SRM medium contained per liter of distilled water: 3.5 g sodium lactate, 0.1 g yeast extract, 2.0 g $MgSO_4 \cdot 7H_2O$, 1.0 g Na_2SO_4, 0.05 g $FeSO_4$, 0.1 g Na thioglycolate, 0.1 g ascorbic acid, 0.015 g KH_2PO_4, 2.1 g MOPS, 0.5 g KCl, 1.0 g NH_4Cl, and 10.0 g NaCl. The pH was set at 7.5. Three replicates, plus a growth control lacking added tin, of each media type were inoculated with each isolate. Uninoculated media controls included medium alone, added tin, added Na_2S (1mM), and both Na_2S and tin. Cultures and uninoculated controls were incubated under O_2-free N_2 (SRM) or CO_2 (GAM) for two weeks, and then analyzed for methyltins.

Methylation in sediment cores

In order to measure _in situ_ methyltin production from inorganic tin and to study the interactions between the sulfur cycle and tin methylation, methyltin production and variety of other parameters, including sulfide concentration and bacterial numbers, were measured with depth in relatively undisturbed sediments. Seven replicate cores (5 cm I.D.) were taken from each of three sites along a mid-salinity transect of Chesapeake Bay. Station D was in the deep channel at 42 m water depth. This sediment was black silty-clay, and contained methane gas bubbles. Station S was on the more shallow western shelf at 14 m depth. Sediments at station S were dense, black clay, and contained no gas bubbles. Station G was near western shore in 7.5 m of water. Station G sediments were silty-sand and extremely bioturbated. Mini-cores (10 ml vol.) were removed from each of the larger cores through ports in the core tube at 2.5 cm intervals. Mini-cores were sealed and kept at _in situ_ temperature. Of the seven replicate cores recovered at each site, one was used to measure background methyltin concentrations; mini-cores from the second were incubated with added tin to determine methyltin production; and the remaining cores were used to estimate bacterial numbers and to measure the following parameters: pH, Eh, wet and dry weights, sulfate and sulfide concentrations, and bacterial enzyme activities. Mini-cores used to measure methyltin production were line-injected with 100 ug Sn as anhydrous $SnCl_4$, and incubated at room temperature (about 22°C) inside an anaerobic glove box.

Methyltin analysis was performed using the cryogenic trapping method as described above, with 1 to 3 grams of sediment as the sample. Sediments spiked with methyltins indicated near total recovery of amended organotins by this procedure.

Bacteria were enumerated using a most probable number technique [39]. Five growth media were used to estimate numbers of aerobic heterotrophs, sulfide-oxidizing autotrophs, anaerobic heterotrophs, and SRB which utilize acetate or lactate as carbon and energy sources. The growth medium for aerobic heterotrophs consisted of modified half strength Nelson's medium [40]. Sulfide-oxidizing bacteria were counted in medium consisting of sulfide containing agar slants overlayed with broth. The broth contained, per liter of distilled water, 0.2 g $NaHCO_3$, 0.15 g $CaSO_4$, 0.1 g KCl, 2.1 g MOPS, 0.01 g phenol red, 0.1 g KH_2PO_4, 0.5 g NH_4Cl, 0.25 g $MgSO_4 \cdot 7H_2O$, 10 g NaCl, 5.0 g $Na_2S_2O_3 \cdot 5H_2O$, 1 ml trace metal solution, and 1 ml vitamin solution at pH 7.2. The trace metal solution consisted of 2 g $FeSO_4 \cdot 7H_2O$, 80 mg $ZnSO_4 \cdot 7H_2O$, 100 mg $MnCl_2 \cdot 4H_2O$, 60 mgH3BO4, 190 mg $CoCl_2 \cdot 6H_2O$, 20 mg $CuSO_4$, 20 mg $NiCl_2 \cdot 6H_2O$ and 40 mg $Na_2MoO_4 \cdot 2H_2O$ in 1 liter of distilled water adjusted to pH 1.0 with HCl. The vitamin solution contained 20 mg cobalamine, 2 mg biotin, and 20 mg thiamine per liter of distilled water. The vitamin solution was filter sterilized and added separately to the other components after autoclaving and cooling. Thiosulfate was autoclaved separately and added with the vitamins. The media of Laanboek and Pfennig [41] were used to enumerate acetate and lactate utilizing SRB. Anaerobic heterotrophs were enumerated in thioglycolate medium (Difco) amended with estuarine salts. Anaerobic dilution cultures were incubated at room temperature for three weeks.

Redox potential measurements were made using a hand-made Pt micro-electrode with a calomel reference electrode. To measure pH, mini-cores were extruded in an O_2-free atmosphere into polypropylene centrifuge tubes, mixed with 5 ml boiled out distilled water, and allowed to settle. A pH electrode plus a calomel reference electrode were used to determine pH in the water above the settled sediment.

Mini-cores were extruded into sealed centrifuge tubes as above and centrifuged at 3000 X g to obtain porewater. Sulfate concentrations in porewater were determined according to Howarth [42]. H_2S was removed prior to sulfate analysis by bubbling with N_2. Sulfide concentrations were measured using a modified Pachmeyer determination [43].

To measure assimilation of CO_2 by sediment microorganisms, mini-cores were injected with $NaH^{14}CO_2$ and incubated in the dark at <u>in situ</u> temperature for 24 hrs. Ten ul of bicarbonate (10 uCi, specific activity of 1 uCi/umol) was line-injected into mini-cores. Microbial activity was terminated by freezing the cores. To determine assimilated radioactivity, thawed cores were extruded into flasks containing 10 ml 5% trichloroacetic acid, and the mixture was stirred and purged with N_2 for 10 min to drive off unassmilated CO_2. Base (20 ml 4N NaOH) was added and the sample boiled gently for 1 hr. After boiling, the volume was readjusted to 40 ml with distilled water and the samples allowed to settle overnight. Radioactivity in the supernatant, representative of incorporated CO_2 was assayed by liquid scintillation counting (LSC). Quench was corrected by external standard.

Ribulose 1,5 bisphosphate carboxylase (RuBPCase) and phosphoenol pyruvate carboxylase (PEPCase) were assayed according to Tuttle [44] with an amended preassay sample preparation designed for determination of enzyme activity in sediment. For each assay, one mini-core was extruded into a flask containing 100 ml of phosphate buffered saline and the mixture was shaken for 15 min. The mixture was decanted into a graduated cylinder and allowed to settle for 1hr. Portions of supernatant were filtered (0.2 u Nuclepore filter) and the filters then used in enzyme assays [44]. Radioactivity incorporated into organic material was assayed by LSC as described above.

RESULTS

Anaerobic vs. Aerobic Methylation by Sediment Slurries

Sediment slurries from sites 1 and 2 were tested for their ability to methylate added inorganic tin. After six weeks, none of the microaerophilic slurries produced methyltins; one aerobic culture from site 1 produced monomethyltin (MMT) and dimethyltin (DMT) in 41 days; but all anaerobic slurries produced MMT and DMT after only 14 days in site 1 slurries and 22 days in site 2 slurries (Table 1). Trimethyltin (TMT) was also detected in anaerobic cultures at 38 days. Carbon dioxide, methane, and hydrogen sulfide were found in the headspace gases of the anoxic slurries indicating active metabolism of microbes, including methanogens and sulfate reducers.

Time Course of Anaerobic Methylation

Anaerobic slurry cultures from sites 1 and 2 produced MMT and DMT over 61 days, while heat-killed slurries exhibited background levels of MMT, DMT, and TMT which remained nearly constant during the incubation period. MMT and DMT concentrations over time in one Baltimore Harbor slurry culture and control are shown in Figure 2. MMT was the predominant product in all cultures, with DMT production averaging 50 times lower than MMT production. Baltimore Harbor cultures produced MMT at levels up to 9.4 ng MMT/ml slurry, with an average of 6.3 ng at 61 days, and DMT at a maximum concentration of 0.5 ng/ml slurry (Table 2). DMT production decreased in the cultures after 41 days, while MMT levels increased throughout the incubation. TMT was found initially in some slurries, but did not increase with time. Methyltin concentrations in Baltimore Harbor controls

Table 1. Production of methyltins in sediment slurries incubated with 10 mg Sn/L as $SnCl_4$ under different redox conditions at 30°C.

Sample	Redox	Incubation Time (Days) 8	14	22	34	41
Baltimore Harbor (Site 1)	Aerobic	-	-	-	-	M,D
	Microaerophilic	-	-	-	-	-
	Anaerobic	-	-	M,D	M,D	M,D
Patuxent River (Site 2)	Aerobic	-	-	-	-	-
	Microaerophilic	-	-	-	-	-
	Anaerobic	-	M,D	M,D	M,D	M,D

M = Production of monomethyltin
D = Production of dimethyltin

Table 2. Amounts of Methyltins Formed in Sediment Slurry Cultures Incubated at 27°C

	ug/L (as Sn)		% Conversion[a] to:	
Sample	MMT[c]	DMT[b]	MMT	DMT
Baltimore Harbor #1	9.35	0.503	0.0145	0.00096
Baltimore Harbor #2	3.31	0.162	0.0024	0.00028
Patuxent River #1	1.42	0.143	0.0072	0.00088
Patuxent River #2	1.75	0.098	0.0105	0.00043

[a]Baltimore Harbor and Patuxent River cultures were amended with 50 and 10 mg Sn/L as $SnCl_4$.

[b]After 41 days

[c]After 61 days

stabilized below 1 ng MMT/ml and 0.1 ng DMT/ml. Methyltin concentrations in both cultures and controls were much lower in slurries from the relatively pristine Patuxent River, although the pattern of methyltin production was similar to that in Baltimore Harbor sediment slurries. MMT averaged 1.6 ng/ml at 61 days in spiked site 2 slurries, while control concentrations remained constant at about 0.1 ng/ml slurry. DMT in cultures increased from an initial level of 0.01 ng/ml to a maximum of 0.14 ng/ml at 41 days. Although the percent conversion of added $SnCl_4$ was very low (0.002 to 0.015%), slurries from both sites showed at least one order of magnitude increases in MMT and DMT concentrations during the two month incubation.

Methylation by Isolated Microorganisms

Isolated organisms were tested for their ability to methylate inorganic tin in GAM and SRM medium without sediment. Of 15 cultures tested, 8 were sulfate-reducing Gram-negative rods and vibrios, 4 were Gram-negative non-sulfate-reducing rods, and 3 were large Gram-positive, non-sulfate-reducing rods. All were obligately anaerobic except one facultative Gram-negative non-sulfate-reducer. Following incubation at 27°C for 2 weeks, five of the isolates had produced up to 4 ug methyltins/l from 10 mg/l tin (Table 3). Four of these cultures were sulfate-reducing bacteria (SRB), the other was the facultatively anaerobic organism. One of the SRB cultures produced 0.3 ug DMT/l. The methylating SRB (tentatively identified as *Desulfovibio* spp.) produced methyltins only on SRM medium, although they grew well and produced sulfide on GAM medium. The facultative anaerobe did not grow on SRM, and therefore only methylated tin in GAM medium. Uninoculated culture medium controls amended with tin, sulfide or both produced no detectable methyltins.

Methylation in Sediment Cores

Sub-samples of sediment cores incubated with inorganic tin for 3 weeks produced MMT and DMT in every instance. Initial methyltin concentrations and concentrations after incubation are shown with depth for the three cores examined (Fig. 3a-c). The average MMT concentration for the three sites was 33 pg/gm wet weight sediment, while the average DMT concentration was lower, 22 pg/gm wet weight sediment. Most of the mini-core samples contained no detectable TMT, although a few contained TMT at up to 21 pg/gm, with an overall average of 4 pg/gm wet weight. Tin amended sediments from each of the three sites produced MMT, DMT, and possibly TMT in 21 days. MMT was the predominant methyltin product in all three sediments, averaging 1547 pg/gm wet weight for the three sites after 21 days. DMT production was nearly ten times lower, averaging 239 pg/gm, and TMT concentrations were lowest with an average of 13 pg/gm. In general, methyltin concentrations decreased with depth.

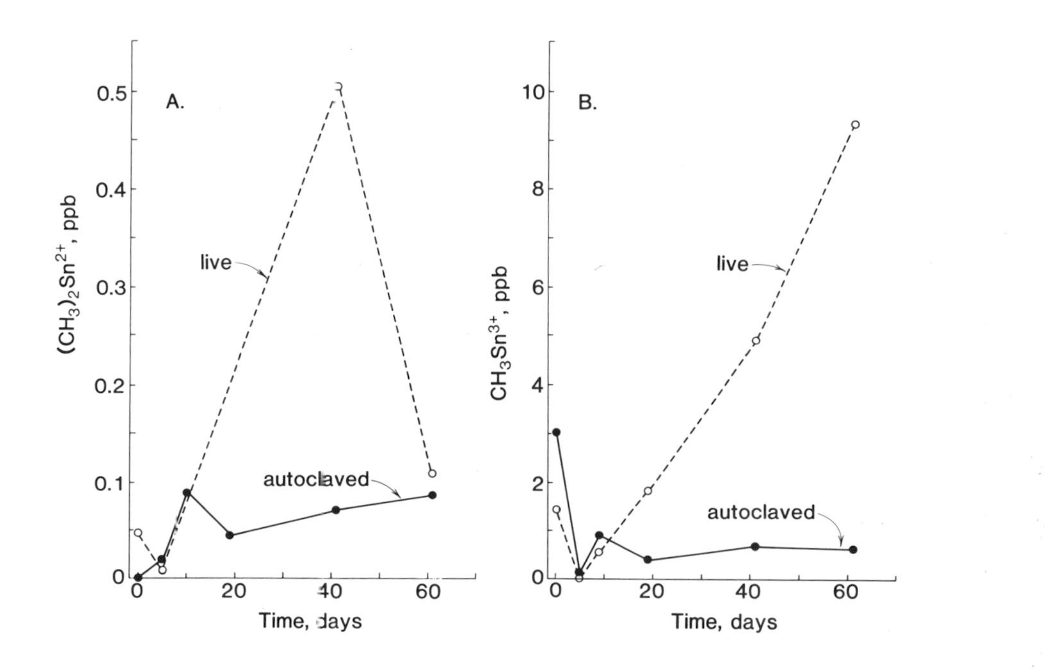

Figure 2. Methyltin concentrations in Baltimore Harbor sediment slurries amended with 10 ug/l Sn as $SnCl_4$ over 61 days. Live (0), killed (●). Left - MMT; Right - DMT.

Table 3. Production of Methyltins from Inorganic Sn, 10 mg/L, by Isolated Bacteria in Anaerobic Culture Media at 27°C

	Methyltin Concentration, ug/L as Sn				
	SRM		GAM		
Isolate	MMT	DMT	MMT	DMT	Bacterial Type
1	0	0	0	0	Sulfate Reducing Bacteria
2	2.4	0.3	0	0	"
3	3.0	0	0	0	"
4	4.0	0	0	0	"
5	1.7	0	0	0	"
6	0	0	0	0	"
7	0	0	0	0	"
8	0	0	0	0	"
9			0	0	Anaerobic Heterotrophs
10			0	0	"
11			0	0	"
12			1.5	0.1	Facultative Heterotrophs
13			0	0	Clostridia
14			0	0	"
15			0	0	"

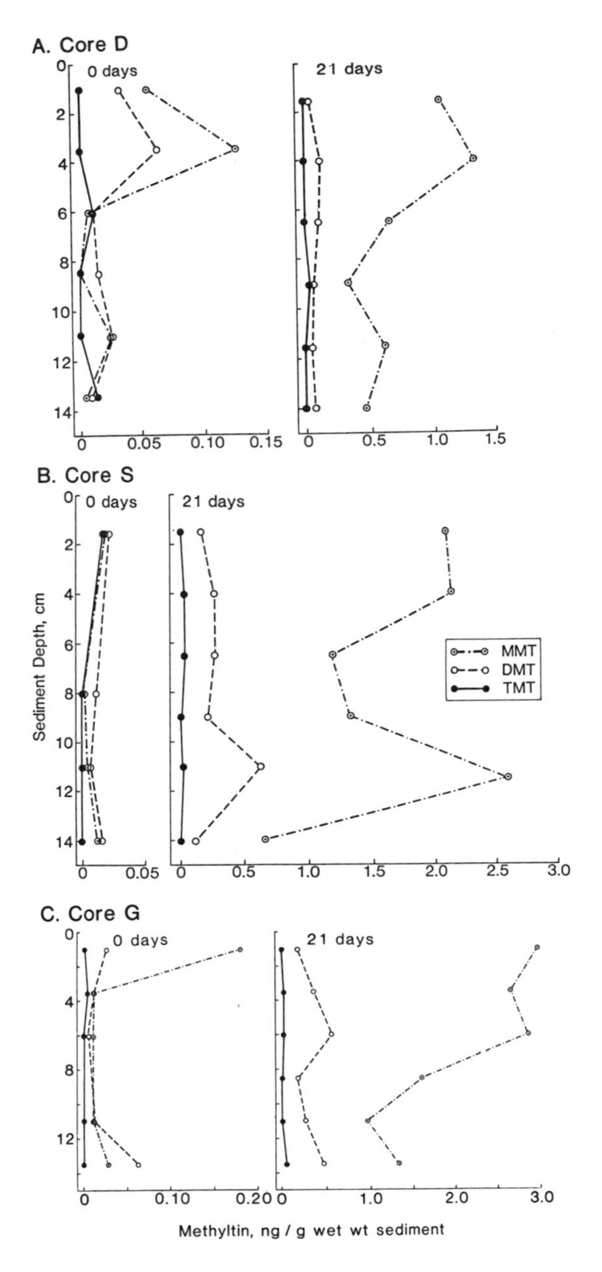

Figure 3. Methyltin concentrations with depth in sediments from three sites on a transect of Chesapeake Bay; background values (0 days) and concentrations after 3-week incubation with 10 ug Sn as Sn Cl_4/l (21 days). Top - 42 m station D; middle - 14 m station S; bottom - 7.5 m station G.

Redox potential and pH were nearly constant with depth at all three sites. Sediments were anoxic throughout, except site G which was heavily bioturbated and, although predominantly anoxic, contained regions with Eh up to +200 mV. A nearly constant pH of about 8 was found in all samples. Thus there were no significant correlations between redox potential or pH and methyltin production.

Average sulfide and sulfate concentrations varied considerably between sites (Table 4). Comparisons between sites showed a strong negative correlation between average sulfide concentration and methyltin production. Pore water sulfate concentrations showed no relationship to methylation either within or between cores. Overlying water at site D, the sediment exhibiting lowest methyltin production and highest sulfide concentration, had 23 ppt salinity, higher than the 16 ppt levels at the other two sites.

Table 4. Average methyltin, sulfide, & sulfate concentrations and salinity for Sites D, S, and G.

	Average Site Value		
Sediment	D	S	G
MMT production ng/gm wet weight as Sn	0.73	1.81	2.06
pore water sulfide, mM	0.52	0.19	0.15
pore water sulfate, mM	8.50	4.15	10.86
salinity, ppt	23	16	16

Correlations between bacterial numbers, enzyme activities, dark CO_2 fixation and methyltin production were made within each sediment and with pooled data from all sites (Table 5). Numbers of aerobic and anaerobic heterotrophs showed only a gradual decrease with depth and positive but insignificant ($p > 0.05$) correlations with tin methylation. Numbers of acetate utilizing SRB showed strong correlations with methyltin production both within and between cores while lactate utilizing SRB did not. It is interesting to note that the ratio of acetate to lactate utilizing SRB increased with decreasing water depths and

Table 5. Average Bacterial Numbers, Enzyme Activities, & CO_2 Fixation Rates for Sites D, S, & G

Bacterial Type	log average MPN/gm wet weight sediment			correlation with MMT production
	D	S	G	r
aerobic heterotrophs	5.99	6.06	6.90	0.451
anaerobic heterotrophs	5.17	4.85	4.79	0.199
sulfide oxidizers	0.00	0.00	2.73	0.397
sulfate-reducing bacteria:				
lactate oxidizers	3.79	3.80	3.46	0.008
acetate oxidizers	2.75	3.25	3.14	0.579**
Parameter	**Average Value**			
RuBPCase enzyme units/g wet wt. sed.	1.49×10^{-5}	7.72×10^{-5}	1.78×10^{-5}	0.209
PEPCase enzyme units/g wet wt. sed.	1.48×10^{-5}	1.76×10^{-5}	1.24×10^{-5}	0.033
Dark CO_2 fix. umoles CO_2 fixed/m2/day	406.7	1136	5902	0.492*

** $p < 0.01$
* $p < 0.05$

decreasing sulfide levels, and that methylation rates also increased with this ratio. Sulfide-oxidizing bacteria were found only in sediment G. The rate of sediment CO_2 assimilation was also significantly correlated with methylation, although PEPCase activity and RuBPCase activity, the latter indicative of carbon fixation in autotrophic organisms, were not.

DISCUSSION

Our results demonstrate that tin methylation in estuarine sediments is a microbially mediated process which is favored under anoxic conditions. Monomethyltin was always the predominant product of biomethylation, representing more than 90% of all methyltin produced in the majority of trials. Heat inactivated sediment slurries did not methylate added tin and evidence for methylation without bacterial activity was not found in our study. Bacterial enrichments and isolated cultures of anaerobes isolated from methylating sediments retained their ability to produce methyltins in culture medium without sediment.

The chemistry and bacteriology of the sulfur cycle seem to be intimately related to tin methylation in estuarine sediments. After finding that methylation was favored under anaerobic conditions, testing of anaerobic organisms isolated from methylating sediments showed that sulfate-reducing bacteria (SRB) in particular are capable of tin methylation. The amounts and rates of methyltin production by several axenic cultures of sulfate-reducing bacteria were similar to amounts produced by sediment slurries. It is noteworthy that methylation of tin by SRB cultures proceeded only in a medium with a relatively high iron concentration, even though the microorganisms grew well on other media. Although sulfide is known to catalyze the methylation and transmethylation of other metals and organometallics [22-26], sulfide added to culture medium blanks did not catalyze tin methylation.

In agreement with pure culture studies, organisms of the sulfur cycle were related to tin methylation by intact cores. Although sulfide produced by bacterial activity seems to have had an inhibitory effect on tin methylation, numbers of acetate-utlizing sulfate-reducing bacteria showed significant correlation with methyltin production. It may also be significant that sulfide-oxidizing bacteria were found only at the site with the highest methyltin production. These organisms could assist tin methylation by removing sulfide as sulfate reduction continues. Sulfide most likely inhibits tin methylation by the formation of insoluble tin sulfides. The optimal conditions for transfer of methyl groups from biological methylating agents and the methylation mechanism itself, enzymatic or not, have not yet been elucidated. Closer investigation of the solution chemistry of tin in methylating systems, especially in relation to sulfide, iron, and oxygen, is needed, along with investigation of the interactions between microorganisms and metal sulfides.

In all of the experimental methylating systems described above, the amount of methyltin production was very similar, reaching a maximum of 10 ppb MMT, 1 ppb DMT and 0.05 ppb TMT. These low methyltin production rates are consistent with sediment methyltin concentrations observed in Chesapeake Bay [12]. The *in situ* ratios of methyltin species in estuarine sediments also match the product distribution of sediment methylation of inorganic tin observed here. Although methylation rates and percent conversion of inorganic tin seem to be low in the environments studied, the products formed, especially TMT, can be toxic to marine organisms [45] and to mammals [2-4] at ppb and even sub-ppb levels. Methyltins may bioaccumulate [46], although no evidence of food poisoning has been described. Our data suggests that methyltin production could be a hazard in anoxic aqueous systems such as drinking water aquifers and in above ground basins with anaerobic sediment pore waters.

ACKNOWLEDGEMENTS

Contribution No. 1603, Center for Environmental and Estuarine Studies of the University of Maryland. From a dissertation to the Graduate School, University of Maryland, by Cynthia C. Gilmour in partial fulfillment of the requirements for the Ph.D. degree in Marine Estuarine and Environmental Studies. CCG is supported by a Chesapeake Biological Laboratory Graduate Research Assistantship.

REFERENCES

1. Chau, Y. K. and P. T. S. Wong. "Occurrence of Biological Methylation of Elements in the Environment," in Organometals and Organometallics in the Environment, F. E. Brinckman and J. M. Bellama, Eds. (American Chemical Society, 1978), pp. 39-51.
2. Doctor, S. V., and D. A. Fox. "Effects of Organotin Compounds on Maximal Electroshock Seizure Responsiveness in Mice. 1. Tri(n-alkyl) Tin Compounds," J. Tox. Environ. Hlth. 10:43-52 (1982).
3. Chang, L. W., T. M. Tiemeyer, G. R. Wenger, D. E. McMillan, and K. R. Reuhl. "Neuropathology of Trimethyltin Intoxication I. Light Microscopy Study," Environ. Res. 29:435-444 (1982).
4. Chang, L. W., T. M. Tiemeyer, G. R. Wenger, and D. E. McMillan. "Neuropathology of Trimethyltin Intoxication III. Changes in the Brain Stem Neurons," Environ. Res. 30:399-411 (1983).

5. Noland, E. A., P. T. McCarthy, and R. J. Bull. "Dimethyltin Dichloride: Investigations into its Gastrointestinal Absorption and Transplacental Transfer," J. Tox. and Environ. Hlth. 12:89-98 (1983).
6. Hallas, L. E., and J. J. Cooney. "Effects of Stannic Chloride and Organotin Compounds on Estuarine Microorganisms," Developments in Industrial Microbiology, 22:529-535 (1981).
7. Zuckerman, J. J. "Organotins in Biology and the Environment" in Organometals and Organometallics in the Environment, F. E. Brinckman and J. M. Bellama, Eds. (American Chemical Society, 1978), pp. 388-422.
8. Jonas, R. B., C. C. Gilmour, D. L. Stoner, M. M. Weir and J. H. Tuttle. "Comparison of Methods to Measure Acute Metal and Organometal Toxicity to Natural Aquatic Microbial Communities," Appl. Environ. Microbiol. 47:1005-1011 (1984).
9. Wong, P. T. S., O. Kramer and G. A. Bengert. "Biological Methylation of Tin Compounds in the Aquatic Environment," paper presented at the Third International Conference on Organometallic Coordination Chemistry of Germanium, Tin and Lead, University of Dortmund, Fed. Rep. Germany, June, 1980.
10. Hallas, L. E., J. C. Means and J. J. Cooney. "Methylation of Tin by Estuarine Microorganisms," Science 215:1505-1507 (1982).
11. Byrd, J. T., and M. O. Andreae. "Tin and Methyltin species in Seawater: Concentrations and Fluxes," Science 218:565-569 (1982).
12. Tugrul, S., T. I. Balkas, and E. Goldberg. "Methyltins in the Marine Environment," Marine Poll. Bull. 14:297-303 (1983).
13. Ridley, W. P., L. J. Dizikes, and J. M. Wood. "Biomethylation of Toxic Elements in the Environment," Science 197:329-332 (1977).
14. Imura, N., E. Sukegawa, S.-K. Pan, K. Nagao, J.-Y. Kim, T. Kwan, and T. Ukita. "Chemical Methylation of Inorganic Mercury with Methylcobalamin, a Vitamin B12 Analog," Science 172:1248-1249 (1971).
15. Rodgers, R. D. "Abiological Methylation of Mercury in Soil," EPA Report 600/3-77-007 (1977).
16. Bertlisson, L., and H. Y. Neujahr. 1971. "Methylation of mercury compounds by methylcobalamin," Biochem. 10:2805-2808.
17. Bisogni, J. J., Jr., and A. W. Lawrence. 1975. Kinetics of mercury methylation in aerobic and anaerobic aquatic environments. J. Water Poll. Control. Fed. 47:135-152.
18. Kudo, A., D. R. Miller, H. Akagi, D. C. Mortimer, A. S. DeFreitas, H. Nagase, D. R. Townsend, and R. G. Warnock. "The Role of Sediments on Mercury Transport (Total- and Methyl-) in a River System," Prog. Water Tech. 10:329-339 (1978).

19. Compeau, G. and R. Bartha. "Methylation and demethylation of mercury under controlled redox, pH, and salinity conditions," Appl. Environ. Microbiol. 48:1203-1207 (1984).
20. Yamada, M., and K. Tonomura. "Formation of Methylmercury Compounds from Inorganic Mercury by Clostridium cochlearium," J. Ferment. Tech. 50:159-166 (1972).
21. Olson, B. H., and R. C. Cooper. In Situ Methylation of Mercury in Sediment," Nature 252:682-683 (1974).
22. Jarvie, A. W. P., A. P. Whitmore, R. N. Markall, and H. R. Potter. "The Sulphide Assisted Decomposition of Trimethyl Lead Chloride in Aqueous Systems," Environ. Pollut. (Series B) 6:69-79 (1983).
23. Reisinger, K., M. Stoeppler, and H. W. Nurnberg. "Evidence for the Absence of Biological Methylation of Lead in the Environment," Nature 291:228-230 (1981).
24. Snyder, L. J., and J. M. Bentz. "Alkylation of Lead(II) Salts to Tetraalkyllead in Aqueous Solution," Nature 296:228-229 (1982).
25. Craig, P. J. and P. D. Bartlett. "The Role of Hydrogen Sulphide in Environmental Transport of Mercury," Nature 275:635-637 (1978).
26. Rowland, I. R., M. J. Davies and P. Grasso. "Volatilisation of Methylmercuric Chloride by Hydrogen Sulphide," Nature 265:718-719 (1977).
27. Manders, W. F., G. J. Olson, F. E. Brinckman, and J. M. Bellama. "A Novel Synthesis of Methyltin Tri-Iodide with Environmental Implications," J. Chem. Soc., Chem. Comm. 726-729 (1984).
28. Bertlisson, L., and H. Y. Neujahr. "Methylation of Mercury Compounds by Methylcobalamin," Biochem. 10:2805-2808 (1971).
29. Craig, P. J., and S. Rapsomanikis. "A New Route to Tris(Dimethyltin Sulphide) with Tetramethyltin as Co-product; the Wider Implications of This and Some Other Reactions Leading to Tetramethyl-Tin and Lead from CH_3I," J. Chem. Soc., Chem. Commun. 1982, p.114.
30. Jarvie, A. W. P., R. N. Markall, and H. R. Potter. "Chemical Alkylation of Lead," Nature 255:217-218 (1975).
31. Craig, P. J. "Methylation of Trimethyl Lead Species in the Environment; an Abiotic Process?," Environ. Tech. Lett. 1:17-20 (1980).
32. Guard, H. E., A. B. Cobet, and W. M. Coleman, III. "Methylation of Trimethyltin Compounds by Estuarine Sediments," Science 213:770-771 (1981).
33. Fanchiang, Y. T. and J. M. Wood. "Alkylation of tin by alkylcobalamins. Kinetics and mechanism," J. Am. Chem. Soc. 103:5100-5103 (1981).
34. Jackson, J. A., W. R. Blair, F. E. Brinckman, and W. P. Iverson. "Gas-Chromatographic Speciation of Methylstannanes in the Chesapeake Bay using Purge and Trap Sampling with a Tin-Selective Detector," Environ. Sci. Tech. 16:110-119 (1982).

35. Means, J. C., and K. L. Hulebak. "A Methodology for Speciation of Methyltins in Mammalian Tissue," NeuroToxicology 4:37-44 (1983).
36. Postgate, J. R. "Enrichment and Isolation of Sulfate Reducing Bacteria," in Anreicherungskultur und Mutantenauslese, H. G. Schlegel and E. Kroger, Eds. (Stuttgart, Fed. Rep. Germany: Fischer Verlag, 1965).
37. Hungate, R. L. in Methods in Microbiology, J. R. Norris and D. W. Ribbons, Eds. (New York, N. Y.: Academic Press, 1969), vol. 3B, p. 117.
38. Buchanan, R. E. and N. E. Gibbons, Eds. Bergey's Manual of Determinative Bacteriology, 8th. ed. (Baltimore, MD: The Williams and Wilkins Co., 1974).
39. Koch, A. L. "Growth Measurement," in Manual of Methods for General Bacteriology (Washington, D. C., American Society for Microbiology, 1981), pp. 179-207.
40. Nelson, J. D. and R. R. Colwell. "The Ecology of Mercury-Resistant Bacteria in Chesapeake Bay," Microb. Ecol. 1:191-218 (1975).
41. Laanboek, H. J. and N. Pfennig "Oxidation of Short Chain Fatty Acids by Sulfate-Reducing Bacteria in Freshwater and Marine Sediments," Arch. Microbiol. 128:330-335 (1981).
42. Howarth, R. W. "A Rapid and Precise Method for Determining Sulfate in Seawater, Estuarine Water, and Sediment Pore Water," Limnol. and Oceanog. 23:1066-1069 (1978).
43. Trupper, H.G. and H.G. Schlegel. "Sulphur metabolism in Thiorhodaceae. 1. Antonie van Leeuwenhoek," J. Microbiol. Serol. 30:225-238 (1964).
44. Tuttle, J. H. "The Role of Sulfer-Oxidizing Bacteria at Deep Sea Hydrothermal Vents," Proc. Biol. Soc. Washington, D. C. (in press).
45. Ishii, T. "Tin in Marine Algae," Bull. Soc. Jap. Sci. Fisheries 48:1609-1615 (1982).
46. Blair, W. R., G. J. Olson, F. E. Brinckman and W. P. Iverson. "Accumulation and Fate of Tri-n-Butyltin Cation in Estuarine Bacteria," Microb. Ecol. 8:241-251 (1982).

CHAPTER 18

TOTAL AND ORGANIC MERCURY IN SEA WATER IN THE WESTERN NORTH PACIFIC

Yoshimi Suzuki and Yukio Sugimura
Geochemical Laboratory
Meteorological Research Institute
Nagamine 1-1, Yatabe, Tsukuba,
Ibaraki 305, Japan

ABSTRACT

Mercury dissolved in sea water can be separated into inorganic and organic species by adsorption on XAD-2 resin with and without a chelating agent, respectively. The results of an extensive study on mercury concentration in sea water in the western North Pacific over depth revealed that the concentration of total mercury is relatively high in the surface (average 10.9 ± 4.1 ng l^{-1}) and lower in the intermediate and deep water (average 5.1 ± 0.4 ng l^{-1}). Organic species with molecular weights of approximately 9000 daltons account for 30 to 60 % of the total. The concentration of inorganic mercury species are nearly constant with depth. Tests for alkyl mercury in sea water gave negative results. The previous work on alkyl mercury may be drived by deoxymercuration of macromolecular mercury compounds during the analytical procedure. The results of the present work are higher than most previous reports, since the present work represents the total and the previous was concerned with inorganic or labile species.

INTRODUCTION

Many studies have been done to develop the analytical methods and to determine concentrations of mercury in environmental samples, such as natural water, ambient air, soil, sediments and biological samples [1,2].

Following these studies, model calculations on the global mercury cycle were developed [3-5]. These models in the study are pre-industrial and contempory. In the pre-industrial model, the flux of mercury balances between input and output in the marine environment, but in the contemporary, a non-steady state model is proposed on the excess input of mercury in the marine environment. In all of the models, the role of alkyl mercury compounds is emphasized in evasion process through ocean/air interface although there are

MARINE AND ESTUARINE GEOCHEMISTRY, Sigleo, A. C., and A. Hattori (Editors)

little evidences on the occurrence of alkyl mercury in the open ocean water.

In order to understand the geochemical cycle of mercury in the ocean, it is important to clarify the chemical nature of mercury species dissolved in sea water.

Most of the studies on mercury concentrations in sea water have been performed using the cold vapor atomic absorption spectrophotometry method without preconcentration [6,7], and the results, therefore include only inorganic or labile organic mercury species. The organic mercury species are determined as the increase in mercury concentration after persulfate or UV oxidation of the sample [8,9]. However, the nature of the organic mercury species is not well known.

Results of recent studies on several transition or biophile elements dissolved in sea water revealed that many of these elements are complexed to macromolecular organic matter with molecular weights ranging from 10^3 to 10^5 daltons [10-14]. Based on the data on biologically essential elements, we hypothesized that a significant amount of mercury, a non-essential elements, in sea water might be associated with such macromolecular organic matter.

The study on the chemical form of mercury revealed that the mercury dissolved in sea water can be separated into inorganic and organic species by using adsorption on XAD-2 resin with and without a chelating agent, respectively. The present study presents the results of determinations of total and organic mercury in the ocean waters of the western North Pacific with some examination of the chemical nature of the macromolecular organic mercury compounds. The significance of these data in the geochemical balance of mercury in the marine environments also is discussed.

METHODS AND SAMPLES

After preconcentration by adsorption on XAD-2 resin from two aliquots of the same sample water with the presence of oxine (8-hydroxyquinoline) and without the organic reagent, the determination of mercury was done by cold vapor atomic absorption spectrophotometry. In this study, the difference of total and organic is defined as inorganic or labile mercury.

Cold vapor atomic absorption spectrophotometry was performed using a Shimazu AA-625 spectrophotometer with deuterium background correction. The mercury compounds were reduced to metallic mercury with 10 % stannous chloride in 6N HCl. Throughout the work, the 253.7 nm mercury absorption line was used with a 1 nm band width. The mercury vapor generator (Shimazu MVU-1A) was operated with nitrogen as a carrier gas at a flow rate of 300 ml min^{-1} and an optical cell (10 mm i.d. and 20 cm long) was used.

For the identification and determination of alkyl mercury compound, a Shimazu 5A gas-chromatograph equipped with an electron capture detector (^{63}Ni) was used. A column (1 m x 3 mm i.d.)packed with 15 % DEGS on 80/100 mesh chromosorb W was used to separate the alkyl mercury compounds.

Unless otherwise stated, all reagents were ultrapure reagent grade. A stock mercury(II) solution 1000 $\mu g\ ml^{-1}$ was prepared by dissolving mercury(II) sulfate in 50 % v/v sulfuric acid. A methyl mercury chloride solution (Hg 1000 $\mu g\ ml^{-1}$) was prepared in 1N HCl. Standard for lower concentrations were obtained by serial dilution. Macroreticular resin beads (Amberlite XAD-2, Rohm and Haas Co., Philadelphia, USA) were washed with methyl alcohol, dilute sulfuric acid and deionized water (Milli-R/Q water) before use.

A 3 % solution of 8-hydroxyquinoline (oxine) was prepared in 50 % acetic acid. For the gel-filtration chromatography, Sephadex gel (G-50) was slurry packed in a 1 cm by 50 cm glass column. The gel bed was washed and equilibrated with a mixture of 3 % NaCl and 1 M Na_2HPO_4, which was also used as the eluant. Fractions 2 ml were collected in glass test tubes using a fraction collector. Samples were eluted at a flow rate of 0.2 ml min^{-1} and passed through a UV monitor (254 nm).

Precaution were taken to avoid the contamination of glass or plastic, reagents, and analytical instruments as suggested by the previous researchers [15-18].

For total mercury, sea water samples (1 to 5 liter) were filtered (0.45 µm), mixed with 5 ml of 3 % oxine solution, and then was passed through a glass column containing 10 g of precleaned XAD 2 resin at a flow rate of 1 ml min^{-1}. Mercury compounds were eluted from the resin columns with a mixed solution of 20 ml of 2M HNO_3 and 2 ml of conc. H_2SO_4 at a flow rate of 5 ml min^{-1}. The organic matter in the eluate was digested and decomposed by adding 1g of $K_2S_2O_8$ to the eluates and heating on a hot plate for one hour in a glass flask with a reflux condensor. The digested solution was then analyzed for mercury as described below.

Mercury organic compounds were analyzed in sea water samples (10 liter) which were filtered (0.45 µm) and then passed through an XAD-2 resin column containing 30 g of the resin at a flow rate of 5 ml min^{-1}. The effluent of these columns was subjected to analysis of inorganic mercury compounds adsorbed onto the resin was eluted with 20 ml of methanol at a flow rate of 1 ml min^{-1}. Ten ml of 2N H_2SO_4 was added to the eluate and the methanol was removed under vacuum. Organic matter in the solution was decomposed by the digestion with 1 g of $K_2S_2O_8$ on a hot plate for one hour in a glass flask with a reflux condensor.

"Inorganic" or labile mercury: Aliquots (1-5 liter) of the effluent solution for the resin column after removal of organic Hg compounds were used for the determination of inorganic or labile Hg. The inorganic mercury was concentrated after addition of 3% oxine as described in the total mercury procedure.

For alkyl mercury, a group of sea water samples were prepared as shown in Fig. 1. Alkyl mercury compounds were extracted from the samples using the benzene-cysteine extraction method of Fujita and Iwashima (1981) [19] and quantitation was performed with gas-chromatography with ECD.

Owing to the volatility of mercury compounds, contamination from the laboratory atmosphere is a major risk during the handling of water sample under the normal conditions. In order to minimize such contamination, and to avoid changes of chemical forms during the storage of sample, the following procedures were used for ship board work.

The sample water was filtered through the membrane filter

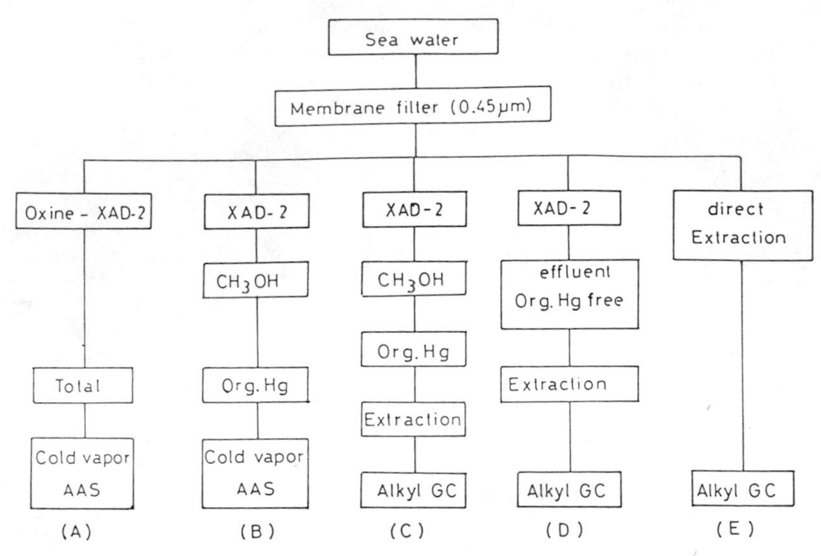

Fig. 1 The schematic procedure for an examination of the occurrence of alkyl mercury in sea water

(Millipore HA, or Nuclepore 0.45 µm of pore size) immediately after collection, with minimum exposure to laboratory air, using a pressurized filtration system which consists of a flexible sample container and filtrate reservoir (low density polyethylene bottle), water pump of glass made and inline filter.

The preconcentration of sample for total, organic mercury and inorganic mercury compounds was done after collection on board. An aliquot of the filtered sea water was mixed with 5 ml of 3% oxine solution and 10 g of XAD-2 resin and shaken. Mercury oxinate together with natural organic mercury compounds in the solution were thus adsorbed on the resin by the batch equilibrium method within the sample bottle. Further separation and analysis was done in the laboratory.

Another aliquot of filtered sea water was prepared for determination of the organic mercury compounds by adsorption on 30 g of XAD-2 resin within the sample bottle. Elution and determination of the organic mercury compounds from the resin was performed on land as described in previous section.

All laboratory work was performed under HEPA filtered air. The blank value for the whole procedure was 0.93 ± 0.22 ng Hg based on the average of 12 duplicate tests. The absolute detection limit of the cold vapor atomic absorption spectrophotometry was 0.2 ng Hg and that of gas-chromatograph-ECD system was 0.03 ng Hg.

Water samples were collected during the years 1981 to 1984 using a non-metallic sampler at depths ranging from surface to bottom in the western North Pacific aboard the M.S. Ryofu-maru (Japan Meteorological Agency). Samples were covered from subarctic to equatorial regions as shown in Fig. 2.

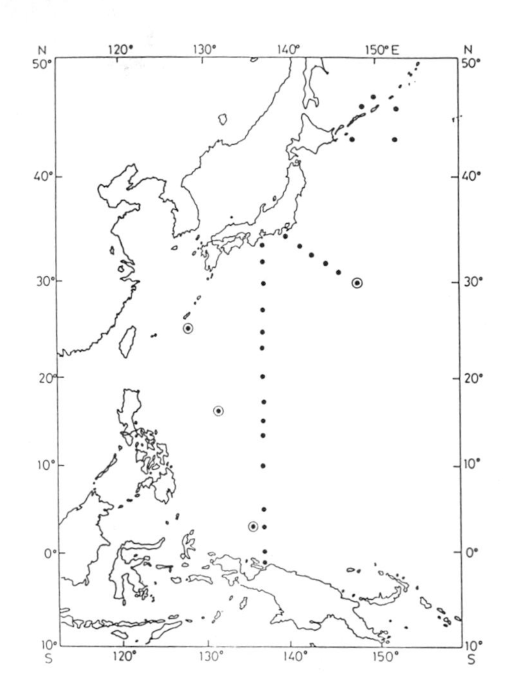

Fig. 2 The location of water sampling sites in the western North Pacific (• :surface, ◉:deep)

RESULTS AND DISCUSSION

The concentration of mercury in sea water

The concentrations of total mercury ranged from 6.7 to 23.8 ng l^{-1} with an average of 10.9 ± 4.1 ng l^{-1} (Table I). Organic mercury concentrations ranged from 2.3 to 11 ng l^{-1} with an average of 4.9 ± 2.3 ng l^{-1}. The average concentration of inorganic or labile mercury was 6.1 ± 2.0 ng l^{-1} and ranged from 3.8 to 12.8 ng l^{-1}. In the water near the coast of Japan relatively higher concentrations

Table I The summary of concentrations of total, organic and inorganic mercury in surface water in the western North Pacific

(ng l^{-1})

	Range	Average
Total	6.7 - 23.8	10.9 ± 4.1
Organic	2.3 - 11.0	4.9 ± 2.3
Inorganic	3.8 - 12.8	6.1 ± 2.0
Organic/Total	30 - 60%	
Particulate	0.2 - 0.9	0.5 ± 0.2
Particulate/Total	3 - 10%	

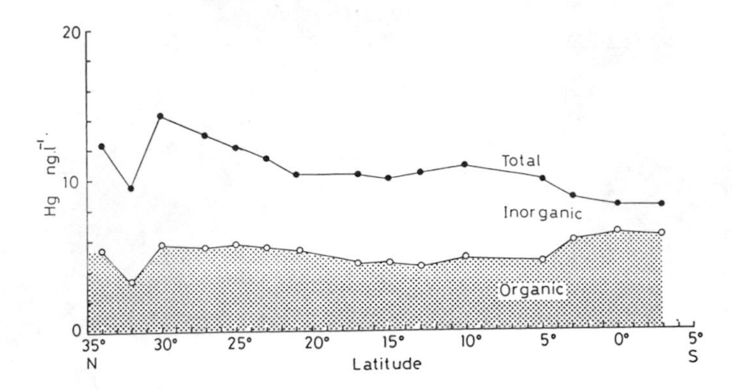

Fig. 3 The meridional distribution of total, inorganic and organic mercury in surface water along 137°E in the western North Pacific (Jan. - Feb., 1984)

of both total and organic mercury were observed and compared to concentration in open sea area. Approximately 3 to 10 % of the total was determined as particulate mercury[21].

The meridional distribution of total, inorganic and organic mercury are given in Fig. 3. As seen in the figure, the concentrations of total and inorganic mercury are slightly higher in temperate zone and lower in tropical regions. Organic mercury shows a fairly uniform distribution of all locations.

The vertical distribution of mercury

Vertical distribution of total and organic mercury at five stations in the western North Pacific was obtained. Two examples of the vertical distribution of total and organic mercury at stations in the temperate zone and the tropical zone in the western North Pacific are given in Fig. 4. Profiles of water temperature, salinity and dissolved oxygen at these stations are also given.

At the temperate zone station, the concentrations of total mercury in the surface layer (0-100 m) ranged from 14 to 20 ng l^{-1}. These were relatively higher than those observed at the tropical region station where the concentrations ranged from 7 to 11 ng l^{-1}. The concentration decreased rapidly with depth to about 4 to 8 ng l^{-1} at 500 m depth. In the intermediate ($>$ 500 m) and deep layer, the concentration is nearly constant with depth except in the near bottom layer where the concentration increased up to 13 ng l^{-1} at the temperate station (30°N, 147°E).

The concentrations of organic mercury in the surface layer ranged from 5 to 9 ng l^{-1} and were also low in the tropical water. In the intermediate and deep water, the concentration decreased rapidly, and the distribution nearly uniform with horizontal and vertical directions from 500 m to 5,000 m (about 2 ng l^{-1}). At the station in the temperate zone where near bottom increase of total mercury was observed, the concentration of organic mercury also in-

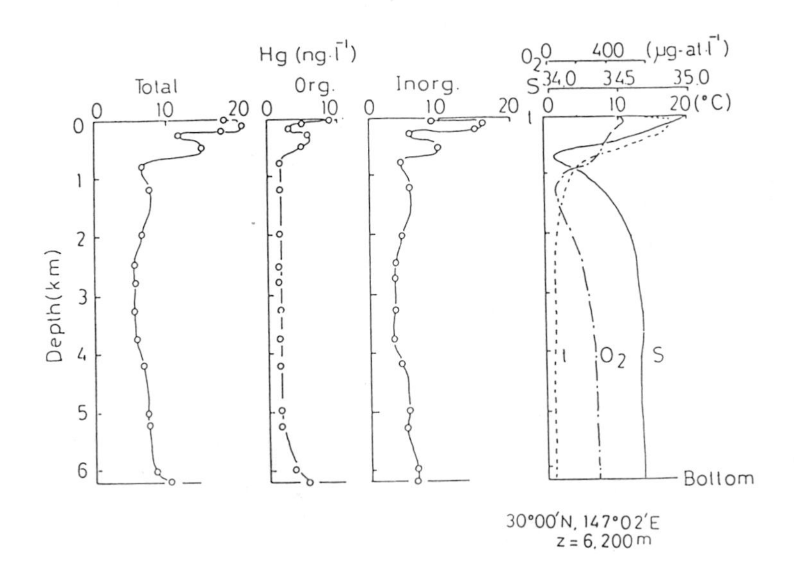

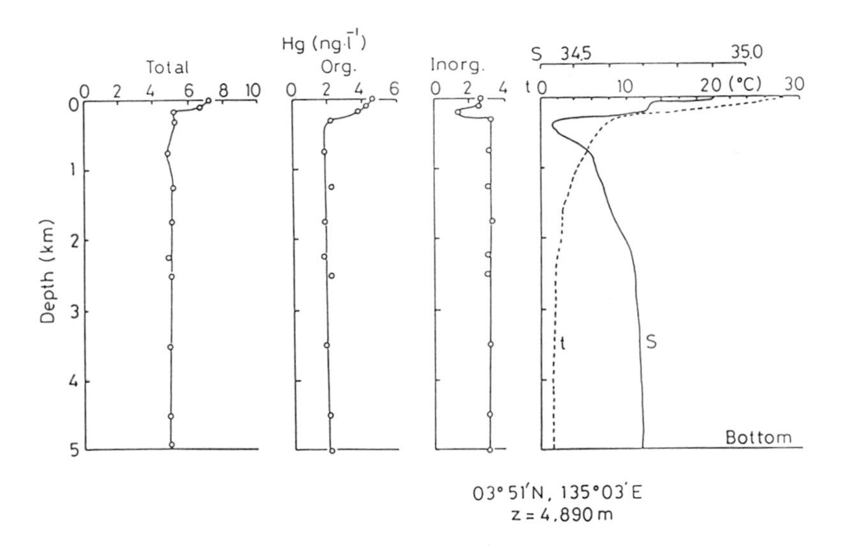

Fig. 4 The vertical distributions of total, organic and inorganic mercury at temperate and tropical zone stations in the western North Pacific

creased in the near bottom layer.

The differences between total mercury concentrations in temperate and tropical waters were mainly due to the differences in inorganic or labile mercury, which was higher in the temperate zone (5 ng l^{-1}) and low in the tropical zone(about 2 ng l^{-1}). Causes of the high concentration in the temperate zone is a problem for further study, but it may be due to greater release rates of inorganic mercury in the middle latitude zone from natural and anthropogenic sources relative to other areas.

Therefore, it may be concluded that in the western North Pacific, the concentrations of total mercury are higher than previously reported due to the inclusion of organic mercury which was not detected in the method used previously[6].

Nature of organic mercury compounds

Alkyl mercury compounds

Although extensive work has been done on the concentrations of mercury in sea water, there are still some uncertainties about the presence of alkyl mercury in sea water. There is some evidences that the determined levels of alkyl mercury compounds in sea water [19,20] may be an artifact or a secondary product produced at some stage of the analytical procedure by mechanisms such as deoxymercuration of some other organic mercury compounds.

Therefore it is important to establish whether alkyl mercury exists as free ion or compound in marine environment, prior to the

Table II The results of an examination of the occurrence of alkyl mercury in surface water (ng l^{-1})

Location	Total Hg (A)	macromolecular Hg (B)	Alkyl Hg (C)	(D)	(E)
33°30'N 141°30'E	9.4	3.7	0.43	< 0.03	0.07
38°31'N 145°29'E	8.7	2.6	0.59	< 0.03	0.19
Sagami Bay	12.3	4.9	0.78	< 0.03	0.10

(A) Inorganic + Organic (XAD-Oxine method)

(B) Macromolecular organic Hg adsorbed on XAD-2 resin

(C) Alkyl Hg in a sample of (B)

(D) Alkyl Hg in sea water of Macromolecular organic Hg free

(E) Alkyl Hg in original sea water

presentation of organic mercury concentrations in the western North Pacific. A detailed report on the examination of alkyl mercury in sea water will be given elsewhere[22], but the results are summarised briefly here (Table II).

The concentration of total mercury in the samples ranged from 9 to 12 ng l^{-1} of which organic mercury was 3 to 5 ng l^{-1}. Although the alkyl mercury ion does not adsorb on the XAD-2 resin, the detected concentration of alkyl mercury in organic macromolecular fraction (C) ranged from 0.4 to 0.8 ng l^{-1}. After removal of the organic mercury fraction by adsorption on XAD-2 resin, alkyl mercury was not detected in the effluent from the resin (D). A trace amount of alkyl mercury is detected by direct extraction of sea water.

Therefore, these data suggest that alkyl mercury compounds in sea water reported in the open and coastal oceans[19, 20] may be a secondary product derived from deoxymercuration of some other macromolecular organic mercury compounds during the analytical procedure, and that if they exist in sea water, the concentrations of alkyl mercury compounds will be very low (less than 0.03 ng l^{-1}).

Macromolecular organic mercury compounds

The concentrations of organic mercury are highly correlated with the total amino acids adsorbed on the XAD-2 resin (Fig. 5). Another approach for elucidate the nature of organic mercury is to examine the molecular size distribution of these compounds. Using gel-chromatography, the molecular size distribution of the organic mercury was determined. Although several size groups of organic matter were observed with molecular weights ranging from 10^3 to 10^5 daltons, the mercury was detected only in the limited fraction with molecular weight of 9×10^3 daltons[21]. Based on these findings, further studies have been performed on samples from the surface and deep waters in the western North Pacific.

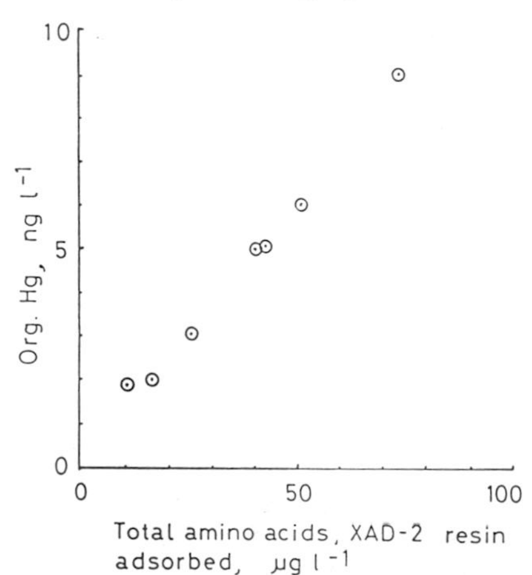

Fig. 5 The relation between the concentration of organic mercury and that of total amino acids adsorbed on the XAD-2 resin

Surface water : The surface distributions of molecular sizes for organic mercury compounds along a transect at 137°E are given in Fig. 6, together with those of organic matter.

As seen in Fig. 6, several size groups of organic matter were observed with molecular weights ranging from 10^3 to 10^5 daltons.

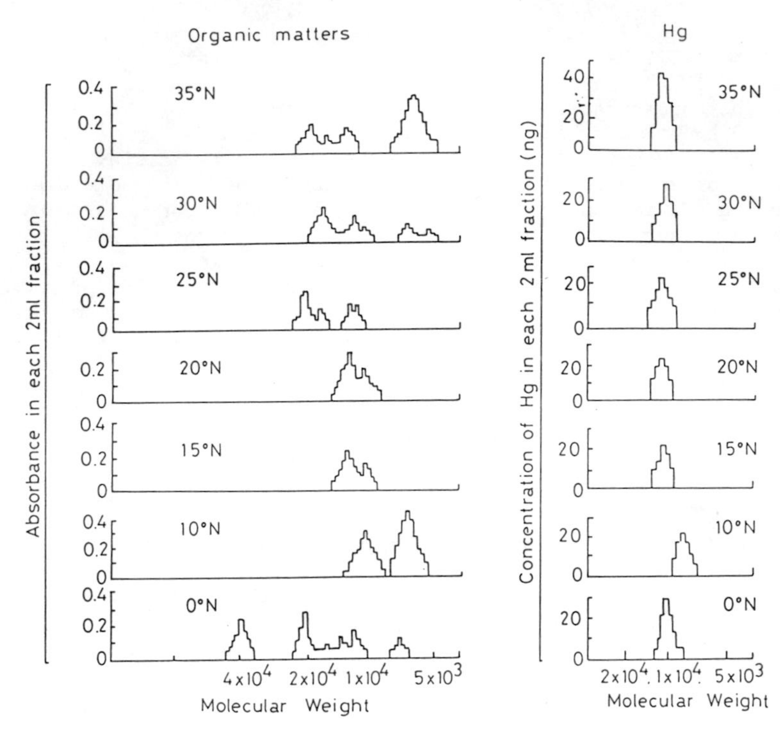

Fig. 6 The surface distribution of molecular sizes of organic mercury compounds along 137°E in the western North Pacific

Mercury was detected only in the fractions with molecular weights of 9×10^3 daltons except at the 10°N 137°E station, in the western North Pacific, where the mercury was associated with a lower molecular weight fraction.

Deep water: An example of the vertical distributions of selected molecular weight fractions of organic mercury compounds are given in Fig. 7. Profiles of water temperature, salinity, dissolved oxygen and dissolved amino acids are also given.

As seen in Fig. 7, the dominant species of organic mercury compounds had molecular weights of 9×10^3 daltons, and the concentrations are high in the surface layer. From 500 m to very close to the bottom, this fraction is low and fairly uniform in distribution. In the near bottom water, the concentrations of organic mercury increased. This increment is associated with the appearence of low molecular weight mercury organic compounds with molecular weights of 3×10^3 daltons. The pattern with depth near the bottom strongly suggests that the some kind of the regeneration process from deep sea sediment involving a low molecular weight organic mercury complex.

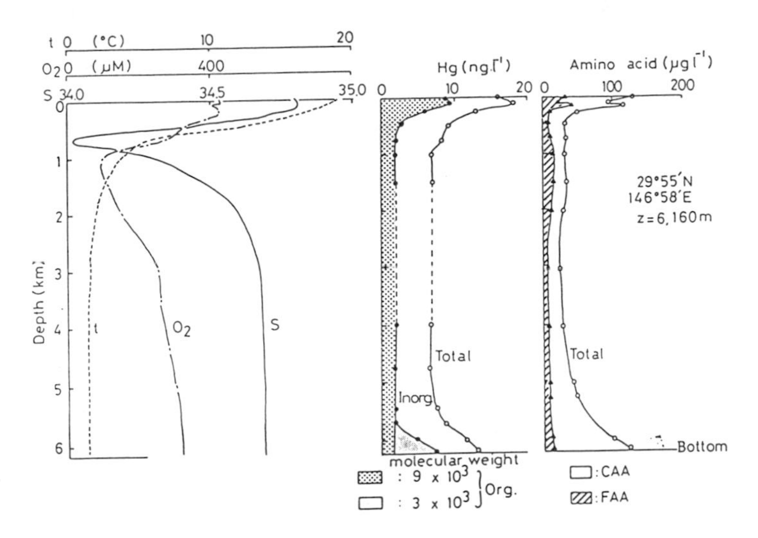

Fig. 7 The vertical distributions of selected molecular weight fractions organic mercury compounds, dissolved amino acids, water temperature, salinity and dissolved oxygen

Geochemical balance of mercury in the ocean

Three different models which describe the different vertical distribution of mercury are given in Fig. 8.

In general, the material balance of a certain element in marine environment can be described in the following equation:

$$\frac{dC}{dt} = R + F - D - E$$

Where R: entry to the ocean through river water

F: entry from wet and dry deposition over the ocean

D: Deposition associated with sediment particles

E: evasion through ocean/air interface

In the steady state condition, it is known that the distribution of total concentrations of a specific biophile element (inorganic plus organic) with a sufficiently long residence time in the ocean is nearly uniform with depth as seen in that of total nitrogen in the ocean[23,24]. In this case, organic species are dominant in surface waters, while inorganic species dominate in deep waters.

In the transient, non-steady state model, because of relatively slow rate of mixing between surface and deep waters, a newly added element entering through the air/sea interface may exist in the surface of the ocean at concentration in excess of the steady state

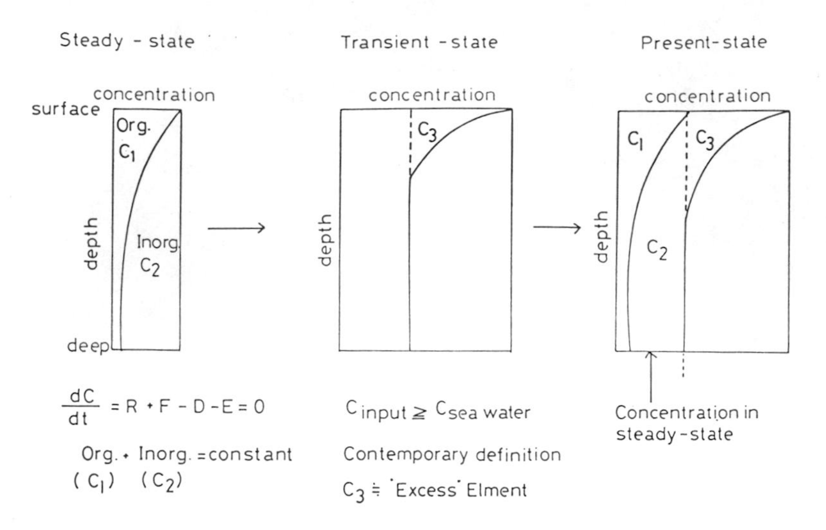

Fig.8 Three models of different vertical distributions of mercury

concentrations, when the concentrations of that element in rain or river water are nearly the same or higher than that of sea water as seen in H-3 distributions. In this case, the concentrations of the element are high in the surface, and low and uniform in the intermediate and deep water.

The vertical distribution of total mercury, as given in the previous section, is high in the surface, and low and uniform in the deeper layers. The vertical distributions of organic mercury species are similar to that of organic nitrogen, while the distributions of total mercury concentration follow a H-3 or Cs-137 type profile (Fig. 9) [25,26].

Considering the chemical form ($HgCl_4^{3-}$ and organic species) and biophilic nature of mercury (biological concentration factor = 10^5), it is quite natural to assume that the steady state distribution of mercury must be uniform with depth as seen in total nitrogen. Therefore, the observed surface maximum in mercury distribution may represent the amount of newly added species of mercury entering the ocean.

Because the concentrations of mercury at the steady state condition are not known yet, at the first approximation, the steady state concentration was chosen to be the value at uniform and low concentration in the deeper layers. We defined the difference between surface and deep values as "excess" mercury.

In Fig. 10, the amount of "excess" mercury is high in temperate zone (3 mg m^{-2}) and low in equatorial (less than 1 mg m^{-2}), the spatial distribution of "excess" mercury may be explained by the fact that there are concentrated human activities and a higher ratio of land to sea area in the temperate zone of the northern hemisphere.

In this connection, a similar pattern of the meridional distri-

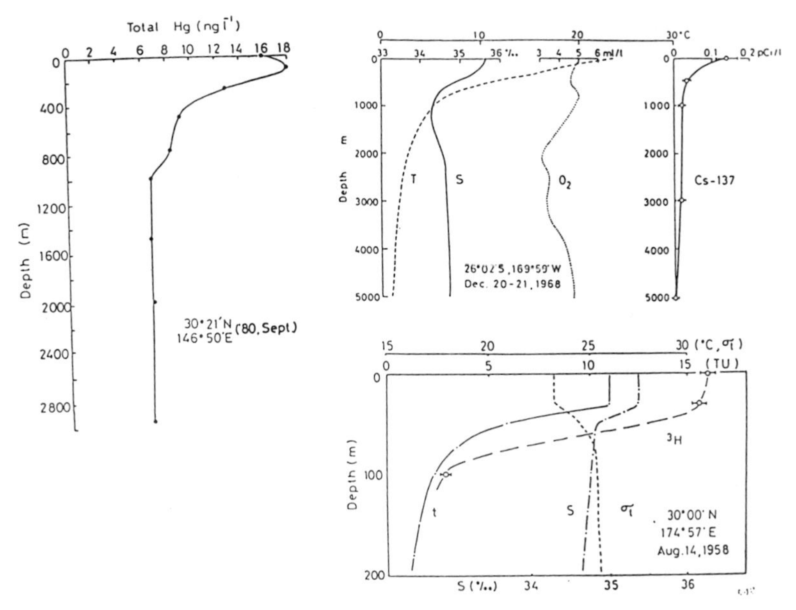

Fig. 9 The similarity of the vertical distribution of total mercury to that of H-3 and Cs-137

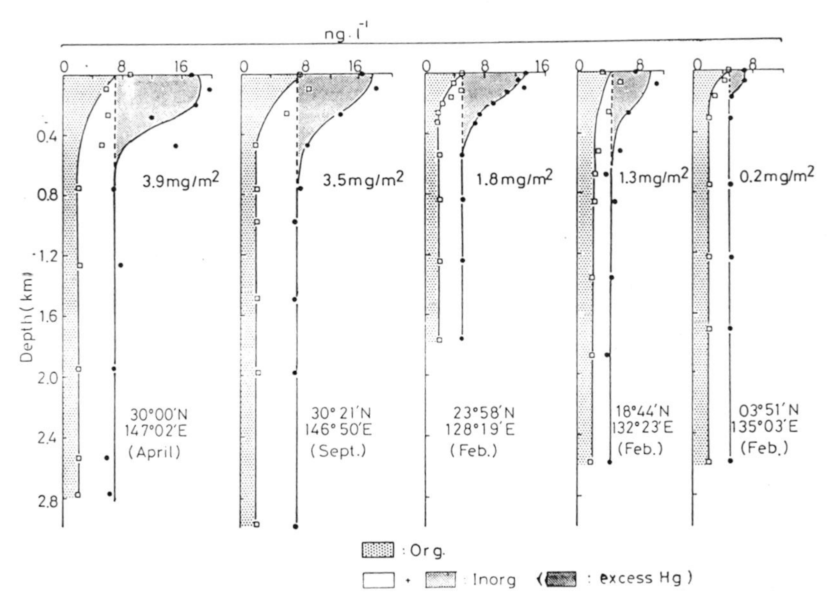

Fig. 10 The vertical distributions of mercury at five stations in the western North Pacific with the amount of "excess" mercury calculated and plotted

butions of mercury (mainly inorganic or labile species) was previously reported [7] in the western North Pacific.

Therefore. it seems that the ocean as a whole is in a transient and non-steady state with respect to mercury at present, and that "excess" mercury over the steady state concentration exist in the surface layers of the ocean as inorganic or labile species.

CONCLUSION

Results of the extensive study on the concentrations and chemical forms of mercury dissolved in sea water revealed that the average total concentration of mercury is 14 ng l^{-1}, of which 30 to 60% is in organic complex with molecular weights of 9×10^3 daltons. In the near bottom layer, the possibility of regeneration of a low molecular weight organic mercury compound (3×10^3 daltons) was suggested. Alkyl mercury compounds were not detected. The vertical distributions of mercury indicate that the recent increased release of mercury by human activity may be reflected in increased concentrations of mercury in the surface layer of the ocean.

Further studies are required on the chemical forms of mercury which may be involved in evasion processes through air/sea interface.

REFERENCES

1. National Academy of Sciences(NAS). "An Assessment of Mercury in the Environment" (Washington, D.C., 1978), pp. 185.
2. Nriagu, J.O.,Ed. "The Biogeochemistry of Mercury in the Environment", (Elsevier/North Holland Biomedical Press, New York 1979).
3. Mackenzie, F.T. and R. Wollast. "Sedimentary Cycling Models of Global Processes", In *The Sea(Marine modeling,* vol.6), E.D. Goldberg, I.N. McCave, J.J.O. Brine and J.H. Steele, Ed. (John Wiley, New York, 1977), pp. 765-777.
4. Lantzy, R.J. and F.T. Mackenzie. "Atmospheric Trace Metals: Global Cycles and Assessment of Man' s Impact", Geochim. Cosmochim. Acta 43: 511-525 (1978).
5. Brosset, C.. "The Mercury Cycle", Water, Air and Soil Pollution 16: 253-255 (1981).
6. Matsunaga, K.. "Estimation of Variation of Mercury Concentration in the Oceans during the Last Several Decades", J. Oceanogr. Soc. Japan 32: 51-52 (1976).
7. Nishimura, M., S. Konishi, K. Matsunaga, K. Hata and T. Kosuga. "Mercury Concentration in the Ocean", J. Oceanogr. Soc. Japan 39: 295-300 (1983).
8. Fitzgerald, W.F. and C.D. Lyons. "Organic Hg Compounds in Coastal Waters", Nature 241: 526-527 (1973).
9. Williams, P.M., K. Chew and H.V. Weiss. "Mercury in the South Polar Seas and in the northeast Pacific Ocean", Mar. Chem. 2: 287-299 (1974).

10. Sugimura, Y., Y. Suzuki and Y. Miyake. "Chemical Form of Minor Metallic Elements in the Oceans" , J. Oceanogr. Soc. Japan 34: 93-96 (1978).
11. Sugimura, Y., Y. Suzuki and Y. Miyake. "The Dissolved Organic Iron in Sea Water", Deep-Sea Res. 25: 306-314 (1978).
12. Sugimura, Y., Y. Suzuki and Y. Miyake. "The Behavior and the Chemical Forms of Metallic Elements Dissolved in Ocean Waters", In *Marine Radioecology* (OECD-NEA, Paris, 1980), pp. 131-141.
13. Sugimura, Y. and Y. Suzuki. "A Method of Determination of Metal Organic Compounds Dissolved in Sea Water using XAD-2 resin", Mar. Chem.,(in press).
14. Sugimura, Y. and Y. Suzuki. "Amino Acids Dissolved in the Western North Pacific Waters", Pap. Met. Geophys. 34: 267-289 (1983).
15. Shimomura, Y., Y. Nishihara and Y. Tanase. "Decrease of Mercury in the diluted Mercury(II) solutions", Bunsekikagaku 17: 1148-1149 (1968).
16. Toribara, Y., C.P. Shields and L. Koval. "The Behavior of Dilute Solutions of Mercury", Talanta 17: 1025-1028 (1970).
17. Feldeman, C.. "Preservation of Dilute Hg Solutions", Anal. Chem. 46: 99-102 (1974).
18. Heiden, R.W. and D.A. Aikens. "Pretreatment of Polyolefine Bottles with Chloroform and Aqua Regia Vapor to Prevent Losses from Stored Trace Mercury(II) Solutions", Anal. Chem. 51: 151-156 (1979).
19. Fujita, M. and K. Iwashima. "Estimation of Organic and Total Mercury in Sea Water around the Japanese Archipelago", Environ. Sci. Technol. 15: 929-933 (1981).
20. Chiba, K., K. Yoshida, T. Tanabe, H. Haraguchi and K. Fuwa. "Determination of Alkylmercury in Sea Water at the Nanogram per Liter Level by GC/Atmospheric Pressure Helium Microwave Induced Plasma Emission Spectrometry", Anal. Chem. 55: 450-453 (1983).
21. Miyake, Y. and Y. Suzuki. "The Concentration and Chemical Forms of Mercury in Waters of the Western North Pacific", Deep-Sea Res. 30: 615-627 (1983).
22. Suzuki, Y. and Y. Sugimura. "A New Method of Determination of Total and Organic Mercury Dissolved in Sea Water", Anal. Chim. Acta (in press).
23. Suzuki, Y., Y. Sugimura and T. Itoh. "A Catalytic Oxidation Method for the Determination of Total and Organic Nitrogen Dissolved in Sea Water", Mar. Chem. 15: (in press) (1985).
24. Sagi, T., Y. Miyake and K. Saruhashi. "On Nitrogen and Phosphorous in Sea Water", Bull. Soc. Sea Water Sci. Japan 36: 253-262 (1982)(in Japanese).
25. Miyake, Y., T. Shimada, K. Kawamura, Y. Sugimura, K. Shigehara and K. Saruhashi. "Distribution of Tritium in the Pacific Ocean", Rec. Oceanogr. Works Japan 13: 17-32 (1975).
26. Saruhashi, K., Y. Katsuragi, T. Kanazawa, Y. Sugimura and Y. Miyake. "Sr-90 and Cs-137 in the Pacific Waters", Rec. Oceanogr. Works Japan 13: 1-15 (1975).

CHAPTER 19

SPECIATION OF DISSOLVED SELENIUM IN THE UPPER ST.LAWRENCE ESTUARY

Kazufumi Takayanagi and Daniel Cossa

Champlain Centre for Marine Science and Surveys
Department of Fisheries and Oceans
P.O. Box 15500, 901 Cap Diamant
Quebec (Que.), G1K 7Y7
Canada

ABSTRACT

Water samples collected from the upper St.Lawrence Estuary were analyzed for selenite, dissolved inorganic selenium and dissolved total selenium (ΣSe). The concentrations ranged from 0.25 to 1.71 nmole/kg for selenite, from 0.41 to 2.08 nmole/kg for inorganic Se and from 0.62 to 2.34 nmole/kg for ΣSe over a salinity range of 0.05 to 31.2 $^{o}/_{oo}$. Selenite, inorganic Se and ΣSe were found to behave conservatively at salinities above 0.2 $^{o}/_{oo}$, with the concentration of each species decreasing with increasing salinity. Selenate, calculated as the difference between inorganic Se and selenite, and organic Se, calculated as the difference between ΣSe and inorganic Se, were also conservative, with no significant interconversion between selenium species observed in these salinity ranges. Although selenite appeared to be rapidly removed at salinities less than 0.2 $^{o}/_{oo}$, a corresponding decrease of inorganic Se or ΣSe, or a corresponding increase of selenate was not observed in this salinity range. Selenite was the predominant species in the river endmember, while selenate, selenite and organic Se each shared significant fractions of the total selenium in the ocean endmember.

INTRODUCTION

Although selenate is the thermodynamically stable species of selenium in oxic river and estuarine waters, as well as in seawater, at a pE of 12.5 [1,2], the co-existence of selenite and selenate in natural waters has been reported by many workers [e.g. 3,4,5, 6]. The presence of selenite in the ocean may be explained by reaction kinetics. A slow oxidation rate of selenite to selenate may result in the presence of thermodynamically unfavorable selenite [6,7,8].

However, no general consensus is available on the behaviour and speciation of selenium in estuaries where a variety of biogeochemical processes take place. Laboratory experiments of Kharkar et al. [9] have shown that dissolved selenium is produced

MARINE AND ESTUARINE GEOCHEMISTRY, Sigleo, A. C., and A. Hattori (Editors)

during estuarine mixing (no speciation studies were done). However, a production of dissolved selenium was not observed in the field studies of Measures and Burton [4] or Takayanagi and Wong [5]. Measures and Burton [4] reported that both dissolved total selenium and selenite were conservative during estuarine mixing in the River Test at salinities above 7 $^{o}/_{oo}$ and that selenite was present as a minor fraction (less than 10%) of the dissolved total selenium. Takayanagi and Wong [5] found that selenite was the major species in the James River Estuary, along with a removal of selenite during estuarine mixing, possibly as a result of oxidation to selenate. In the present study, we have investigated the distributions of the selenium species in the Upper St.Lawrence Estuary over a salinity range of 0.05 to 31.2 $^{o}/_{oo}$.

STUDY AREA

The upper St.Lawrence estuary is classified as a partially mixed estuary extending from Quebec City to Tadoussac at the mouth of the Saguenay Fjord (Fig. 1). The depth ranges from 100 m, near Tadoussac, to 20 m at Quebec City. There are no major tributaries in the upper estuary. The main source of fresh water is the St.Lawrence River, with an average annual river flow rate of 1.10 x 10^4 m^3/sec [10]. The St.Lawrence River drains through the Canadian Shield and St.Lawrence Lowlands. The former consists almost entirely of mixed crystalline rocks of Precambrian age while the latter is made up of carbonate-rich sedimentary rock [11]. The drainage basin is 1.13 x 10^6 km^2 with 29% of the area being lakes developed on the glaciated topography [12]. El-Sabh [13] estimated the flushing time of fresh water in the region between Quebec City and Rimouski to be approximately one month. Therefore, the flushing time of the upper estuary is probably slightly less than one month.

SAMPLING AND ANALYTICAL METHODS

Water samples for selenium analysis were collected in 5 liter Niskin bottles at various depths at 10 stations in the Upper St.Lawrence Estuary during a cruise in October, 1983 (Fig. 1). At station 115, subsurface water samples at ~ 0.5 m were collected every two hours over one tidal cycle (~ 14 hours). The samples were pressure filtered (N_2) through pre-combusted Gelman micro quart filters (1.0 μm pore size) directly into 2 liter polyethy lene bottles. The filtered samples were acidified with HCl to pH ~ 2.0 and stored at ~ 4°C until analysed. Analyses for sele nite and total Se were performed using the fluorometric method of Takayanagi and Wong [14]. Subsamples were irradiated with ultra violet (UV) light in the presence of hydrogen peroxide at pH ~ 2.0 for five hours utilizing a 1200 watt mercury lamp. These UV-oxidized subsamples were also analyzed fluorometrically for total Se, following the pre-concentration step by co-precipitation with

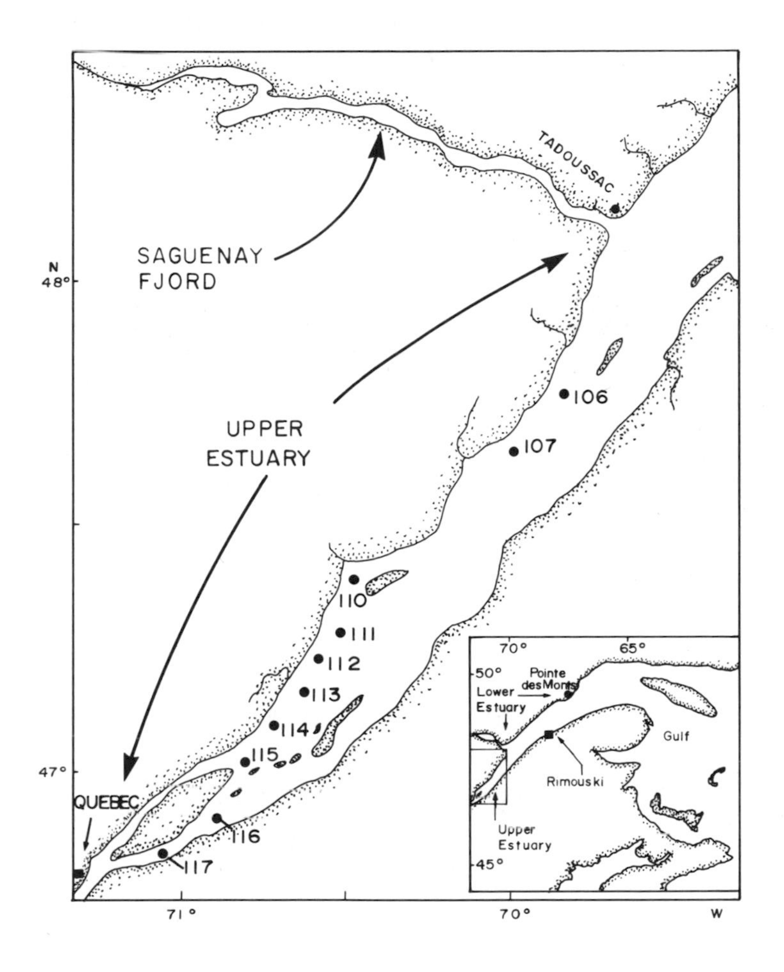

Fig. 1 Sampling stations in the upper St.Lawrence Estuary.

tellurium. Organic Se was calculated as the difference between the total Se, in a sample before (inorganic Se) and after UV-oxidation (ΣSe). Selenate was calculated as the difference between selenite and total Se before UV-oxidation. Inorganic Se and ΣSe represent dissolved inorganic selenium and dissolved total selenium, respectively. The analytical uncertainties were ± 0.02 nmole/kg and ± 0.04 nmole/kg for the determination of Se(IV) and total Se, respectively. Salinities were determined by conductivity measurements with a Guideline Instrument model 8400 Salinometer.

RESULTS AND DISCUSSION

Behaviour of selenium species

Salinity was constant (0.05 $^{o}/_{oo}$) with depth at stations 116 and 117, where three depths were sampled. At these stations, no vertical trend of inorganic Se was observed and there was no significant difference in the concentration of inorganic Se between these two stations. The average concentration of six samples obtained from various depths at these stations was 1.96 ± 0.08 nmole/kg. This value represents the concentration of the river endmember. The concentration decreased with increasing salinity from the river endmember to 0.38 nmole/kg in the ocean endmember of 31.2 $^{o}/_{oo}$ (Sta. 106). A linear correlation with salinity was observed with a slope of -0.045 and a correlation coefficient of 0.97, implying conservative behaviour for inorganic Se (Fig. 2).

The composite plot of ΣSe vs salinity (Fig. 3) also indicates conservative behaviour in the estuary. The average concentration of five samples in the river endmember was 2.12 ± 0.12 nmole/kg with the concentration decreasing to 0.78 nmole/kg at 31.2 $^{o}/_{oo}$. The slope of the linear regression line is -0.045, indistinguishable from that obtained from inorganic Se vs salinity (Fig. 2). These results suggest that organic Se is also conservative and its concentration is constant within the entire estuary. The actual concentrations of organic Se are highly variable, ranging from undetectable (<0.06 nmole/kg) to 0.64 nmole/kg, and organic Se shows only a weak relationship with salinity (Fig. 4). However, the uncertainty in the concentration of organic Se needs to be kept in mind. It was calculated as the difference between two large numbers, the concentrations of ΣSe and inorganic Se. The uncertainty of an individual point is approximately ±0.06 nmole/kg. This relatively large uncertainty may mask a relationship with salinity, even if it exists. Although organic Se was a minor species in the river endmember, only 10% of ΣSe, its contribution to ΣSe increased with increasing salinity and it represented about 30% of ΣSe in the ocean endmember.

At salinities between 0.2 and 31.2 $^{o}/_{oo}$, the concentra-

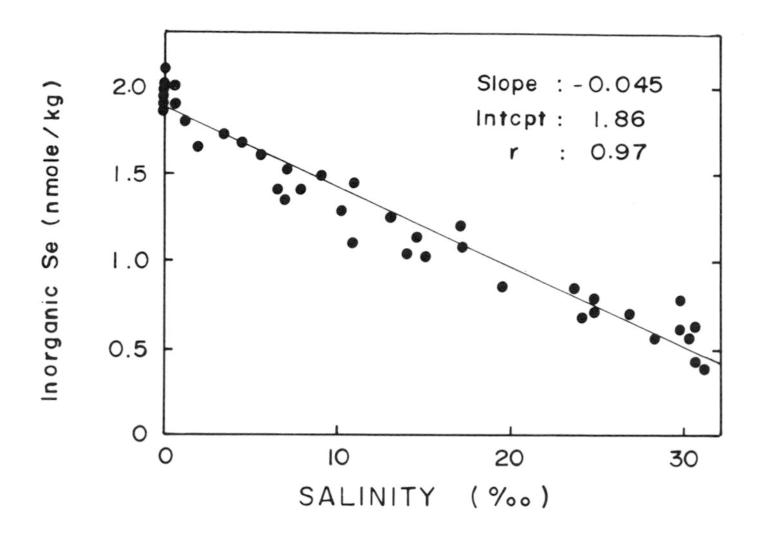

Fig. 2 Inorganic Se vs salinity. The linear regression line is also shown.

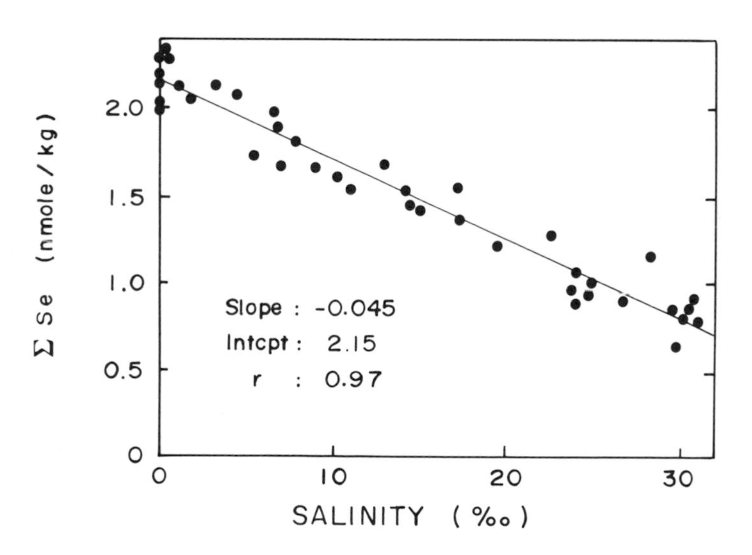

Fig. 3 ΣSe vs salinity. The linear regression line is also shown.

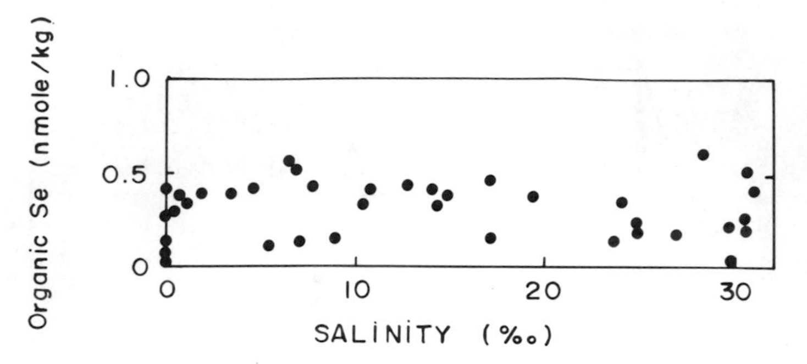

Fig. 4 Organic Se vs salinity.

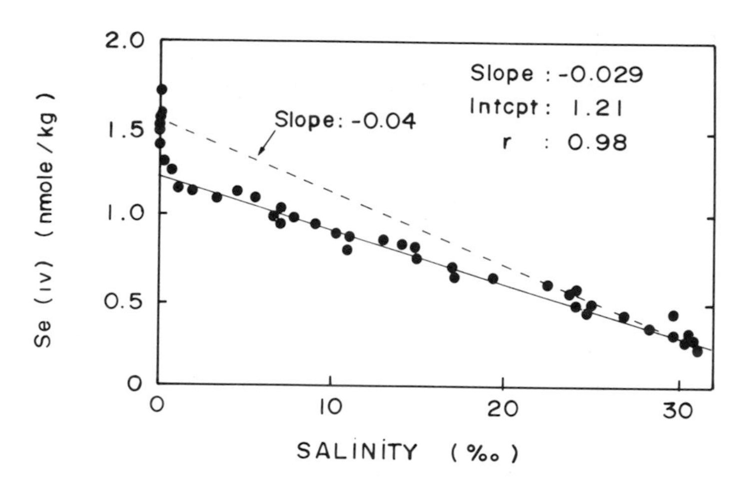

Fig. 5 Selenite vs salinity. The linear regression line (solid line) for the points with salinities above 0.2 $^{o}/oo$ and theoretical dilution line (dashed line) are also shown.

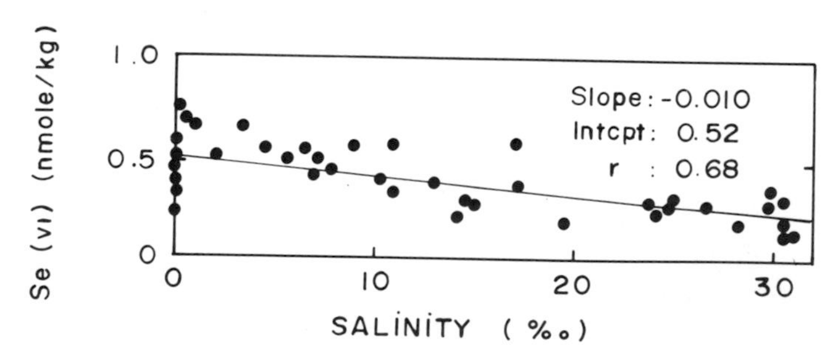

Fig. 6 Selenate vs salinity. The linear regression line is also shown.

tion of selenite decreased from 1.28 to 0.38 nmole/kg. A linear relationship between the concentration of selenite and salinity was observed with a slope of -0.029 and a correlation coefficient of 0.98 (Fig. 5). This also implies conservative behaviour for selenite. However, the concentration of six samples collected at various depths at stations 116 and 117 ranged from 1.40 to 1.71 nmole/kg with no systematic vertical or horizontal gradients. If the average of these samples, 1.55 ± 0.10 nmole/kg, is taken as the concentration of the river endmember, all the observed points lie below the theoretical dilution line connecting the concentrations between the river and ocean endmembers (the slope of this dilution line, -0.04, is similar to those obtained from inorganic Se vs salinity and ΣSe vs salinity). Some removal mechanism for selenite would have to be suggested in this case, especially in the very low salinity region of the estuary where the sharp drop in concentration was observed. An increase in the river discharge rate prior to our sampling could cause this type of profile [15, 16]. However, during two months prior to the sampling, the daily discharge rate measured near Montreal, 250 km upstream of Quebec City, was reasonably constant, ranging from 8030 to 8680 m^3/sec (personnal communication, Water Resources Branch, Environment Canada). Furthermore, if an increase in the river discharge rate is the cause of this profile, a similar feature should be seen in all selenium species. However, a sharp drop in the concentration was not observed for other selenium species.

If selenite was indeed removed, an increase in the concentration of selenate might have been expected in the low salinity region due to the oxidation of selenite to selenate, as suggested by Takayanagi and Wong [5] in the James River Estuary. However, the observed concentrations are variable and a corresponding increase in the concentration of selenate is not recognizable in the composite plot of selenate vs salinity (Fig. 6). The concentration of selenate was, in general, higher (up to 0.73 nmole/kg) in the low salinity region and decreased with increasing salinity to a minimum of 0.15 nmole/kg. Alternatively, selenite may have been transformed to another form of dissolved selenium which was not measured by our analytical method. Since the average concentration of selenite in the river endmember is 1.55 nmole/kg and the y intercept of the linear regression line based on the points for salinities above 0.2 $^{o}/_{oo}$ (Fig. 4) is 1.21 nmole/kg, the calculated amount of selenite removed is 0.34 nmole/kg, or 22 % of the river concentration.

Dissolved selenium may have been transformed to a particulate form. If this is the case, and the amount of selenite and dissolved selenium which are removed are the same, then the removal of selenite could be explained by the reduction of selenite to selenium in the elemental state, which may exist as particulate. Or, a preferential adsorption of selenite onto ferric hydroxides [9,17], major carrier phases in estuaries, is another possible removal mechanism since dissolved iron is known to be removed in the low salinity waters of the St.Lawrence Estuary [18,19].

River flux of selenium

The average river concentration of ΣSe (2.12 ± 0.12 nmole/kg) is similar to values determined in 7 US rivers and the Rhone and Amazon Rivers reported by Kharker et al. [9] and is slightly lower than those in 4 rivers in England reported by Measures and Burton [4]. Cossa and Tremblay [10] reported that the average concentration of sulfate in the St.Lawrence River is 25.4 mg/1, which leads to a sulfur to selenium weight ratio of 5×10^4. If selenium in the earth's crust is assumed to exist as selenides within sulfide minerals at a sulfur to selenium weight ratio of 6×10^3 [20,21] and it is then separated from sulfur during weathering, selenium is mobilized 10 times less than sulfur in the St.Lawrence River drainage basin. However, this weight ratio in the river water is much lower than the ratio of 5×10^6 in the ocean.

CONCLUSIONS

Within our analytical uncertainties, all the selenium species analyzed behaved conservatively at salinities above 0.2 $^o/_{oo}$ and no significant interconversion among the species was observed within these salinity ranges during the period sampled. However, selenite appeared to be removed at salinities less than 0.2 $^o/_{oo}$. Possible removal mechanisms of selenite include a conversion of selenite to a particulate form or a conversion of selenite to other dissolved species which were not determined by our analytical method. Oxidation of selenite to selenate seems to be unlikely. Selenite was the major species of dissolved total selenium in the river endmember constituting up to 60 % of ΣSe. However, selenite, selenate and organic Se each shared significant fractions of the dissolved total selenium in the ocean endmember.

ACKNOWLEDGEMENTS

We wish to thank the officers and crew of "NAVIMAR UN" for their assistance in the collection of the samples. We also thank C. Gobeil and T.J. Oatts for their comments and discussions. One of us (K.T.) was supported by the Natural Science and Engineering Research Council of Canada.

REFERENCES

1. Sillen, L.G. "The physical chemistry of sea water" In Oceanography, M. Sears, Ed. (AAAS, Washington, D.C., 1961), pp. 549-581.
2. Turner, D.R., M. Whitfield and A.G. Dickson. "The equilibrium speciation of dissolved components in freshwater and seawater at 25°C and 1 atm pressure," Geochim. Cosmochim.

Acta 45: 855-881 (1981).
3. Sugimura, Y., Y. Suzuki and Y. Miyake. "The content of selenium and its chemical form in sea water," J. Oceanogr. Soc. Japan 32: 235-241 (1976).
4. Measures, C.I. and J.D. Burton. "Behaviour and speciation of dissolved selenium in estuarine waters," Nature 273: 293-295 (1978).
5. Takayanagi, K. and G.T.F. Wong. "Total selenium and selenium (IV) in the James River Estuary and southern Chesapeake Bay," Estuarine Coastal Shelf Sci. 18: 113-119 (1984).
6. Cutter, G.A. and K.W. Bruland. "The marine biogeochemistry of selenium: a reevaluation," Limnol. Oceanogr. 29: 1179-1192 (1984).
7. Measures, C.I., B.C. Grant, B.J. Mangum and J.M. Edmond. "The relationship of the distribution of dissolved selenium IV and VI in three oceans to physical and biological processes," In Trace Metals in Sea Water, C.S. Wong, E. Boyle, K.W. Bruland, J.D. Burton and E.D. Goldberg, Eds (Plenum Press, New York, 1983), pp. 73-83.
8. Measures, C.I., R.E. McDuff and J.M. Edmond. "Selenium redox chemistry at GEOSECS I re-occupation," Earth Planet. Sci. Lett. 49: 102-108 (1980).
9. Kharkar, D.P., K.K. Turekian and K.K. Bertine. "Stream supply of dissolved silver, molybdenum, antimony, selenium, chromium, cobalt, rubidium and cesium to the ocean," Geochim. Cosmochim. Acta 32: 285-298 (1968).
10. Cossa, D. and G. Tremblay. "Major ions composition of the St.Lawrence River: Seasonal variability and fluxes. In Transport of Carbon and Minerals in Major World Rivers, Part II, E.T. Degens, Ed, (University of Hamburg, Hamburg, 1983), pp. 253-259.
11. Loring, D.H. and D.J.G. Nota. Morphology and Sediments of the Gulf of St.Lawrence. Fish. Res. Board Can., Bull., 182. 146 pp. (1973).
12. Pardé, M. "Hydrologie du Saint-Laurent et de ses affluents," Rev. Can. Géogr. 2: 35-83 (1948).
13. El-Sabh, M.I. "The lower St.Lawrence Estuary as a physical oceanographic system," Naturaliste Can. 106: 55-73 (1979).
14. Takayanagi, K. and G.T.F. Wong. "Fluorimetric determination of selenium (IV) and total selenium in natural waters," Anal. Chim. Acta 148: 263-269 (1983).
15. Loder, T.C. and R.P. Reichard. "The dynamics of conservative mixing in estuaries," Estuaries 4: 64-69 (1981).
16. Officer, C.B. and D.R. Lynch. "Dynamics of mixing in estuaries," Estuarine Coastal Shelf Sci. 12: 525-533 (1981).
17. Geering, H.R., E.E. Cary, L.H.P. Jones and W.H. Allaway. "Solubility and redox criteria for the possible forms of selenium in soils," Soil. Sci. Soc. Amer. Proc. 32: 35-40 (1968).
18. Subramanian, V. and B. d'Anglejan. "Water Chemistry of the

St.Lawrence Estuary," J. Hydrol. 29: 341-354 (1976).
19. Bewers, J.M. and P.A. Yeats. "The behavior of trace metals in estuaries of the St.Lawrence Basin," Naturaliste Can. 106: 149-161 (1979).
20. Goldshmidt, V.M. and L.M. Strock. "Zur Geochimie des Selens II," Nachr. Ges. Wiss. Göttingen, Math-Phys. Kl. Fachgruppe IV, 1: 123-142 (1935).
21. Howard, J.H. III. "Geochemistry of selenium: formation of ferroselite and selenium behavior in the vicinity of oxidizing sulfide and uranium deposits," Geochim. Cosmochim. Acta 41: 1665-1678 (1977).

CHAPTER 20

SILVER SPECIATION IN SEAWATER: THE IMPORTANCE OF SULFIDE AND ORGANIC COMPLEXATION

Christina E. Cowan
Everett A. Jenne
Eric A. Crecelius
Pacific Northwest Laboratory
P.O. Box 999
Richland, Washington 99352

ABSTRACT

The effect of variations in pH, salinity, sulfide and dissolved organic carbon (DOC) concentrations on the speciation of dissolved Ag were calculated using the MINTEQ geochemical model. Molecular-weight-averaged conditional equilibrium constants for natural Ag-organic complexes were estimated from the equilibrium constants for specific metal-organic complexes and the concentration of the organic compounds, reported to be present in seawater, summed by principal functional group. These calculations show that the major Ag complexes in seawater are $AgHS^0$, for S[-II] concentrations above 0.01 µg L^{-1}, $AgCl^0$, $AgCl_2^-$, $AgCl_3^-$ and $AgCl_4^0$. If the $AgCl^0$ complex is the primary bioavailable or toxic chemical species as has been reported, the toxicity of silver will decrease significantly with increased S[-II] concentration, pH, and with decreased salinity. Organically complexed silver was calculated to represent <1% of the total dissolved Ag, which suggests that changes in Ag speciation previously attributed to DOC complexation may actually be due to formation of S[-II] or sulfhydral complexes.

INTRODUCTION

Silver, one of the most toxic of the heavy metals in aquatic ecosystems, has an $^{96}LC_{50}$ level (concentration resulting in 50% mortality in 96 hours) for the American oyster, *Crassostrea virginica*, of about one-hundredth of that for zinc or copper [1]. The toxicity of silver to aquatic organisms has been of little concern because concentrations in the environment were considered to be too low to produce adverse effects. However, the amounts of silver that continue to enter surface waters from industry and municipal sewage may result in increased accumulation of silver by various aquatic organisms in local environs [2,3] and could result in toxic effects.

Studies with other heavy metals show that their toxicity is often more dependent upon the concentration of complexing

MARINE AND ESTUARINE GEOCHEMISTRY, Sigleo, A. C., and A. Hattori (Editors)

ligands than on the total dissolved concentration of the metal [4,5]. Assuming that this also applies to silver, its distribution among the various aqueous complexes and the thermodynamic activity of its bioavailable complex(es) will determine its bioaccumulation in and toxicity to aquatic organisms. The limited research on silver suggests, based on a plot of toxicity versus the aqueous complex activity, that the neutral complex, $AgCl^0$, may be the primary bioavailable species for yeast [6,7]. Recently, the data of Andrew et al. [8] on Cu toxicity to _Daphnia magna_ has been reexamined using geochemical modeling and multivariant statistical testing [9]. The results suggest that the hydroxide species of Cu are the most toxic species whereas the uncomplexed Cu^{2+}, and other highly correlated species, showed little toxicity in contrast to previous statements based on plotting that the Cu^{2+} ion is the most toxic species. These results indicate that although $AgCl^0$ may be a toxic species, determining whether it is the primary toxic species will require consideration of all the aqueous species and application of a quantitation technique as in [9].

The inorganic composition of open ocean water shows minimal spatial and temporal variation in chemical constituents except in upwelling areas. For example, the pH of open ocean water varies over a very narrow range of 8.0 ± 0.7 [10]. In contrast, fresh water inflow may cause significant variations of dissolved constituents in coastal waters. For a small group of constituents of importance to silver speciation, concentrations may vary notably both spatially and temporally even in the open ocean. For example, sulfide concentrations of 1.7 to >100 μg L^{-1} have been measured in estuarine water columns and above anoxic coastal sediments [11,12,13]. The disequilibrium coexistence of sulfide and oxygen is apparently due to a kinetic constraint on the oxidation of S[-II] [11,14].

Dissolved organic carbon (DOC) concentration in seawater varies spatially and temporally. For open ocean water, Duursma [15] reported a DOC range of 1.32 to 3.32 mg L^{-1} of C, Dawson [16] gave 3.40 mg L^{-1} of C as the average concentration and Stumm and Morgan [14] state that the average dissolved organic carbon concentration in seawater is 5.00 mg L^{-1} of C. Organic carbon concentrations of estuarine water compiled by Jenne [17] range from 1.9 to 17 mg L^{-1}, although most values fall between 2 and 6 mg L^{-1}. Some of the dissolved organic compounds in the sea may complex some heavy metals [18]. However, the importance of this complexation for silver is unknown because of the difficulty of measuring complexation capacity at the observed concentrations of silver. Jenne et al. [19] suggested that the calculated oversaturation of San Francisco Bay water with respect to $Ag_2S(s)$ might be due, at least in part, to organic complexation. Humic material from estuarine sediment has been observed to decrease the rate of uptake of silver by the excised gills of the little neck clam, _Protothaca staminea_ and to decrease the rate of silver plating onto gold when this technique is used in the analysis of dissolved silver (E. A. Crecelius, unpublished data). These studies suggest that the organic complexation of silver may be important in marine waters and needs to be evaluated.

Our immediate objective was to investigate the effect of several variables on the equilibrium distribution of silver complexes in coastal waters via sensitivity analysis using a comprehensive geochemical model. The longer range objective was to provide rationale for the design of laboratory experiments for the evaluation of silver bioavailability in coastal marine waters including the identification of any additional analytical parameters needed.

METHODS

Speciation and saturation indices were calculated with the MINTEQ (version K1) geochemical model [20], which includes the following Ag species in the data base: Ag^+, $AgBr^0$, $AgBr_2^-$, $AgCl^0$, $AgCl_2^-$, $AgCl_3^{2-}$, $AgCl_4^{3-}$, AgF^0, $AgHS^0$, $Ag(HS)_2^-$, AgI^0, AgI_2^-, $AgOH^0$, $Ag(OH)_2^-$, $AgSO_4^-$, $AgNO_3^0$, $Ag(NO_2)_2^-$, $AgBr_3^{2-}$, AgI_3^{2-}, AgI_4^{3-}, $Ag(S_4)_2^-$, $AgS_4S_5^-$, $Ag(HS)S_4^-$. The thermodynamic data are given by Ball et al. [21]. The following complexes were added temporarily to the thermodynamic data base for this study: Ag-amino acid, Ag-2 amino acid, Ag-carb, 2Ag-carb, Ag-2 carb, Ag-phenol. Silver concentrations of 0.5 and 1.0 ng L^{-1}, at pH values of 7.0, 7.5 and 8.2, S[-II] concentrations of 0.0, 0.1, 1.0 and 10.0 μg L^{-1} and DOC concentrations of 1.0 and 5.0 mg L^{-1} for salinities of 27.0, 29.5 and 32.0 °/oo were used. The concentrations of other dissolved constituents at the indicated salinity levels were computed[a] from their concentration at 35 °/oo given in Nordstrom et al. [22]. In these sensitivity analyses, the concentration of one constituent was varied while the remaining constituents were held at their base concentrations (e.g., Ag = 0.5 μg L^{-1}, salinity = 30 °/oo, pH = 8.2, DOC = 1 mg L^{-1}, S[-II] = 1 μg L^{-1}).

RESULTS

Solid Phases of Silver

One of the first steps in designing or interpreting bioaccumulation or toxicity experiments is to check for possible oversaturation of any solid phases that might occur. This is important because the formation of a solid phase(s) would significantly reduce the activities of bioavailable species. The

---

(a) Constituent concentration at indicated salinity in mg L^{-1} = (indicated salinity °/oo/35.05 °/oo) concentration at 35 °/oo where chlorinity = Σ Cl + Br + I = 19.42 g kg and salinity = 0.03 + 1.805 chlorinity.

formation of a silver precipitate would, at best, require frequent analytical determinations but could well invalidate the experimental results.

Acanthite, Ag_2S, calculates to be one-half to two orders of magnitude oversaturated at a sulfide and silver concentration of 0.1 to 10 μg L^{-1} and 0.05 ng L^{-}, respectively. All other silver solids calculate to be undersaturated.

Inorganic Complexes

The sensitivity analyses show that the aqueous species that constitute greater than 1% of the total dissolved Ag are $AgHS^0$, $AgCl_2^-$, $AgCl_3^{2-}$ and $AgCl_4^{3-}$. The activities of the Cl complexes increase with salinity as the Cl^- activity increases (Figure 1). The most abundant aqueous species is the silver bisulfide complex, $AgHS^0$, even at S[-II] concentrations only slightly above 0.01 μg L^{-1}. Only when the S[-II] concentration is reduced to below 0.01 μg L^{-1} is the activity of this complex of the same order of magnitude as the most abundant Cl complex. Decreasing the sulfide concentration results in an increase in the activity of both the Ag^+ and the Cl complexes (Figure 2). The activity of individual silver chloride species increases by approximately one order of magnitude for a two log unit decrease in the concentration of dissolved S[-II].

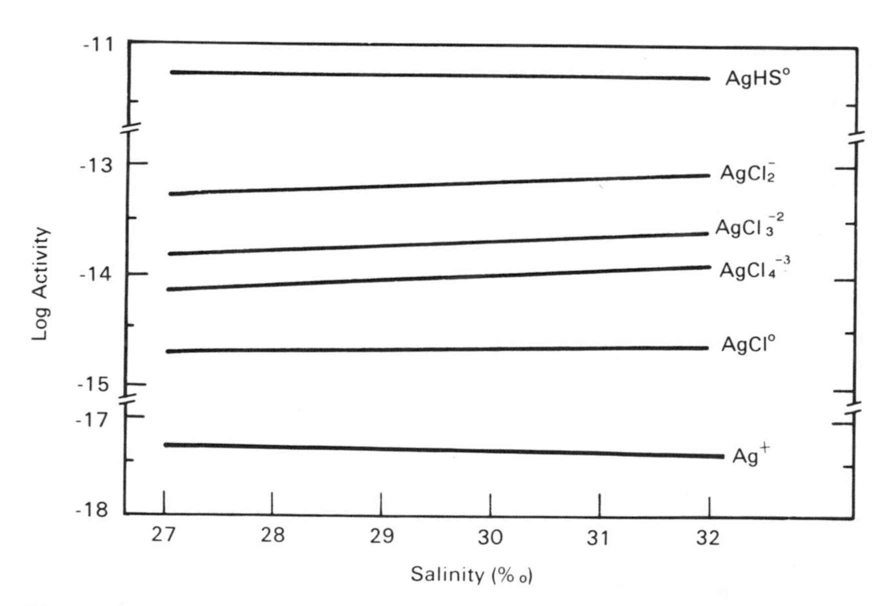

Figure 1. Variation in the Activity of Selected Silver Species with Salinity [pH = 8.2, S(-II) = 1 μg L^{-1}, Ag = 0.5 ng L^{-1}, DOC = 1 mg L^{-1}]

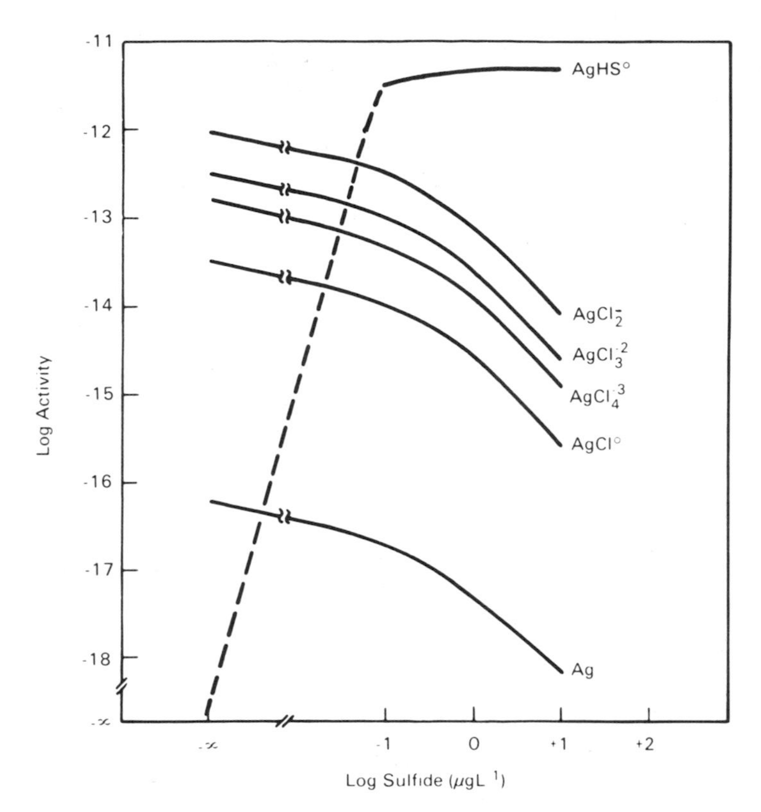

Figure 2. Variation in the Activity of Selected Silver Species Sulfide Concentration [pH = 8.2, S = 30 ^{o}/oo, Ag = 0.5 ng L^{-1}, DOC = 1 mg L^{-1}]

The activities of the Cl complexes ($AgCl^0$, $AgCl_2^-$, $AgCl_3^{2-}$, and $AgCl_4^{3-}$) are sensitive to changes in salinity as indicated by an increase of >5% for the change in salinity from S = 29.5 ^{o}/oo to the base value of S = 30 ^{o}/oo although the activity of the neutral Cl complex increases the least for given changes in salinity (Table I). In contrast, the concentration of Ag^+ is relatively constant for a given S[-II] concentration (S[-II] = 1 μg L^{-1}) regardless of variations in salinity (Table I).

The activities of the chloride complexes are highly dependent on pH (Figure 3) decreasing by about 104% with a pH decrease from 8.2 to 7.0. The silver bisulfide complex activity changes less but is still significantly affected (8.2%) by a pH decrease of 0.7 log units. The activity of Ag^+ exhibits the same magnitude of variability with pH change as the Cl complexes.

Table I. Change in Activity of Major Aqueous Species of Silver with Salinity [base case is S = 30 $^{o}/oo$; DOC = 1 mg L^{-1}, HS = 1 μg L^{-1}, Ag = 0.5 ng L^{-1}, pH = 8.2]

Species	Salinity ($^{o}/oo$) 27	29.5	30	32	27	29.5	32
	- - - - - - - Log Activity			- - - - - -	% Change from Base Case		
$AgHS^{o}$	-11.292	-11.292	-11.292	-11.293	0	0	0.2
$AgCl^{o}$	-14.712	-14.669	-14.649	-14.628	13.5	4.5	5.0
$AgCl_2^{-}$	-13.285	-13.207	-13.171	-13.132	13.1	8.0	9.4
$AgCl_3^{-2}$	-13.838	-13.724	-13.674	-13.617	31.4	10.9	14.0
$AgCl_4^{-3}$	-14.190	-14.042	-13.976	-13.902	38.9	14.1	18.6
Ag^{+}	-17.410	-17.401	-17.397	-17.393	2.9	0.9	0.9

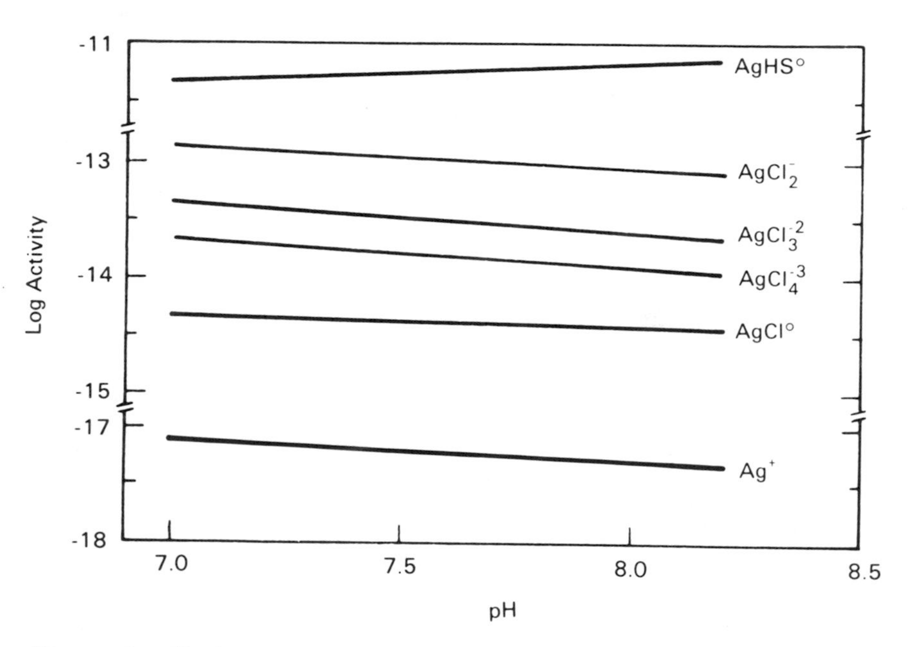

Figure 3. Variation in the Activity of Selected Silver Species with pH [S = 30 $^{o}/oo$, S(-II) = 1 ug L^{-1}, Ag = 0.5 ng L^{-1}, DOC = 1 mg L^{-1}]

Organic Complexes

No conditional stability constants for natural dissolved organic matter complexes of Ag were found. As a first approximation of the content of the organic matter, we used the available compilations of specific organic compound concentrations in seawater and compilations of stability constants for complexes of silver with organic compounds to obtain a series of representative reactions and stability constants. Stability constants for Ag [23,24] are available (Table II) for only a few of the many organic compounds whose concentrations have been determined in seawater [15,16,25]. Because these compounds represent only a small fraction of the total dissolved organic matter [15,16, 25,26], some method was needed to extend this limited database to represent the bulk of the organic ligands.

Calculation of Molecular Weight Averaged Conditional Stability Constants

The first approach was to assume that the available silver-organic compound stability constants (log K_r^0) can be used to represent all other organic compounds possessing similar functional groups. Differences in pK values for Ag with specific organic compounds occurs due to variations in chain length and the presence of additional functional groups. However, for the straight-chained carboxylic acids, the proton dissociation constant (pK_a) shows little variation with chain length [27]; therefore, in the approach used here it is assumed that within a class of compounds the effect of chain length on the stability constant is minimal. This assumption was evaluated by plotting the log K_r^0 values for silver-amino acid complexes versus the number of carbon groups in the radical attached to the CNH_3COOH group (data in Table II) to determine if the log K_r^0 changes with chain length (Figure 4). The few available log K values were all very similar, which agrees with the relationship between chain length and pK_a observed for carboxylic acids [27]. No trend in log K_r^0 with chain length could be discerned.

The functional groups represented by the organic compounds in Table II are CH_2OOH by acetic acid, COH_2COOH by glycollic acid, CNH_3COOH by methionine, glycine and alanine and phenolic-OH by phenol. Based on metal ion classification (i.e., Type A and B metal cations) [14], the most stable complexes of Ag with organic compounds are expected to be those with the CNH_3COOH functional group because of the strong preference of silver for compounds containing nitrogen. This is observed in Table II. The second strongest complex is that with the COH_2COOH functional group and the weakest with the CH_2OOH and OH functional groups. This ordering of stability constants for the functional groups with silver is representative of the general reactivity of these type of compounds [28]. Generally, the order of reactivity and pK_a [27] for these functional groups is: COOH > OH > COH > C=0.

Table II. Organic Complexes of Silver and Their Stability Constants

Compound	Representation[a]	HL	H_2L	AgL	AgL_2	Ag_2L	Composition	Functional Group
		Log K[b]						
Acetic Acid	HL	4.757 ± 0.002		0.73	0.64	1.14	CH_3COOH	-COOH
Glycollic Acid	HL	3.831 ± 0.001		2.90 ± 0.02	4.66		$HOCH_2COOH$	-COOH and -OH
Methionine	HL	9.05 ± 0.02 (25°, 0.1)	2.20 ± 0.04 (25°, 0.1)	3.17 (25°, 0.1)			$CH_3SCH_2CH_2CH$	$-NH_2COOH$
Glycine	HL	9.778 ± 0.0	2.350	3.51	6.89		CH_2NH_2COOH	$-NH_2COOH$
Alanine	H_2L	9.867	2.348	3.64	7.18		$CH_3CH(NH_2)COOH$	$-NH_2COOH$
Phenol	HL	9.98		0.42			C_6H_5OH	-OH

(a) HL, H_2L signify reactions with acid; for example, HL means $H^+ + L \rightarrow HL$. AgL, AgL_2 and Ag_2L signify reactions of silver ion with ligand; for example, AgL means $Ag^+ + L \rightarrow AgL$. Source: Martell and Smith [23,24].

(b) Unless otherwise noted, constants are for 25° and 0 ionic strength.

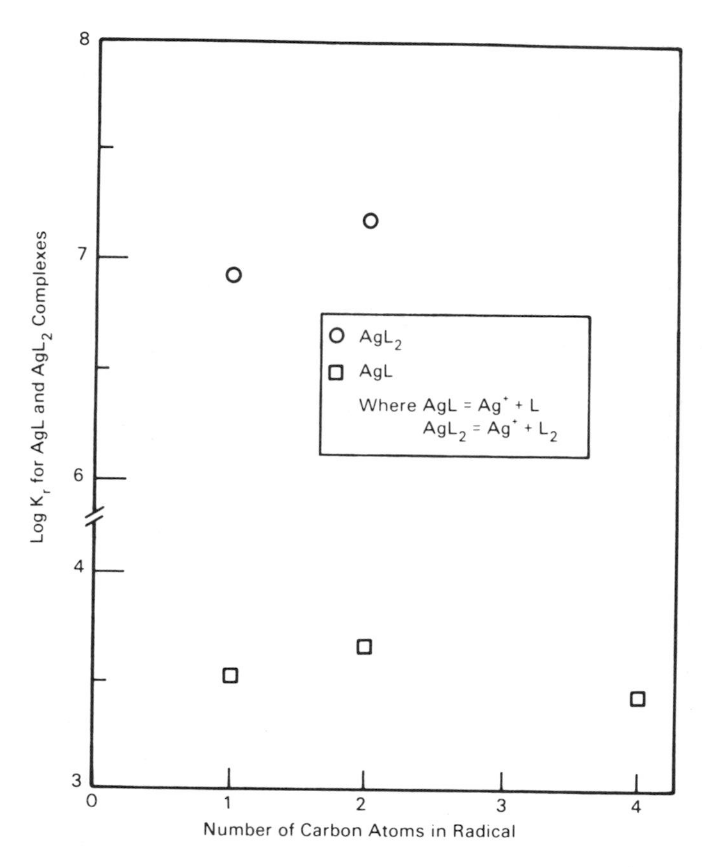

Figure 4. Stability Constants as a Function of the Number of Carbons in the Radical for Amino Acids Using the Data in Table II

The log K_r^0 values chosen to represent the CH_2OOH functional group, the COH_2COOH functional group and the phenolic-OH functional group were those for acetic acid, glycollic acid, and phenol, respectively. The log K_r^0 for the CNH_3COOH functional group was chosen as the average of the reported equilibrium constants, i.e., log K = 3.5 for the AgL complex and 7.0 for AgL_2 complex (where L is some organic ligand). Although Ag^+ forms organic solid complexes with coordination numbers up to 3 with certain organic ligands (i.e., olefins, aromatic compounds and acetylenes) [39], there is no evidence [23,24] that tri-dentate aqueous silver-organic complexes are formed.

A comparison of the two major compilations of organic compounds in seawater (Table III) shows that they are similar except that Dawson [16] lists additional compound groups. These

Table III. Composition of the Soluble Organic Fraction in Seawater by Compound Group and Functional Group

Compound Group	Functional Group	Duursma[a] μg L^{-1}	Duursma[a] %	Duursma[a] % (combined)	Dawson[b] μg L^{-1}	Dawson[b] %	Dawson[b] % (combined)
Free sugars	-OH	14.6 to 36.9	11	11	20	12.5	12.6
Sterol	-OH				0.2	0.1	
Phenol	-OH	3 to 9	3		2	1.2	
Combined amino acids	$-NH_2COOH$	6.9 to 39.1	12	47	50	31.2	43.7
Free amino acids	$-NH_2COOH$	8.0 to 117.4	35		20	12.5	
Fatty acids	-COOH	130	39	39	10	6.2	17.4
Uronic acid	-COOH				18	11.2	
Ketones	C=O				10	6.2	9.3
Aldehydes	C=O				5	3.1	
Urea	$(NH_2)_2C=O$				20	12.5	
Hydrocarbons	-				5	3.1	

(a) Duursma [15]. Data was compiled for the Pacific Ocean by Degens et al. [34], Williams [35] and Lewis and Rakestraw [36] for 0.45 μ filtered seawater.

(b) Dawson [16].

data were composited by functional groups in Table IV. The COH_2COOH functional group is represented by such a small fraction (<<1%) of the organic compounds that, even though it forms the second strongest complex, it is not included in this table and is not considered further.

No stability constant data was found during the compilation of the available stability constants data (Table II) that could be used to represent the sugar-OH or C=O functional groups. The lack of stability constant data for these two functional groups can be handled in several ways. First, assuming the fraction of the organic carbon represented by the sugar-OH and C=O groups does not complex with silver, then 89% of the Duursma organics or 62.3% of the Dawson organics could be represented by the existing data. An alternative approach is to estimate a stability constant for these two functional groups. The complexation of silver with organic compounds containing sugar-OH or C=O functional groups was assumed to be zero because sugars are very unreactive and any metal complexes formed are expected to be very

Table IV. Composition of the Soluble Organic Fraction of Seawater by Functional Group

Functional Group	Duursma %	Dawson %
OH	11	12.6
COOH	39	17.4
CNH_2COOH	47	43.7
C=O	-	9.3
Other	-	15.6
Phenol-OH	3	1.2
	82.9	99.8

weak. The ketone (C=O) and aldehyde (C=O) complexes would probably form even weaker complexes with silver than the phenols or carboxylic acids [28]. Initial simulations indicated that the inclusion of sugars, ketones and aldehydes would not significantly increase the mass of organically complexed Ag where these log K_r^0 values were assumed to be equal to that of carboxylic acid - OH groups.

To convert from the published weight concentration of dissolved organic carbon to molar units, an average molecular weight for the functional group categories had to be estimated. A weighted-average molecular weight of 115 grams for the CNH_2COOH functional group category, 250 grams for the COOH functional group category and 168 grams for phenols was obtained by averaging the formula weights for the compounds given by Duursma [15].

Using the above data, the organic complexation with silver can be calculated by: 1) assuming three functional group categories for the organics, which form silver complexes and can be represented by the stability constants given in Table II; 2) using the average molecular weight; and 3) using the percent distribution of the organic compounds by class calculated from Duursma [15] and alternatively by Dawson [16]. The resulting reactions, weighted average molecular weights, stability constants and percent distribution of functional group classes are given in Table V.

Alternative Approaches

Given the number of assumptions that were made in order to derive the stability constants in Table V, alternate approaches to estimating the stability constants were evaluated.

Table V. Organic Complexes of Silver, the Stability Constants, and Amount in Each Functional Group[a] Used in MINTEQ Geochemical Model

Functional Group	Molecular Weight	Reaction	Log Stability Constant	Amount in Functional Group
-COOH	250.0	$R\text{-}COO^- + H^+ \leftrightarrows R\text{-}COOH^o$	4.76	18.6 to 39
		$R\text{-}COO^- + Ag^+ \leftrightarrows R\text{-}COOAg^o$	0.73	
		$2R\text{-}COO^- + Ag^+ \leftrightarrows (R\text{-}COO)_2Ag^-$	0.64	
		$R\text{-}COO^- + 2Ag^+ \leftrightarrows R\text{-}COO\ Ag_2^+$	1.14	
$-CNH_3COOH$	115.7	$R\text{-}CNH_3COO^- + H^+ \leftrightarrows R\text{-}CNH_3COOH^o$	9.68	43.7 to 47
		$R\text{-}CNH_3COOH + H^+ \leftrightarrows R\text{-}CNH_3COOH_2^+$	2.3	
		$R\text{-}CNH_3COO^- + Ag \leftrightarrows R\text{-}CNH_3COOAg^o$	3.5	
		$2R\text{-}CNH_3COO^- + Ag \leftrightarrows (R\text{-}CNH_3COO)_2Ag^-$	7.0	
Phenol-OH	168.15	$R\text{-}O^- + H^+ \leftrightarrows R\text{-}OH$	9.98	1.2 to 3.0
		$R\text{-}O^- + Ag^+ \leftrightarrows R\text{-}OAg^o$	0.42	

(a) Method of calculating the molecular-weight-averaged stability constant is explained in the text. Functional group concentration is from Table III.

The first alternative method tested was to plot the log K_r^0 values versus the electronegativity of the metals for a large group of metals (Figure 5). Electronegativity can often be used as a measure of the reactivity of an element with ligands, thus, measured stability constants can sometimes be related to electronegativity [29]. However, the scatter in the data was so great ($r^2 \simeq 0.5$) that prediction was not feasible.

The second alternative method used was based on the work of Nieboer and McBryde [30] who showed that stability constants for inorganic complexes could be predicted from plots of log K_r^0

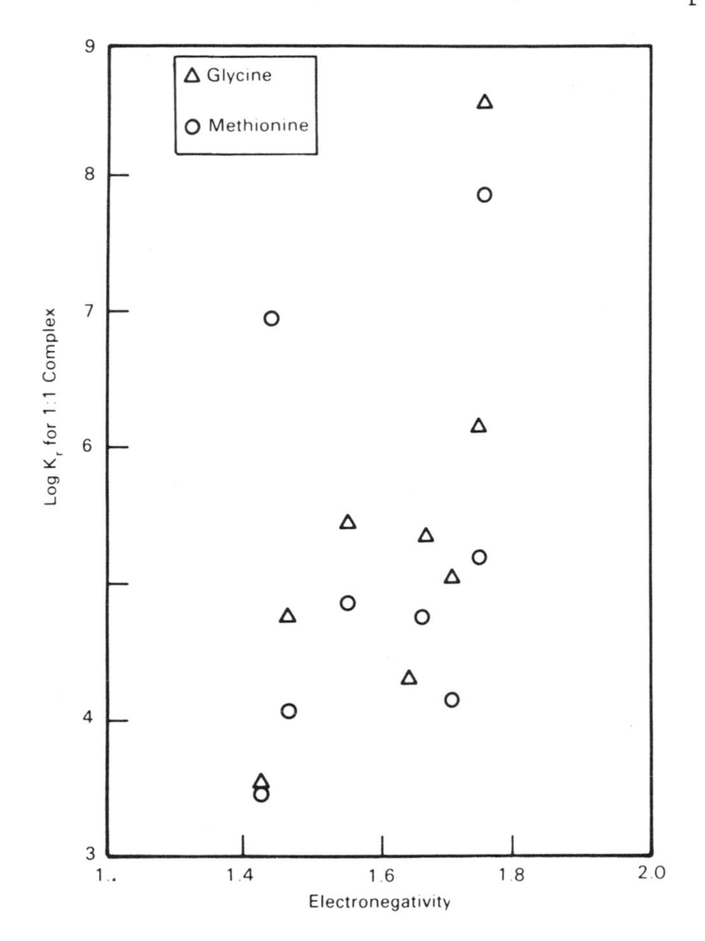

Figure 5. Log K of Reaction Versus Electronegativity of Complexing Metals for the Amino Acids, Glycine and Methionine [Log K values from Smith and Martell [37]; electronegativity from CRC Handbook of Chemistry and Physics, 47th Edition, [38]]

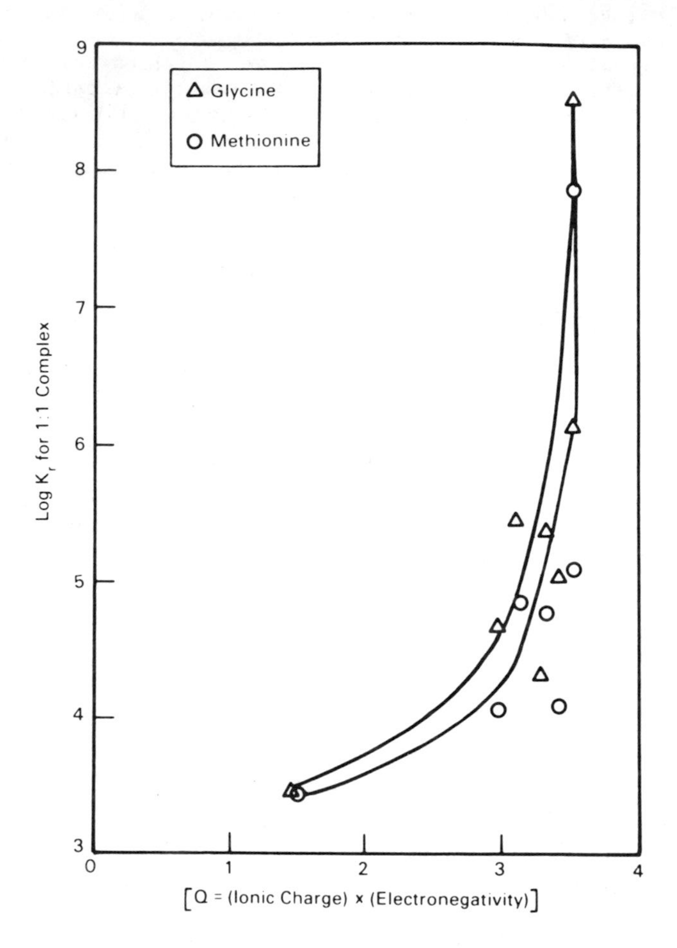

Figure 6. Log K of Reaction Versus Metal Ion Properties for the Amino Acids, Glycine and Methionine [Log K values from Smith and Martell [37]; electronegativity from CRC Handbook of Chemistry and Physics, 47th Edition, [38]]

versus Z•Xm, where Z is the ionic charge and Xm is the electronegativity of the metal ions. Figure 6 is a plot of this data for the amino acids, glycine and methionine. As with Figure 5, the scatter was too great to make estimation of log K_r^0 values for other amino acids feasible.

Sensitivity Analysis

The common-functional-group average stability constants, common-functional-group average-compound molecular weight and total common-functional-group class-compound concentrations given in Table V were used to obtain best estimates of the magnitude and importance of organic ligand complexation to silver speciation in coastal and marine waters. For this simulation study, the amino acid functional group, CNH_3COOH, was assumed to represent 45% of the DOC; the carbonic acid functional group, 42% of the DOC; and phenolic-OH functional group, 2.0% of the DOC. The remaining 11% was assumed to be represented by non-reactive organic matter, mostly the sugars identified by Duursma [15] and Dawson [16].

The activities of the silver-organic complexes were very small, comprising much less than 1% of the total dissolved Ag. A five times increase in the total DOC from 1 mg L^{-1} to 5 mg L^{-1} resulted in a significant increase in the activities of silver-organic complexes but did not change the activity of the free metal ion or the chloride complexes. A dissolved organic matter concentration of several hundreds of milligrams per liter would be required to decrease the neutral Cl complex activity by even as much as 5% at S[-II] = 0 and a much higher concentration of DOC would be required at S[-II] concentrations above zero. Because of the overwhelming importance of S[-II] complexation of Ag as compared to the calculated organic complexation, the observed effects of organic matter on silver uptake and plating (E. A. Crecelius, unpublished results) may actually be due to dissolved S[-II] in the media or to a sulfhydryl group in the organic matter not accounted for in this analysis.

DISCUSSION AND CONCLUSIONS

The importance of the variables examined on the speciation of aqueous Ag decreased in the order S[-II] (0.0 to 10.0 µg L^{-1}), pH (7.0 to 8.2), salinity (27 to 32 $^{o}/oo$) and dissolved organic carbon (1 to 5 mg L^{-1}). The overwhelming effect of even low S[-II] concentrations on speciation indicates the importance of accurate analytical determination of sulfide concentrations in order to reliably compute the Ag speciation. The aqueous speciation of Ag is also dependent on the pH and Cl concentration; predictably, the activity of the silver chloride complexes is dependent on the salinity of the solution. Assuming that $AgCl^0$ is the principal toxic species, then any decrease in pH to below 8.2 would result in a significant increase in toxicity for silver, because of the decrease in the silver hydroxide species and corresponding increase in the silver chloride species with decreasing pH. This suggests that pH must be closely monitored in toxicity studies with an accuracy of 0.05 pH units clearly required but an accuracy of 0.01 pH units desirable. Assuming that the $AgCl^0$ complex is the one that is bioavailable, then the accumulation and toxicity of the Ag will decrease significantly with increases in S[-II] and will decrease significantly with decreases in salinity. We combined the limited available data on organic carbon content in marine waters with the available data on the stability constants of silver with organic compounds to obtain a series of reactions to represent the complexation of Ag and DOC in seawater. Using these reactions and constants, organic complexation of Ag was found to be insignificant.

Acanthite ($Ag_2S(s)$) was calculated to be oversaturated by 0.5 to ~2 orders of magnitude in waters containing S[-II] in the 0.1 to 10 $\mu g\ L^{-1}$ range. Jenne et al. [19] similarly reported acanthite to be oversaturated by many orders of magnitude in Sacramento River water assuming a sulfide concentration of 0.01 $\mu g\ L^{-1}$. Although acanthite is a common silver mineral and known to form in hydrothermal veins [31,32], it is not clear if it forms at ambient temperatures. The calculated oversaturation could result from use of incorrect or inappropriate thermodynamic data, kinetic constraints on precipitation or silver complexation reactions not included in the model. The thermodynamic data used here as well as by Jenne et al. [19] is from studies [33] presumably for a well crystallized sample. This study indicates that organic complexation of silver in seawater is unlikely to be of any consequence. Therefore, we suggest that the most likely explanation for the calculated oversaturation is the occurrence of an amorphic form of silver sulfide. Clearly, thermodynamic data on the amorphic Ag_2S is required to determine if it is a solubility control for silver and to evaluate if an amorphic silver sulfide will form at silver and sulfide levels encountered in estuarine and coastal waters. Although it is not known if acanthite or amorphic Ag_2S precipitates from seawater at ambient temperatures, the potential for acanthite formation further emphasizes the importance of determining the sulfide concentrations of waters used in studies of silver speciation and toxicity.

ACKNOWLEDGMENTS

This work was supported by the U.S. Department of Energy under Contract DE-AC06-76RLO 1830.

REFERENCES

1. Calabrese, A., R. S. Collier, D. A. Nelson, and J. B. MacInnes. "The Toxicity of Heavy Metals to Embryos of the American Oyster Crassostrea virginica," *Marine Biology* 18:162-166 (1973).

2. Luoma, S. N., and G. W. Bryan. "A Statistical Assessment of the Form of Trace Metals in Oxidized Estuarine Sediment Employing Chemical Extractants," *Sci. Total Environ.* 17:165-196 (1981).
3. Pavlou, S. P., R. F. Shokes, W. Hom, P. Hamilton, J. T. Gunn, R. D. Muench, J. Vinelli, and E. A. Crecelius. "Dynamics and Biological Impacts of Toxicants in the Main Basin of Puget Sound and Lake Washington," Volume I, Draft Final Report, Metro Toxicant Pretreatment Planning Study, Draft Task C.1 Report, Municipality of Metropolitan Seattle, Seattle, WA. (in press)
4. Pagenkopf, G. K., R. C. Russo, and R. V. Thurston. "Effect of Complexation on Toxicity of Copper to Fishes," *J. Fish. Res. Bd. Can.* 31:462-465 (1974).
5. Sunda, W. G., and R. R. L. Guillard. "The Relationship Between Cupric Ion Activity and the Toxicity of Copper to Phytoplankton. *J. of Marine Research* 34:511-529 (1976).
6. Engel, D. W., W. G. Sunda, and B. A. Fowler. "Biological Monitoring of Marine Pollutants," in *Biological Monitoring of Marine Pollutants*, F. J. Vernberg, A. Calabrese, F. P. Thurberg, and W. B. Vernberg, Ed. (NY: Academic Press, 1981).
7. Palumbo, A. V., and W. G. Sunda. "Trace Metal Accumulation by Marine Microorganisms." (in press)
8. Andrew, R. W., K. E. Biesinger, and G. F. Glass. "Effects of Inorganic Complexing on the Toxicity of Copper to Daphnia magna." *Water Res.* 11:309-315 (1977).
9. Cowan, Christina E., Everett A. Jenne, and Robert R. Kinnison. "A Methodology for Determining the Toxic Chemical Species of Copper in Toxicity Experiments and Natural Waters," in *Trace Substances in Environmental Health-XVIII*, University of Missouri, Columbia, MO, June 4-7, 1984, D. D. Hemphill, Ed. (in press).
10. Riley, J. P., and R. Chester. *Introduction to Marine Chemistry*, (NY: Academic Press, 1971).
11. Bella, D. A. "Tidal Flats in Estuarine Water Quality Analysis," U.S. Environmental Protection Agency EPA-660/3-75-025 (1975).
12. Karl, D. M. "Microbial Transformations of Organic Matter at Oceanic Interfaces: A Review and Prospectus," *EOS* 63(5):138-140 (1982).
13. Vivit, D. V., J. W. Ball, and E. A. Jenne. "Specific-Ion Electrode Determinations of Sulfide Preconcentrated from San Francisco Bay Waters," *Environ. Geol.* (in press).
14. Stumm, W., and J. J. Morgan. *Aquatic Chemistry: An Introduction Emphasizing Chemical Equilibria in Natural Waters*, (New York, NY: John Wiley and Sons, Inc., 1981).
15. Duursma, E. K. "The Dissolved Organic Constituents of Sea Water," in *Chemical Oceanography*, V. 1, Ch. 11, J. B. Riley and G. Shirrow, Ed., (1965), pp. 433-475.

16. Dawson, R. "Water Soluble Organic Compounds in Seawater," (The Characterized Fraction), in *Symposium on Concepts in Marine Oceanic Chemistry,* (Edinbergh 1976).
17. Jenne, E. A. "Trace Element Sorption by Sediments and Soil--Sites and Processes," in *Symposium on Molybdenum in the Environment,* Volume 2, W. Chappell and K. Petersen, Ed. (NY: M. Dekker, Inc., 1977) pp. 425-553.
18. Mantouro, R. F. C. "Organo-Metallic Interactions in Natural Waters," in *Marine Organic Chemistry: Evolution, Composition, Interactions, and Chemistry of Organic Matter in Seawater,* E. K. Duursma and R. Dawson, Ed., Elsevier Oceanography Serv., (NY: Elsevier Scientific Publishing Company, 1981).
19. Jenne, E. A., D. C. Girvin, J. W. Ball, and J. M. Burchard. "Inorganic Speciation of Silver in Natural Waters - Fresh to Marine," in *Environmental Impacts of Artificial Ice Nucleating Agents,* D. A. Klein, Ed. (Stroudburg, PA: Dowden, Hutchinson and Ross, 1978).
20. Felmy, A. R., D. C. Girvin, and E. A. Jenne. "MINTEQ - A Computer Program for Calculating Aqueous Equilibria," NTIS PB 84157148 (1984).
21. Ball, J. W., E. A. Jenne, and D. K. Nordstrom. "Additional and Revised Thermochemical Data and Computer Code for WATEQ2--A Computerized Chemical Model for Trace and Major Element Speciation and Mineral Equilibria of Natural Waters," U.S. Geol. Survey Water Res. Invest. 78-116 (NTIS PB-80 224 140) (1980).
22. Nordstrom, D. K., et al. "A Comparison of Computerized Chemical Models for Equilibrium Calculations in Aqueous Systems," in *Chemical Modeling in Aqueous Systems,* ACS Symposium Series 93, E. A. Jenne, Ed. (Washington, DC: American Chemical Society, 1979).
23. Martell, A. E., and R. M. Smith. *Critical Stability Constants. V. 1: Amino Acids,* (New York, NY: Plenum Press, 1974).
24. Martell, A. E., and R. M. Smith. *Critical Stability Constants. V. 3: Other Organic Ligands,* (New York, NY: Plenum Press, 1977).
25. Gagosian, R. B. "Review of Marine Organic Geochemistry," *Reviews of Geophysics and Space Physics* 21:1245-1258 (1983).
26. Ehrhardt, M. "Organic Substances in Seawater," *Mar. Chem.* 5:307-316 (1977).
27. Perrin, D. D., B. Dempsey, and E. P. Serjent. *pKa Reduction for Organic Acids and Bases,* (New York, NY: Chapman and Hall, 1981).
28. Morrison, R. T., and R. N. Boyd. *Organic Chemistry,* 3rd Edition, (Boston, MA: Allyn and Bacon, Inc., 1973).
29. Langmuir, D. "Techniques of Estimating Thermodynamic Properties for Some Aqueous Complexes of Geochemical Interest," in *Chemical Modeling in Aqueous Systems,* ACS Symposium Series 93, E. A. Jenne, Ed. (Washington, DC: American Chemical Society, 1979).

30. Nieboer, E., and W. A. E. McBryde. "Free-Energy Relationships in Coordination Chemistry. III. A Comprehensive Index to Complex Stability," *Can. J. Chem.* 51:2512-2524 (1973).
31. Correns, C. W. *Introduction to Mineralogy,* (New York, NY: Springer-Verlag, 1969).
32. Bates, R. L., and J. A. Jackson (Ed). *Glossary of Geology,* (Falls Church, VA: American Geological Institute, 1980).
33. Robie, R. A., and D. R. Waldbaum. "Thermodynamic Properties of Minerals and Related Substances at 298.15°K (25°C) and One Atmosphere (1.013 Bars) Pressure and at Higher Temperatures," U.S. Geol. Survey Bulletin 1259, (1968) 256 p.
34. Degens, E. T., J. H. Reuter, and K. N. F. Shaw. "Biochemical Compounds in Offshore California Sediments and Seawaters," *Geochemica et Cosmochimica Acta* 28:45-66 (1964).
35. Williams, P. M. "Organic Acids in Pacific Ocean Waters," *Nature* 189:219-220 (1961).
36. Lewis, G. J., and N. W. Rakestraw. "Carbohydrates in Seawater," *J. Mar. Res.* 14:253 (1955).
37. Smith, R. M., and A. E. Martell. *Critical Stability Constants Vol. 2: Amines,* (New York, NY: Plenum Press, 1976).
38. Weast, Robert C., and Samuel M. Selby, Ed. *CRC Handbook of Chemistry and Physics,* (Cleveland, OH: The Chemical Rubber Co., 1966-1967).
39. Cotton, F. A., and G. Wilkinson. *Advanced Inorganic Chemistry: A Comprehensive Text,* Third Edition, (New York, NY: Interscience Publishers, New York, 1972).

CHAPTER 21

SPARTINA ALTERNIFLORA LITTER IN SALT MARSH GEOCHEMISTRY

Robert E. Pellenbarg
Combustion and Fuels Branch, Code 6180
Chemistry Division
The Naval Research Laboratory
Washington, D. C.

ABSTRACT

Abiotic interactions between *Spartina alterniflora* litter and the waters and sediments of a salt marsh are examined in this review paper. The litter is shown to be capable of scavenging the aqueous surface microlayer and associated trace metals from marsh tidal waters. Such litter-water interactions can lead to an eight-fold increase in litter zinc content in the short term. Litter sediment interchanges are affected by redox processes in the anoxic sediments of the marsh which seasonally release dissolved, reduced iron. As the reduced iron is oxidized at the sediment surface, and retained in part as a coating on litter there, other metals such as copper and zinc are enriched in the litter by coprecipitation.

INTRODUCTION

Salt marshes are ecosystem-level interfaces between fresh and salt water regimes. Major components of this interface are sediments, tidal saline waters, and biota. The marsh as a whole processes matter and energy (1) requiring the major elements of the marsh to exchange matter and energy (2-4). A driving force for this movement of matter and energy is tidal flow, which alternately fills and empties the marsh, its major channels, and the smallest feeder channels. Tidal waters transport dissolved material (inorganic salts, and a rich organic "soup"), entrained particles (seston), and an enriched microlayer supported on the water itself (2). These waters impact and influence both the marsh sediments, and the biota of the marsh, but identifying and quantifying such interactions is difficult.

A major component of marsh biota is the vascular plant *Spartina alterniflora*, which densely occupies salt marshes of the East and Gulf Coasts of the United States. In addressing marsh component interactions, one can examine how *Spartina* can interact with the waters or sediments of the salt marsh. Seasonal growth in any ecosystem yields discarded organic matter which is more or less stable on a short time scale. *Spartina* plants shed leaves, stems, stalks and other physical matter which, to a first approximation, are non-biotic, non-respiring. This litter often accumulates in layers and windows in the marsh, and is ultimately converted by microbial activity over months to nutrient and energy

MARINE AND ESTUARINE GEOCHEMISTRY, Sigleo, A. C., and A. Hattori (Editors)

rich particulates for export from the marsh [5-7]. However, in the shorter term, this litter can interact with the tidal waters and sediments of the marsh in an abiotic sense.

As tidal waters fill the salt marsh drainage channels, and percolate across the marsh sediments (especially on spring tides which maximally inundate the marsh), these waters periodically contact recumbent Spartina litter. Associated with the tidal waters is a layer rich in surface active organic compounds, organically coated particulates, trace metals and other components of the aqueous surface microlayer [8-10]. As Spartina litter is an organic substance, one can examine the degree to which the lignin- and cellulose-rich litter could scavenge the hydrophobic microlayer as the litter periodically interact with salt marsh tidal flow. This abiotic micro-layer litter interaction can yield litter greatly enriched in trace metals which have been concentrated in the microlayer by physical and chemical processes beyond the scope of the present research. (See (4) for comments in this regard). Note further that the recumbent litter contacts the marsh sediments, which are anoxic except for the upper few millimeters. Chemical processes, especially those related to redox chemistry of the sediments, have the potential to influence the trace metal content of the litter. This aspect of marsh component interactions has also been examined.

Specifically, three aspects of marsh component interactons will be examined. Data will be presented and discussed which show that over so short a time as a tidal cycle the litter can abiotically scavenge significant amounts of trace metals from the enriched microlayer residing on the waters of the salt marsh. Secondly, results from a longer term field study will document how the litter exhibits seasonal and spatial variability in trace metal content as a result of litter sediment, and litter-water/microlayer interactions. And lastly, analysis of field litter samples will be discussed in an initial explanation of some of the mechanisms, and physio-chemical processes, which influence the interactions between the litter and other major components of the salt marsh.

METHODS

All research discussed in this paper took place in Canary Creek Marsh near Lewes, Delaware U.S.A. (Figure 1). To examine the abiotic microlayer scavenging by Spartina litter, a large portion of the recumbent litter was collected from the marsh, washed/rinsed by hand with distilled water, and oven dried. Aliquots (10 g each) of this processed litter were tied and bundled with nylon monofilament fishing line or PVC webbing stitched in the form of a large pouch. The bundles were deployed below the water surface in Canary Creek (to interact with seston and dissolved components) and to reside at the air/water interface (to interact with seston, dissolved components, and microlayer) for approximately 12 hours. Also, some of the pouches of litter were deployed on the marsh sediments for two weeks so that the pouches and contained litter would interact on a periodic basis with tidal flow in the marsh. All deployed litter portions were rinsed either in the field and/or in the laboratory to dislodge loosely adherent particulates, then freeze dried, and subsampled for trace metal content analysis.

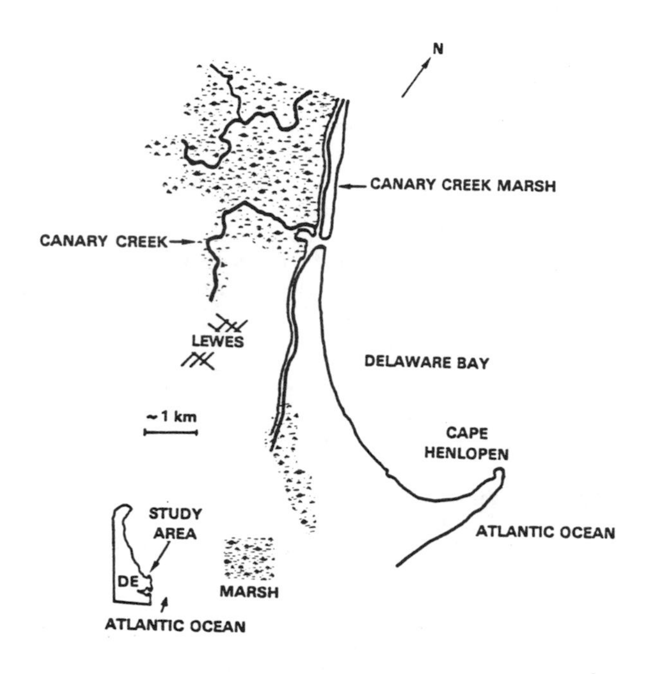

Figure 1: Study area in Delaware, USA.

Small portions (0.2 g each) of the various deployed litter aliquots were wet ashed overnight in covered Teflon crucibles with gentle bottom heat and quartz-distilled nitric acid. Undigested cellulose residues were centrifuged out, and the supernatant transferred to Teflon containers after dilution to 50 ml volume with quartz-distilled water. All aliquots were analyzed with flame and flameless atomic absorption spectrophotometry, using standard analytical procedures. Process blanks gave low- to no-detectable metal content.

Later, in order to evaluate long term litter-microlayer and litter-sediment interactions, samples of recumbent litter were gathered on transects perpendicular to a major water course in Canary Creek (Figure 1). The samples (20 g wet each) were collected from the mat of litter 1, 6, 12, 16, and 24 meters from the creek bank. Litter obviously contaminated with clay materials was rejected, as was standing litter and live plant parts. Collections were in the fall, winter, and late spring of one year (1981-82).

The *Spartina* litter samples were freeze dried in the laboratory, and subsampled in replicate for analysis. Subsamples consisted of 1 g random portions of the stalks, stems and/or leaves which made up each bulk sample. Some replicates were wet ashed with nitric acid to give total metal content analysis, while other replicates were treated to yield data on the physicochemical form of the metals associated with the litter [9, 11]. For these determinations some replicates were extracted with acetone (to yield organically associated metals), exposed to 1 M magnesium chloride (Fisher) solution (to assay exchangeable metals), or interacted with 0.04 M hydroxylamine hydrochloride in 25% (v/v) glacial acetic acid (all Fisher) to quantify metals associated with iron or manganese oxides. (See [12] for a more thorough discussion). The various extracts were dried and wet ashed (acetone residuals) or filtered and diluted prior to analysis by flameless and flame atomic absorption spectrophotometry.

RESULTS AND DISCUSSION

The data in Table 1 clearly show that over short-to-medium term exposure, litter can become significantly enriched in metal content. It should be noted that this situation is essentially non-biotic in nature, and that the mechanism of enrichment in the litter appears to be simple coating of the litter by the intercepted microlayer. Litter deployed in the water column was enriched, but not to the extent displayed by the litter floating at the air-water interface, or residing on the marsh sediment surface. For example, submerged litter was enriched by a factor of 2 in zinc content, whereas litter at the water surface was enriched almost 8-fold in zinc.

Table 1 - Metals* Associated with Control and Deployed *Spartina* Litter

Sample	Copper	Zinc	Iron
Control Litter	2.4	1.7	440
Deployed in Water Column	6.0	4.1	630
Increase	250%	240%	140%
Deployed at Water Surface	8.3	12.9	1640
Increase	350%	760%	370%
Deployed in Marsh	6.0	4.6	1950
Increase	250%	270%	440%

* - All data are on a wt/wt basis, parts per million, adapted from Pellenbarg [13].

A major significance of this litter-metal enrichment mechanism is seen in the following context. Canary Creek marsh yields some 9.4 x 10^8g of dry weight above ground Spartina biomass per year. Much of this biomass becomes litter, a portion of which can decompose to fine detritus for export from the marsh [6-7]. If half of the litter is enriched to the degree indicated by the data in Table 1, some 3 x 10^3g of scavenged zinc could be exported from the marsh. This zinc, and other metals similarly associated with the litter/detritus can be carried rapidly and directly to those components of marsh biota which rely on detritus as a food source. Larval and juvenile forms of crustaceans, fish, and even predators on these elements low in the marsh food chain could be easily impacted by metals sequestered to Spartina litter in the salt marsh.

Previously discussed work had shown that Spartina litter can become enriched in metals upon exposure to the aqueous surface microlayer [13]. The degree of enrichment would be a function of frequency of contact between the litter and metal rich microlayer. Thus, one can speculate that litter closer to a creek as the major source of microlayer covered water would be exposed more often to the microlayer than would litter further from the creek bank. The data in Table 2 pertain in this situation, and

Table 2 - Metals* in Litter as a Function of Distance from Creek Bank, Canary Creek Marsh

Season	Fall			Winter			Spring/Summer		
Distance Inland(m)	Cu	Zn	Fe	Cu	Zn	Fe	Cu	Zn	Fe
1.2	2.3	21	3100	7.5	50	4000	5.0	32	2700
6	3.7	34	3400	10.7	71	8100	6.7	35	5300
12	3.8	35	4200	11.0	77	7900	6.7	47	6000
16	2.9	29	3500	9.6	61	5600	5.1	46	5100
24	3.1	29	3500	10.0	65	6400	6.4	42	5200

* Data are wt/wt parts per million, adapted from Pellenbarg [12].

show that for several metals in Canary Creek Marsh, litter metal content was maximum some 10 meters from the creek bank. Repeated and complete inundation of the litter closest to the creek bank, and interaction with water stripped of its microlayer are offered as explanation to a mid-marsh litter metal maximum: the mid marsh litter would intercept and retain microlayer before it can be carried farther into the marsh.

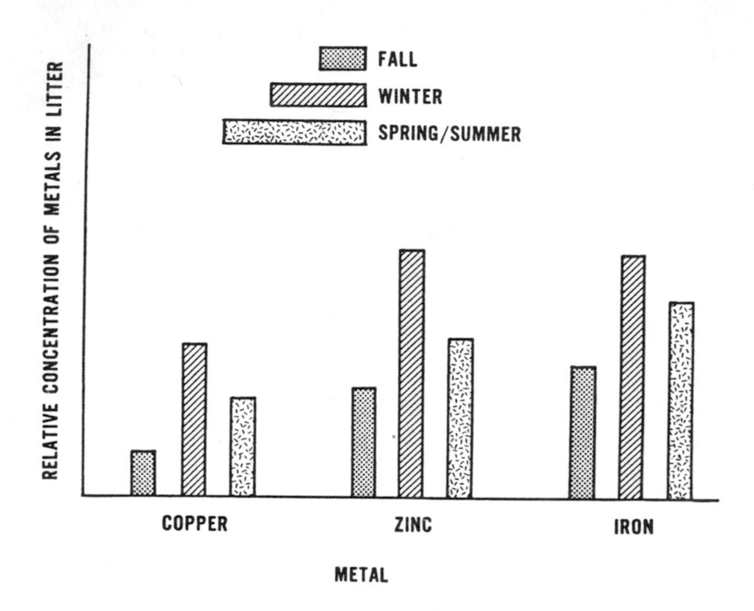

Figure 2: Metals in Spartina litter as a function of season. Copper values range from 3 to 10 ppm; zinc from 30 to 65 ppm; iron from 3500 to 6500 ppm, all on a wt/wt basis.

The data in Table 3 indicate that if one groups the data as a function of season, with the transect stations considered to be portions of a seasonal sample, then litter metals are maximal in the winter (see Figure 2, also). Seasonal sedimentary biological processes will affect litter metal content. Iron and manganese are seasonally released as reduced ions from anoxic marsh sediments [14]. These reduced ions are oxidized at the oxic sediment surface, and, it is hypothesized, can be retained as a coating on recumbent litter there. Biological processes in the sediments release dissolved, reduced iron in the late summer and fall as seasonal organic matter is mineralized with attendant sulfate reduction. In the spring, atmospheric oxygen will react with sedimentary pyrite, yielding hydrogen ion, and sulfate. These seasonal mixes of hydrogen ion, and reduced iron and manganese, will help mobilize iron, and can lead to a net export of dissolved iron from the marsh [2]. However, in the winter, both local oxidation and reduction rates are slowed by biological dormancy, so that previously deposited iron will not be removed by contact with more acidic pore waters exuding from the sediments at the same rate as in biologically active seasons.

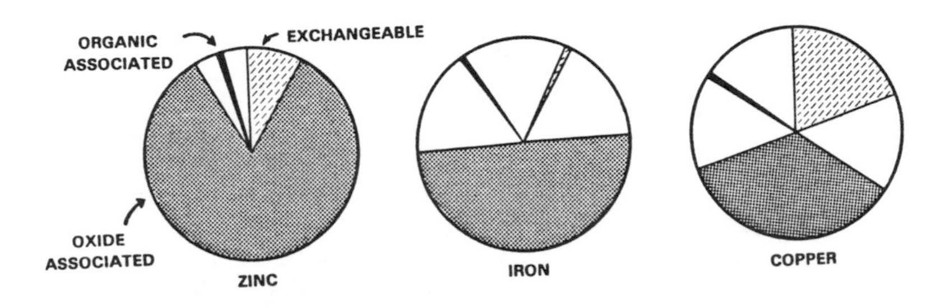

Figure 3: Metal distributions as a function of operationally defined class. Blank sections in the figures are undefined residuals.

Table 3 provides data which indicate that in general, only a very small proportion of the metals can be considered to be associated with organically soluble (acetone extractable) substances (see Figure 3, also). As acetone is a polar solvent, it is reasonable to consider that it would dissolve polar organic materials. Carbohydrates and lignins, and their related degradation products humic and fulvic acids, are polar macromolecules, with affinities for transition metals. It is postulated that such polar macromolecules and associated metals are being collected by the acetone extraction step. However, the data in Table 3 indicate that an appreciable fraction of the copper and zinc are associated with the exchangeable, and especially the oxide class. This fact appears related to the seasonal mobilization of reduced iron which then oxidizes and precipitates on the Spartina litter, co-precipitating quantites of zinc and copper at the same time. Thus, Spartina litter appears as a material which is directly impacted by a major chemical process occurring in the sediments of the salt marsh.

In summary, the trace metal content of Spartina alterniflora litter is directly influenced by salt marsh biogeochemistry. At least two processes can mobilize trace metals which can then be sequestered by the marsh litter. Specifically, the litter can become enriched in trace metals by sorbing metal rich aqueous surface microlayer in what is postulated to be a largely physical process. Furthermore, the litter appears to serve as a collection surface - a substrate - for the deposition of metals,

Table 3 - Litter Metals Proportionality (%) As A Function of Class, Canary Creek Marsh

	FALL			WINTER			SPRING		
	Organic	Exchangeable	Oxide	Organic	Exchangeable	Oxide	Organic	Exchangeable	Oxide
Copper	2	16	45	1	20	35	1.6	14	30
Zinc	0.2	6	82	2	8	82	0.3	6	71
Iron	0	0.1	49	0	0.1	49	0	0.4	29

* Adapted from Pellenbarg [13].

especially iron and manganese which exhibit a redox chemistry under natural environmental conditions. These metals are seasonally mobilized by biological processes in the marsh sediments leaving the sediments as reduced ions, to be deposited on litter in part as oxides. Needed now are data which will show how metals are transferred to the microlayer, and which address the ultimate fate of the metal enriched litter discussed here. Further research should address the impact of metal enriched litter on other biotic components in the marsh ecosystem, and examine the role of the living plant *Spartina alterniflora*, in affecting the redox chemistry of the marsh sediments in which it is rooted.

LITERATURE CITED

1. Woodwell, G. M., Whit[illegible]hton, R. A. 1977. The Flax Po[illegible]of carbon in water between a sa[illegible]. Limnology and Oceanog[illegible]
2. Pellenbarg, R. E., an[illegible]uarine surface microlayer an[illegible]t marsh. Science, 203, 1010-10[illegible].
3. Lion, L., Harvey, R., Young, L., and Leckie, J. 1979. Particulate matter. Its association with microorganisms and trace metals in an estuarine salt marsh microlayer. Environmental Science and Technology, 13, 1522-1526.
4. Lion, L., and Leckie, J. 1982. Accumulation and transport of Cd, Cu, and Pb in an estuarine salt marsh surface micro-layer. Limnology and Oceanography, 27, 111-125.
5. Odum, E. P., and de la Cruz, A. A. 1963. Particulate organic detritus in a Georgia salt marsh-estuarine eco-system. In Estuaries (Lauff, G. H., ed.). American Association for the Advancement of Science, Washington, D. C., Publication #83.
6. Pomeroy, L. R., and Wiegert, R. G. 1981. The Ecology of a Salt Marsh. Springer-Verlag, New York, 271 pp.
7. Banus, M. D., Valiela, I., and Teal, J. M. 1975. Lead, zinc and cadmium budgets in expermentally enriched salt marsh ecosystems. Estuarine and Coastal Marine Science, 3, 421-430.
8. Garrett, W. D., 1965. Collection of slick-forming materials from the sea surface. Limnology and Oceanography, Vol. 10, 602-605.
9. Pellenbarg, R. E. 1981. Trace metal partitioning in the aqueous surface microlayer of a salt marsh Estuarine, Coastal, and Shelf Science, 13, 113-117.
10. Baeir, R. E., Goupil, D. W., Perlmutter, S., and King, R. 1974. Dominant chemical composition of sea-surface films, natural slicks, and foams. Journal de Recherches Atmos-pheriques, 8, 571-600.
11. Tessier, A., Campbell, P. G., and Bisson, M. 1979. Se-quential extraction procedure for speciation of particulate trace metals. Analytical Chemistry, 51, 44-851.
12. Pellenbarg, R. E. 1984. On Spartina alterniflora litter and the trace metal biogeochemistry of a salt marsh. Estuarine, Coastal, and Shelf Science, 18, 331-346.
13. Pellenbarg, R. E. 1978. Spartina alterniflora litter and the aqueous surface microlayer in the salt marsh. Estuarine, Coastal, and Shelf Science, 6, 187-195.
14. Lord, C. J., 1980. The Chemistry and Cycling of Iron, Manganese and Sulphur in Salt Marsh Sediment. Ph. D. Dissertation, University of Delaware, Newark, 176 pp.

CHAPTER 22

THE DISTRIBUTION OF CHEMICAL ELEMENTS IN SELECTED MARINE ORGANISMS: COMPARATIVE BIOGEOCHEMICAL DATA

Toshio Yamamoto, Yukiko Otsuka
Kazumasa Aoyama, Hiroki Tabata and Ken-ichi Okamoto
Department of Chemistry
Kyoto University of Education
Fushimi-ku, Kyoto 612, Japan

ABSTRACT

A Systematic study of 44 elements in various Japanese seaweeds (245 samples) was carried out by chemical and neutron activation analyses. Marine phytoplankton, marine zooplankton and freshwater angiosperms were analyzed for comparative data. A method to compare multi-element data in many samples by the seawater concentration and ocean residence time of the element is proposed. The results of these calculations indicate that Japanese seaweeds have slightly higher contents of elements with long resident times (Fe and Al), whereas phytoplankton contain higher contents of elements with shorter residence times (Na and Mg).

INTRODUCTION

Geochemistry was concerned originally with the chemical composition of the earth. There has been renewed interest in element data on geochemical materials due to the recent awareness of environmental problems. The latest developments in instrumental analysis facilitate environmental measurements and increase the data available on various materials. These data on elemental compositions must be grouped and arranged in a convenient way for comparative studies. Principles governing the distribution of chemical elements in living organisms and in environmental substances may prove useful for the arrangement of element data (1-4). The present study, based on such a principle, was intended for obtaining a general view on the distribution of elements in marine organisms.

EXPERIMENTAL

Samples of Japanese seaweeds, marine phytoplankton, marine zooplankton and freshwater angiosperms were washed thoroughly with distilled water and dried. Before analysis, the samples were dried to constant weight at 105°C and ashed in a muffle furnace at 500°C. The ash samples were analyzed for several elements by neutron activation analysis simultaneously. Additional information on species, collection site and collection data for each sample, and on the analytical procedure for each element can be obtained from a related monograph (4).
For all calculations in this paper the values of ocean residence time and chemical abundance in seawater were taken from

MARINE AND ESTUARINE GEOCHEMISTRY, Sigleo, A. C., and A. Hattori (Editors)

Broecker and Peng (5). The concentration factors were calculated on the basis of wet material assuming that they contained 75% water.

Concentration Ratio to Seawater and Ocean Residence Time

To set up an appropriate reference which can be used for comparing the data on various geochemical substances, the concentration ratio of an element in a sample to its concentration in seawater was fixed as a reference (6), along with the ocean residence time of the element. Because the definition of ocean residence time contains the chemical abundance in seawater as the variable part of the numerator (5), the coefficients result in a negative number.

For this technique of data analysis for each element the log of the concentration factor (y) is inversly proportional to the log of the residence time (x). For nonmarine organisms, the ratio of the level of each element relative to the of seawater can be substituted for the concentration factor (y).

The regression is expressed as the equation:

$$\log y = \log a + b \log x$$

The value of log a is the intercept on the y axis and the slope of the graph is given by the value b. The correlation coefficient r between log x and log y can be calculated from the individual data. From the values of log a, b and r, the distribution of elements in each sample can be obtained. Using the values of log a, b and the equation, the theoretical value of the concentration factor y_o for each element can be calculated.

The values of y/y_o indicate the ratio of the measured concentration factor for each element with respect to the calculated (theoretical) value. Since y and y_o have a common denominator, y/y_o is also the ratio of the measured concentration for each element to the theoretical one. Thus, the value of y/y_o indicates a real excess or deficiency of a given element with respect to the theoretical (calculated) value.

RESULTS AND DISCUSSION

The abundances of up to 44 elements in various Japanese seaweeds (245 samples of 69 species), marine phytoplankton (16 samples composed of mixed species), marine zooplankton (39 samples of 26 species) and freshwater angiosperms (9 samples of 8 species) are summarized in Tables I through IV. The values reported here for seaweeds fall within the ranges reported by other investigators (8-16).

The method of data analysis was applied to the samples for more than seven elements whose ocean residence times are known. Figure 1 shows the relationship for marine phytoplankton (mean values). There was an excellent relationship between concentration factor and residence time for the selected organisms. Table V summarizes the results of the regression analysis for six groups of organisms, including literature values. Within individual biological groups, values for log a and b varied slightly, with an average standard deviation of 7% relative to the mean. All groups of

Table I. The Range of Concentrations of Elements in Japanese Seaweeds

(* mg/Kg dried material)

Element	Number of samples	Minimum*	Maxmum*	Mean*	S.D./Mean
B	102	1.0×10^{1}	3.37×10^{2}	9.9×10^{1}	0.62
Na	90	2.0×10^{2}	6.66×10^{4}	9.6×10^{3}	0.92
Mg	114	3.0×10^{2}	5.34×10^{4}	1.13×10^{4}	0.76
Al	161	2.6×10^{1}	5.7×10^{3}	6.97×10^{2}	1.3
Si	35	4.4×10^{2}	2.18×10^{4}	4.78×10^{3}	1.1
P	53	2.0×10^{2}	2.8×10^{3}	1.05×10^{3}	0.43
Cl	24	2.7×10^{2}	3.4×10^{4}	3.7×10^{3}	1.9
K	87	3.0×10^{2}	4.6×10^{4}	1.25×10^{4}	0.78
Ca	129	2.8×10^{3}	3.16×10^{5}	2.04×10^{4}	1.9
Sc	52	1.1×10^{-2}	6.5	3.15×10^{-1}	2.9
Ti	53	1.8	1.78×10^{2}	3.51×10^{1}	1.1
V	76	2.8×10^{-1}	1.1×10^{1}	3.65	1.2
Cr	88	2.6×10^{-1}	7.3	1.68	0.79
Mn	115	4	1.2×10^{3}	1.04×10^{2}	1.3
Fe	215	2.5×10^{1}	3.41×10^{3}	5.42×10^{2}	1.2
Co	107	4.3×10^{-2}	4.54	7.4×10^{-1}	0.89
Ni	77	1.1×10^{-1}	8.62	2.67	0.72
Cu	66	6.1	2.77×10^{1}	1.39×10^{1}	0.40
Zn	135	1.7×10^{1}	6.8×10^{2}	1.45×10^{2}	0.79
Ga	58	2×10^{-2}	6.4×10^{-1}	1.4×10^{-1}	0.89
As	43	1.2	1.3×10^{2}	2.4×10^{1}	1.4
Se	17	1.5×10^{-2}	3.6×10^{-1}	1.4×10^{-1}	0.74
Br	51	2.7	7.4×10^{2}	1.8×10^{2}	1.0
Rb	52	4.1×10^{-1}	2.6×10^{1}	6.9	0.82
Sr	90	2.0×10^{1}	1.15×10^{4}	1.19×10^{3}	1.4
Mo	73	6×10^{-2}	1.16	3.4×10^{-1}	0.71
Ag	17	6.8×10^{-2}	7.7×10^{-1}	2.6×10^{-1}	0.92
Sb	33	4.0×10^{-2}	6.1	3.6×10^{-1}	2.9
I	14	1.4×10^{1}	2.42×10^{3}	4.34×10^{2}	1.5
Cs	51	1.0×10^{-2}	3.5×10^{-1}	9.3×10^{-2}	1.1
Ba	35	5.8	6.4×10^{1}	3.0×10^{1}	0.47
La	23	8.9×10^{-2}	2.0	7.6×10^{-1}	0.69
Ce	46	9.0×10^{-2}	5.3	1.2	1.0
Sm	27	1.0×10^{-2}	4.3×10^{-1}	1.4×10^{-1}	0.76
Eu	51	1.2×10^{-3}	8.6×10^{-2}	1.8×10^{-2}	1.1
Tb	17	6.4×10^{-3}	7.4×10^{-2}	2.7×10^{-2}	0.78
Yb	12	2.3×10^{-2}	2.1×10^{-1}	8.8×10^{-2}	0.61
Lu	13	6.1×10^{-4}	4.8×10^{-2}	2.0×10^{-2}	0.77
Hf	26	1.2×10^{-2}	5.5×10^{-1}	1.8×10^{-1}	0.90
Ta	7	5.8×10^{-3}	7.7×10^{-2}	3.8×10^{-2}	0.64
Hg	9	5.7×10^{-3}	2.9×10^{-1}	1.1×10^{-1}	1.1
Pb	2	7	1.2×10^{1}	1.0×10^{1}	0.37
Th	43	1.8×10^{-3}	6.9×10^{-1}	1.8×10^{-1}	1.1
U	17	1.6×10^{-1}	7.1	1.1	1.6

Table II. The Range of Concentrations of Elements in Marine Phytoplankton

(* mg/Kg dried material)

Element	Number of samples	Minimum*	Maximum*	Mean*	S.D./Mean
B	2	2.9×10^{1}	6.1×10^{1}	4.5×10^{1}	0.50
Na	16	7.2×10^{2}	3.1×10^{3}	1.43×10^{3}	0.51
Mg	16	4.15×10^{3}	1.1×10^{4}	5.89×10^{3}	0.26
Al	3	3.67×10^{3}	4.51×10^{3}	4.07×10^{3}	0.10
Si	16	9.3×10^{4}	9.1×10^{5}	2.46×10^{5}	0.75
P	16	2.83×10^{3}	4.25×10^{3}	3.45×10^{3}	0.14
K	16	1.98×10^{4}	1.33×10^{5}	6.87×10^{4}	0.53
Ca	16	4.63×10^{3}	8.49×10^{3}	6.41×10^{3}	0.17
Mn	16	7.7×10^{1}	3.93×10^{2}	2.11×10^{2}	0.34
Fe	16	6.24×10^{3}	2.28×10^{4}	1.14×10^{4}	0.40
Zn	10	1.04×10^{2}	1.76×10^{3}	6.6×10^{2}	0.96

Table III. The Range of Concentrations of Elements in Marine Zooplankton

(* mg/Kg dried material)

Element	Number of samples	Minimum*	Maximum*	Mean*	S.D./Mean
B	14	3	2.6×10^{1}	1.1×10^{1}	0.52
Na	39	1.24×10^{3}	2.44×10^{4}	4.37×10^{3}	1.2
Mg	39	6.4×10^{2}	4.68×10^{4}	4.27×10^{3}	1.7
Al	3	1.12×10^{2}	1.26×10^{3}	4.95×10^{2}	1.3
Si	38	6.5×10^{2}	1.42×10^{4}	3.7×10^{3}	0.99
P	39	5.1×10^{2}	1.28×10^{4}	4.94×10^{3}	0.49
K	39	1.92×10^{3}	3.02×10^{4}	5.33×10^{3}	0.97
Ca	39	3.9×10^{2}	1.72×10^{4}	4.36×10^{3}	0.89
Mn	29	2	1.88×10^{2}	2.9×10^{1}	1.6
Fe	39	8.5×10^{1}	5.21×10^{3}	6.2×10^{2}	1.6
Zn	32	2.0×10^{1}	1.07×10^{3}	2.99×10^{2}	0.86

Table IV. The Range of Concentrations of Elements in Freshwater Angiosperms

(*mg/Kg dried material)

Element	Number of samples	Minimum*	Maxmum*	Mean*	S.D./Mean
B	3	1.2×10^{1}	2.6×10^{1}	1.8×10^{1}	0.39
Na	8	6.4×10^{2}	1.07×10^{4}	6.2×10^{3}	0.60
Mg	8	1.16×10^{3}	3.77×10^{3}	2.47×10^{3}	0.37
Al	6	2.43×10^{2}	3.3×10^{3}	1.86×10^{3}	0.72
Si	2	1.02×10^{4}	1.18×10^{4}	1.10×10^{4}	0.10
P	8	1.47×10^{3}	6.18×10^{3}	3.55×10^{3}	0.50
K	8	8.28×10^{3}	7.05×10^{4}	2.53×10^{4}	0.82
Ca	8	2.22×10^{3}	1.68×10^{4}	8.14×10^{3}	0.64
Sc	1	-	-	4.8×10^{-2}	-
Ti	3	2.81×10^{1}	6.99×10^{1}	4.49×10^{1}	0.49
V	1	-	-	1.24	-
Cr	4	6.1×10^{-1}	2.91	1.66	0.66
Fe	9	2.25×10^{2}	8.3×10^{3}	3.36×10^{3}	0.82
Co	2	1.3×10^{-1}	2.86	1.5	1.3
Ni	1	-	-	3.74	-
Zn	4	6.4×10^{1}	5.98×10^{2}	2.61×10^{2}	0.91
As	1	-	-	4.2×10^{-1}	-
Br	1	-	-	1.6×10^{1}	-
Rb	1	-	-	2.9×10^{1}	-
Sr	3	1.8×10^{1}	4.2×10^{1}	3.1×10^{1}	0.39
Mo	6	2.6×10^{-1}	4.6	1.12	1.5
Ag	1	-	-	4.8×10^{-2}	-
Sb	1	-	-	9.4×10^{-2}	-
Cs	1	-	-	5.3×10^{-2}	-
Ba	1	-	-	3.3×10^{1}	-
Ce	1	-	-	3.5×10^{-1}	-
Sm	1	-	-	4.2×10^{-2}	-
Eu	1	-	-	6.4×10^{-3}	-
Tb	1	-	-	1×10^{-2}	-
Yb	1	-	-	2.5×10^{-2}	-
Lu	1	-	-	4.5×10^{-3}	-
Hf	1	-	-	1.8×10^{-2}	-
Ta	1	-	-	6.9×10^{-3}	-
Th	1	-	-	7.4×10^{-1}	-

seaweeds, for example, show a range of similar values for log a and b which might be characteristic of seaweeds. Differences in these values between differenct biological groups were significant (Table V).

Figure 2 shows the comparison of the linear relation drawn by log a and b for Japanese seaweeds (mean values) with the one for marine phytoplankton (mean values). From this figure, it is apparent that Japanese seaweeds have slightly higher contents of elements with long residence times, such as iron and aluminum, while marine phytoplankton have higher contents of elements with shorter residence times such as sodium and magnesium.

The concentration factor of each element (y_o) can be calculated by using the mean values for log a and b given in Table V.

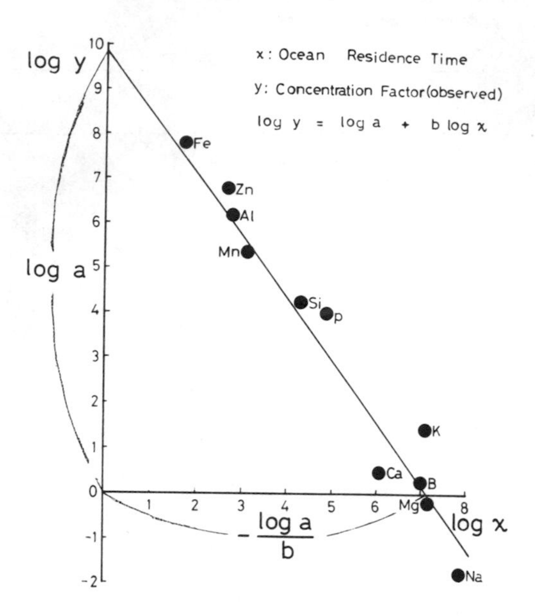

Figure 1. Element Concentrations Relative to Ocean Residence Times for Marine Phytoplankton (log a = 9.8, b = -1.35)

The ratios of the observed concentration factor (y) to the calculated one (y_o) are summarized in Table VI. It is suggested that elements with y/y_o value of greater than 1 might be biologically concentrated, phosphorus and potassium are examples. The elements which y/y_o values are less than 1 are probably more lithophilic. Thus y/y_o values can be used as an index to assess biophilicity and lithophilicity of the chemical elements.

Table V. Relation between element concentration factor (y) and the oceanic residence time (x) for selected organisms.

Biological group	log *a*	*b*	Number of samples	Regression coefficient	Data source
Seaweeds	7.7	-1.05	119	-0.96	This work
Seaweeds	7.7	-1.06	65	-0.93	8-15
Brackish seaweeds	8.0	-1.05	6	-0.96	16
Marine plankton	9.8	-1.35	16	-0.97	This work
Zooplankton	8.1	-1.16	39	-0.95	This work
Angiosperms	8.9	-1.24	8	-0.96	This work

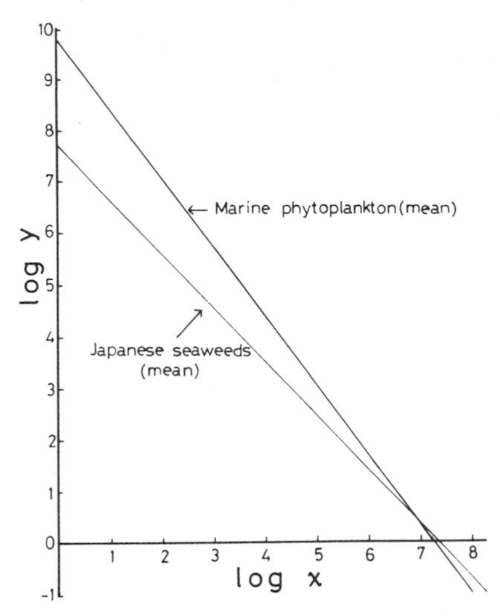

Figure 2. Comparison between linear relations for Japanese seaweeds and marine phytoplankton. Compare with Figure 1 for the position of various elements.

Since the geochemical behavior of the elements is based on atomic properties, general similarities are observed among elemental compositions of many geochemical substances. Table VII summarizes the correlation coefficients for several biological and geological materials with respect to element abundances. In addition to data on seaweeds, marine phytoplankton, and marine zooplankton, data for kale (17), holothuria (18, 19), stream water (20), crustal rock (21), chondrites (22) and shallow water sediments (23) were adapted from several sources. Some values such as those between seaweeds and stream water, show a high degree of correlation, while others such as between holothuria and chondrites indicate a low degree of correlation. In this calculation, if the elemental concentration of the geochemical substances are normalized by those of seawater, the values of the correlation coefficients generally increase (Table VIII). Visual comparison of the results in Table VII and VIII indicate that correlation coefficients between log U/W and log V/W are usually much higher than those between log U and log V. In these formulas, dividing by W means weighting each value of U and V. Since the method for weighting each value is appropriate, higher values of correlation coefficients have resulted. It was also noted that the recent literature data of elemental concentrations of seawater (5) were more appropriate than the older ones (20) for weighting these populations. The correlation seems to be due to the fact that the ocean has an important relationship with all living organisms.

The possibility of using other geochemical abundances as appropriate values of W for U and V also was investigated. When elemental abundances of continental crustal rocks (21) were used as the value of W, there was no good correlation between biochemical materials. The values of the correlation coefficients were in fact much lower than those between log of U and log of V. The

Table VI. The Range of y/y_o Values in Seaweeds, Freshwater Angiosperms, Marine Phytoplankton and Marine Zooplankton, where y/y_o is the ratio of the measured concentration for each element to the calculated theoretical concentration. Values are averages of multiple samples.

Element	Japanese Seaweeds	Seaweeds ref. 8-16	Freshwater Angiosperms	Marine Phytoplankton	Marine Zooplankton
B	2.4	0.83	0.86	1.2	0.57
Na	0.72	2.2	1.2	0.24	0.92
Mg	1.5	1.8	0.41	0.73	0.77
Al	3.0	2.0	2.1	1.2	0.62
Si	0.32	0.05	0.35	2.2	0.19
P	8.4	3.0	17	6.5	51
K	3.1	12	11	22	3.3
Ca	0.47	0.34	0.18	0.09	0.16
V	1.0	1.2	0.24	-	-
Cr	0.48	2.7	0.23	-	-
Mn	2.4	3.2	-	0.49	0.45
Fe	2.5	6.0	1.5	1.7	1.2
Co	0.92	1.4	0.25	-	-
Ni	0.35	0.83	-	-	-
Cu	0.47	0.9	-	-	-
Zn	1.4	0.79	0.55	0.29	1.5
As	5.2	7.8	0.13	-	-
Se	0.2	0.15	-	-	-
Br	2.8	1.2	0.76	-	-
Rb	1.4	17	14	-	-
Sr	6.3	8.0	0.31	-	-
Mo	0.27	0.59	1.5	-	-
Ag	0.39	0.39	0.06	-	-
Sb	0.09	8.5	0.05	-	-
I	24	42	-	-	-
Cs	0.85	1.1	0.97	-	-
Ba	0.16	0.3	0.27	-	-
La	2.7	1.8	-	-	-
Ce	3.2	7.2	2.0	-	-
Sm	3.2	2.3	2.4	-	-
Eu	4.0	3.3	3.0	-	-
Tb	4.7	3.7	4.8	-	-
Yb	3.9	-	3.1	-	-
Lu	2.7	-	2.2	-	-
Hg	0.42	-	-	-	-
Pb	6.4	2.6	-	-	-
U	1.6	0.71	-	-	-

Table VII. Correlation Coefficients between log U and log V

(U :elemental concentration of a geochemical substance
V :elemntal concentration another geochemical substance)

V \ U	1	2	3	4	5	6	7	8	9
1 Seaweeds	1.00								
2 Kale	0.94	1.00							
3 Holothuria	0.94	0.84	1.00						
4 Marine phytoplankton	0.71	0.53	0.26	1.00					
5 Marine zooplankton	0.90	0.83	0.86	0.79	1.00				
6 Stream water	0.95	0.90	0.79	0.72	0.76	1.00			
7 Crustal rock	0.80	0.77	0.45	0.79	0.64	0.81	1.00		
8 Chondrite	0.51	0.66	0.36	0.68	0.52	0.67	0.83	1.00	
9 Shallow water sediments	0.82	0.72	0.67	0.70	0.48	0.87	0.88	0.58	1.00

Table VIII. Correlation Coefficients between log U/W and log V/W

(U :elemental concentration of a geochemical substance
V :elemntal concentration another geochemical substance
W :elemental concentration of seawater)

V \ U	1	2	3	4	5	6	7	8	9
1 Seaweeds	1.00								
2 Kale	0.96	1.00							
3 Holothuria	0.99	0.96	1.00						
4 Marine phytopalnkton	0.98	0.94	0.96	1.00					
5 Marine zooplankton	0.99	0.97	0.99	0.98	1.00				
6 Stream water	0.97	0.91	0.95	0.96	0.95	1.00			
7 Crustal rock	0.95	0.88	0.91	0.96	0.94	0.91	1.00		
8 Chondrite	0.89	0.86	0.86	0.94	0.93	0.87	0.92	1.00	
9 Shallow water sediments	0.97	0.91	0.94	0.97	0.95	0.98	0.97	0.92	1.00

correlation coefficients when condritic abundances of the elements (22) were used were higher than those between log of U and log of V, although seawater abundances provided the highest correlations between log U/W and log V/W.

In the comparison between the results in Table V and Table VIII, there is no difference to be found among the values for each series of correlation coefficients. It seems that every concentration ratio for a biogeochemical substance to seawater can be used as an appropriate reference for comparative data analysis. It means that we will be able to calculate the characteristic values of log a, b and y/y_o for distribution of elements in various organisms and environmental substances by substituting each concentration ratio for ocean residence time. In previous papers (1, 4, 6, 7, 26-29) the values of ocean residence time and chemical abundance in seawater were taken from Goldberg and others (20), while the values in this paper were taken from Broeker and Peng (5). Therefore, this paper reports improved correlations between concentration factors and ocean residence times. The summary of a reassessment on selected organisms (in Table V) shows higher values of correlations coefficients than those values calculated previously.

The magnitude of y/y_o for each individual sample in Table VI is not proportional to the magnitude of the content of the element for every sample, because y/y_o values for each sample are associated with the values of log a and b for the sample. However, since the coefficient of variation for log a and for b in each marine organisms was small, unusually high or low contents of individual samples were easily recognized from y/y_o values as an unusual excess or deficiency of the element with respect to the theoretical value. A specific aspect of y/y_o values is that the magnitude of each element in the sample may be compared with other elements of the same samples. For instance, Japanese seaweeds (Table VI) have high values of y/y_o for iodine, phosphorous and arsenic, whereas they have low values of y/y_o for antimony and silicon. The unusual excess of arsenic with respect to the theoretical value might suggest a biological requirement for this element in seaweeds, while low value of antimony might result from the losses of this element during dry ashing of biological samples. Also, it is observed that Japanese seaweeds have a higher content of strontium, while other seaweeds have higher contents of potassium and iodine.

In summary, we found an excellent linear relationship between the values of log a and b for individual samples through all the organisms. Absolute values of the correlation coefficients (r) are high enough to indicate the general linear relationship. In this way comparative studies of the values of log a and b, r and y/y_o for individual samples provided information on the distribution of chemical elements in seaweeds and related organisms.

ACKNOWLEDGMENT

The authors express a special thanks to Professor Bruce N. Smith (Department of Botany and Range science, Brigham Young University) for his valuable discussions throughout this study.

REFERENCES

1. Yamamoto, T. "The Relations between Concentration Factors in Seaweeds and Residence Time of some Elements in Sea-water," Rec. Oceanogr. Works Japan 11: 65-72 (1972).
2. Cherry, R. D., J.J. Higgo and S. W. Fowler."Zooplankton Feacal Pellets and Element Residence Times in the Ocean," Nature 274: 246-248 (1978).
3. Whitfield, M. and D.R.Turner. "Chemical Periodicity and the Speciation and Cycling of the Elements," in Trace Metals in Sea Water,C.S.Wong, E.Boyle,K.W.Bruland,J.D.Burton and E.D. Goldberg,Eds. (New York, Plenum Press., 1983), PP. 719-750.
4. Yamamoto, T, ED. "Distribution of Trace Elements in Marine Algae-Comparative Biogeochemical Data- ," A report on a Grand-in-Aid for Scientific Research from the Ministry of the Education, Science and Culture, Japan (Grant No. 56117002, 57110003 and 58102003) "Special Project Research : The Ocean Characteristics and their change," (Fushimi-ku, Kyoto, Kyoto University of Education.,1983), PP. 1-499.
5. Broecker, W.C. and T.H. Peng. "Tracers in The Sea," A Publication of the Lamont-Doherty Geological Observatory, Columbia Univercity, (Nem York, Eldigio Press. 1982), PP. 26-27.
6. Yamamoto, T. "Distribution of Chemical Elements in Marine Algae," In Bioaccumlation in the Ocean and Factors affecting it, (NIRS-M-39), Y. Ueda and T. Koyanagi, Eds. (Chiba, Institute of Radiological Science.,1980), PP. 236-253, (in Japanese).
7. Yamamoto, T., Y. Otsuka, K. Aoyama and K. Okamoto. "Character of each Element on its Distribution in Seaweeds," Hydrobiologia 116/117: 510-512 (1973).
8. Black, W.A.P. and R.L. Mitchell. "Trace Elements in the Common Brown Algae and in Seawater," J. Mar. Biol. Ass. U.K. 30: 575-584 (1952).
9. Mita, K. "A Study on the Mucilage of Gloiopeltis furcata-I. On the Imorganic Components," Bull. Japan. Soc. Sci. Fish. 22: 558-560 (1957). (in Japanese).
10. Young, E.G. and W.M. Langille. "The Occurrence of Inorganic Elements in Marine Algae of the Atlantic Provinces of Canada," Can, J. Botany 36: 301-310 (1958).
11. Mita, K. "Chemical Studies on the Green Seaweed-III. On the Inorganic Components of Enteromorpha Compressa, Ulva Pertusa and their Mucilages," Bull. Japan. Soc. Sci. Fish. 27: 239-240 (1961). (in Japanese).
12. Sivalingam, P.M. "Biodeposited Trace Metals and Mineral Content Studies of Some Tropical Marine Algaes," Bot. Mar. 21: 327-330 (1978).
13. Harada, T., K. Oishi, T. Tamaki and M. Koyama. "Radioactivation Analysis of Calcareous Algae and Laminariales," In Proceedings of the 1980 Spring Meeting of the Oceanographical Society of Japan. (preprints., 1980), P.174.

(in Japanese).
14. North, W.J. "Trace Metals in Giant Kelp, Macrocystis," Amer. J. Bot. 67: 1097-1101 (1980).
15. Nakayama, Y., N. Hyakushima, S. Sugihara and Y. Takashima. "Radioactivation Analysis of Ishige okamurai and Sargassum thunbergii," Private communication (1981).
16. Bojanowski, R. "The Occurrence of Major and Minor Chemical Elements in the More Common Baltic Seaweed," Oceanologia NR 2 : 81-152 (1973).
17. Bowen, H.J.M. "Problems in the Elementary Analysis of Standard Biological Materials," J. Raidioanal. Chem. 19: 215-226 (1974).
18. Matsumoto, T., M. Satake and Y. Hirata. "On the Macro Constituent Elements in Marine Invertebrates," J. Oceangr. Soc. Japan 20: 110-116 (1964) (in Japanese).
19. Matsumoto, T., M. Satake, J. Yamamoto and S. Haruna. "On the Micro Constituent Elements in Marine Invertebrates," J. Oceangr. soc. Japan 20: 117-121 (1964), (in Japanese).
20. Goldberg, E.D., W.S. Broecker, M.G. Gross and K.K. Turekian. "Radioactivity in the Marine Enviroment," National Academy of science, Washington, D.C., 137-146 (1971).
21. Taylor, S.R. "Abundance of Chemical Elements in the Continental Crust : A New Table," Geochim. Cosmochim. Acta 28: 1273-1285 (1964).
22. Vinogradov, A.P. "Atomic Abundances of Chemical Elements of the Sun and Stony Meteorites," Geokhimiya 4: 291-295 (1962), (in Russian).
23. Yamamoto, Y., Y. Tanaka and S. Ueda. "The Chemical Composition and Nickel, Cobalt, Copper, Zinc, and lead Contents of Nanao Bay Sediments," J. Oceangr. Soc. Japan 33: 242-246 (1977).
24. Whightfield, M. "The Mean Oceanic Residence Time (MORT) Concept - a Rationalisation," Marine Chem. 8: 101-123 (1979).
25. Ishibashi, M. "Distribution and Regular Relationship of Quantities of Chemical Elements in Sea Water, including so-called Nutrient Elements," in Morning Review Lecture of the 2nd International Oceanographic Congress, Moscow, 1966," (UNESCO, Paris), PP. 83-92.
26. Yamamoto, T., Y. otsuka, M. Okazaki and K. Okamoto. "A Method of Data Analysis on the Distribution of Chemical Elements in the Biosphere," in Analytical Techniques in Environmental Chemistry, J. Albaiges Ed. (Oxford, Pergamon Press., 1980), PP. 401-408.
27. Yamamoto, T., and M. Ishibashi. "The Content of Trace Elements in Seaweeds," in Proceedings of the 7th International Seaweed Symposium, K. Nisizawa, Ed. (Tokyo, University of Tokyo Press., 1972), PP. 511-514.
28. Yamamoto, T., T. Yamaoka, S. Tsuno, R. Tokura, T. Nishimura and H. Hirose. "Microconstituents in Seaweeds," in Proceedings of the 9th International Seaweed Symposium, A. Jensen and J.R. Stein, Eds. (Princeton, Science Press., 1978), PP. 445-450.

29. Yamamoto, T., Y. Otsuka, M. Okazaki and K. Okamoto. "The Distribution of Chemical Elements in Algae," in Marine Algae in Pharmaceutical Science, H.A. Hoppe, T. Levring and Y. Tanaka, Eds. (Berlin, Walter de Gruyter, 1979), PP. 569-607.

SUBJECT INDEX